CHESTER COLLEGE LIBRARY

This book is to be returned on or before the
last date stamped below.

STUDY GUIDE AND SELECTED ANSWER MANUAL

FUNDAMENTALS OF GENERAL, ORGANIC AND BIOLOGICAL CHEMISTRY THIRD EDITION

JOHN R. HOLUM
Augsburg College

JOHN WILEY & SONS
NEW YORK CHICHESTER BRISBANE TORONTO SINGAPORE

ISBN 0 471 81513 6

Printed in the United States of America

10 9 8 7 6 5 4 3 2 1

Preface

Each chapter in this **Study Guide** matches a chapter in **Fundamentals of General, Organic, and Biological Chemistry**. They include five major features:

1. A discussion of the purpose of the chapter and a list of specific objectives.

2. A Glossary of all the important terms in the chapter.

3. Special help units. These may include aids in learning how to work problems, practice in writing formulas or structures, or lists of the important properties of functional groups. Also, several special sets of drill exercises occur in this **Study Guide**, all with answers given.

4. Self-testing questions. Both completion questions and multiple-choice questions are given.

5. Answers to the exercises and self-testing questions.

Answers to selected Review Exercises found in the text are included in the back of this **Guide**.

To June Brady, my typist, and Sandra Olmsted, who checked answers to the questions and exercises, go my special thanks.

Best wishes.

John R. Holum
Augsburg College
Minneapolis, MN 55454

Contents

1 Goals, Methods, and Measurements

STRATEGY FOR STUDY

There's a world of difference between understanding and knowing. Understanding is only the first step, although an essential one, to knowing. This difference means that there are two steps in studying. Neither step may be omitted. The first must occur before the second will work.

1. Study the material until it is understood, until you can say to yourself "that makes sense," or "now I get it," or something similar. With easy material, one reading will get you this far. With the more difficult, you'll need the text, class lectures and discussions, and sometimes extra help.
2. Drill yourself on the understood material until you know it. In some cases, this means memorizing the definitions of new words. In others, you should work problems of a purely drill nature as well as questions that require thought.

What will count from this course throughout your professional life, including later courses and in-service training, is remembered chemistry. "Understood chemistry" is of itself almost never "remembered chemistry." "Understood chemistry," taken alone is chemistry very soon forgotten; it is chemistry that can seldom, if ever, be applied in new situations. It's like a house stopped at the foundation. On the other hand, "rote-memorized chemistry" is almost never "remembered chemistry," either. It is like a house without a foundation and will soon collapse. A complete house requires both a foundation and a superstructure.

This **Study Guide** is designed to help you get from the understanding stage to the knowing stage. The text, lectures, and outside help are your chief aids to understanding. The **Study Guide** can help you by providing examples of worked problems. Sometimes alternative methods are presented. Definitions of important terms used in the text are provided. A number of drill exercises, together with all their answers, are given. You are also given an opportunity to test yourself by two kinds of questions – completion and multiple-choice. As a final check, you can go to the list of objectives for the chapter being studied and see if you can do each of them.

The two main steps for mastering a body of knowledge may be broken down into a series of smaller steps, as follows:

Step 1.   Read the chapter in the text. If at all possible read it once before the class discussion. Use a pencil – not a pen – to place a check in the margin near any part you have trouble understanding. You don't help yourself much by underlining what you understand.[1]

Step 2.   Use the lecture period to clear up as many of the puzzles as possible. Erase each check mark as understanding moves in and puzzlement moves out. If the lecture leaves any points still unclear, see your instructor or a teaching assistant about them as soon as you can. Often another classmate can help you.

Step 3.   After reaching the understanding stage, begin the knowing stage. Learn the definitions of the important terms. They appear in boldface in the text and are defined in the glossaries of each chapter in this **Study Guide.**

---

[1] I owe the formulation of this step and the next to S. Paul Steed of American River College, Sacramento, California. He calls this approach the EYI study technique – "**Eliminate Your Ignorance**" – and it makes great good sense. (Journal of Chemical Education, December 1976, p. 745.)

Step 4.   Where the material involves calculations, rework the examples. Try all of the problems for which answers are provided in order to test yourself. The **Study Guide** has many additional problems for you to solve.

Step 5.   When you are quite sure you have done sufficient studying, test yourself with the Self—Testing Questions in this **Study Guide**. All of the answers are given, and in many instances explanations are also provided. If you do not do very well on these questions, take it as a sign that you need to do more studying.

Step 6.   As a final test, check the list of objectives for the chapter as given in this **Study Guide**. Be sure you can write out a correct response for each objective. Sometimes working in pairs helps.

OBJECTIVES

After you have studied this chapter and have worked the exercises in it, you should be able to do the following. (Other objectives may also be assigned.)

1. State what is meant by a "reproducible fact."
2. Describe the relationships among facts, hypotheses, and theories.
3. Distinguish between a number and a physical quantity.
4. Describe in general terms the difference between a physical and a chemical property.
5. Distinguish between a base quantity, a base unit and a standard of measurement and between a base unit and a derived unit.
6. Give the present name (in the English translation) and the official abbreviation for the successor to the metric system.
7. Name, define, and give the symbols of the SI base units of length, mass, temperature, and time and the derived unit for volume.
8. Describe the advantage of the use of base—10 in the SI.
9. Use the conventions for writing figures and symbols for SI units.
10. Use the prefixes kilo—, centi—, milli—, and micro— with SI units.
11. Convert back and forth between kilograms, grams, milligrams, and micrograms.
12. Convert back and forth between liters, milliliters, and micro-liters.
13. Use the factor label method and conversion factors to change from

one system to another.

14. Tell how many significant figures a number has.
15. Correctly round off numbers obtained by arithmetic operations on physical quantities.
16. State the difference between "precision" and "accuracy," and tell what is meant by an exact number.
17. Convert between the Fahrenheit, Celsius, and Kelvin scales.
18. Do calculations involving densities and specific gravities.
19. Define each term listed in the Glossary.

GLOSSARY

Accuracy. In science, the degree of conformity to some accepted standard or reference; freedom from error or mistake; correctness.

Base Quantity. A fundamental quantity of physical measurement such as mass, length, and time; a quantity used to define derived quantities such as mass/volume for density.

Base Unit. A fundamental unit of measurement for a base quantity – such as the kilogram for mass, the meter for length, the second for time, the kelvin for temperature degree, and the mole for quantity of chemical substance; a unit to which derived units of measurement are related.

Centimeter (cm). A length equal to one-hundreth of the meter. 1 cm = 0.01 m = 0.394 in.

Chemical Property. Any chemical reaction that a substance can undergo and the ability to undergo such a reaction.

Conversion Factor. A fraction that expresses a relationship between quantities that have different units, such as 2.54 cm/in.

Degree Celsius. One-hundreth (1/100) of the interval on a thermometer between the freezing point and the boiling point of water.

Degree Fahrenheit. One-one hundred and eightieth (1/180) of the interval on a thermometer between the freezing point and the boiling point of water.

Density. The ratio of the mass of an object to its volume; the mass per unit volume. Density = mass/volume (usually expressed in g/mL).

Derived Quantity. A quantity based on a relationship that involves one or more base quantities of measurement such as volume (length^3) or density (mass/volume).

Derived Unit. A unit of a derived quantity such as g/mL (density).

Extensive Property. Any property whose value is directly proportional to the size of the sample; for example, the volume or the mass of a sample.

Factor-Label Method. A strategy for solving computational problems that uses conversion factors and the cancellation of the units of physical quantities as an aid in working toward the solution.

Gram (g). A mass equal to one-thousandth of the kilogram mass, the SI standard mass. 1 g = 0.001 kg = 1000 mg; 1 lb = 454 g.

Hypothesis. A conjecture, subject to being disproved, that explains a set of facts in terms of a common cause and that serves as the basis for the design of additional tests or experiments.

Inertia. The resistance of an object to a change in its position or its motion.

Intensive Property. Any property whose value is independent of the size of the sample, such as temperature and density.

International System of Units (SI). The successor to the metric system with new reference standards for the base units but with the same names for the units and the same decimal relationships.

Kelvin. The degree on the Kelvin scale of temperature and equal to the degree Celsius in magnitude.

Kilogram (kg). The SI base unit of mass; 1000 g; 2.205 lb.

Length. The base quantity for expressing distances or how long a thing is.

<u>Liter</u> (<u>L</u>). A volume equal to 1000 cm^3 or 1000 mL or 1.057 liquid
quart.

<u>Mass</u>. A quantitative measure of inertia based on an artifact at
Sevres, France, called the standard kilogram mass; a measure of
the quantity of matter in an object relative to this reference
standard.

<u>Measurement</u>. An operation whereby we compare an unknown physical
quantity such as length or mass with a known physical quantity
such as a meter stick or a gram mass.

<u>Meter</u> (<u>m</u>). The base unit of length in the International System of
Measurements (SI).
1 m = 100 cm = 39.37 in. = 3.280 ft = 1.093 yd

<u>Metric System</u>. A decimal system of weights and measures in which the
conversion of a base unit of measurement into a multiple or a
submultiple is done by moving the decimal point; the predecessor
to the International System of Measurements (SI).

<u>Microgram</u> (μg). A mass equal to one-thousandth of a milligram.
1 μg = 0.001 mg = 1 x 10^{-6} g   (Its symbol is sometimes given
as mcg or as γ in pharmaceutical work.)

<u>Microliter</u> (μL). A volume equal to one-thousandth of a milliliter.
1 μL = 0.001 mL = 1 x 10^{-6} L   (Its symbol is sometimes given as
λ in pharmaceutical work.)

<u>Milligram</u> (<u>mg</u>). A mass equal to one-thousandth of a gram.
1 mg = 0.001 g;   1000 mg = 1 g;   1 grain = 64.8 mg.

<u>Milliliter</u> (<u>mL</u>). A volume equal to one-thousandth of a liter.
1 mL = 0.001 L = 16.23 minim = 1 cm^3;   1 liquid ounce = 29.57 mL
1 liquid quart = 946.4 mL.

<u>Millimeter</u> (<u>mm</u>). A length equal to one-thousandth of a meter.
1 mm = 0.001 m = 0.0394 in.

<u>Physical Property</u>. Any observable characteristic of a substance other

than a chemical property, such as color, density, melting point, boiling point, temperature, and quantity.

Physical Quantity. A property of something to which we assign both a numerical value and a unit, such as mass, volume, or temperature; physical quantity = number x unit.

Precision. The fineness of a measurement or the degree to which successive measurements agree with each other when several are taken one after the other. (See also Accuracy.)

Property. A characteristic of something by means of which we can identify it.

Scientific Method. A method of solving a problem that uses facts to devise a hypothesis to explain the facts and to suggest further tests or experiments designed to discover if the hypothesis is true or false.

Scientific Notation. The method of writing a number as the product of two numbers, one being 10^x, where x is some positive or negative whole number.

Second (s). The SI unit of time; 1/60th minute.

Significant Figures. The number of digits in a numerical measurement or in the result of a calculation that are known with certainty to be accurate plus the first digit whose value is uncertain.

Specific Gravity. The ratio of the density of an object to the density of water.

Standard. A physical description or embodiment of a base unit of measurement, such as the standard meter or the standard kilogram mass.

Temperature. The measure of the hotness or coldness of an object. Degrees of temperature, such as those of the Celsius, Fahrenheit, or Kelvin scales, are intervals of equal separation on the thermometer.

Theory. An explanation for a large number of facts, observations, and hypotheses in terms of one or a few fundamental assumptions of what the world (or some small part of the world) is like.

Time. A period during which something endures, exists, or continues.

Volume. The capacity of an object to occupy space.

Weight. The gravitational force of attraction on an object as compared to that of some reference.

MULTIPLES AND SUBMULTIPLES BY MOVING A DECIMAL POINT

The purpose here is to become skilled in changing a number by units of 10, 100, 1000, or more or by 0.1, 0.01, 0.001, etc., simply by moving the decimal point. We shall practice applying this skill to a set of problems that involves specific quantities stated in metric or SI units.

The most common metric or SI conversions are made by moving the decimal point either three places or multiples of three places. The prefix used indicates the type of change to be made. Consider those prefixes most commonly encountered in the life sciences and in chemistry:

Prefix	Exponential Equivalent
kilo-	10^3
REFERENCE	1 (or 10^0)
milli-	10^{-3}
micro-	10^{-6}

To change a quantity expressed in the unit of the reference (e.g., liters) to the equivalent expressed by the next smaller unit in this list (e.g., milliliter), we simply move the decimal three places to the right.

Example: 3 liters = 3000 milliliters
3.0 liters becomes 3.0.0.0 milliliters

right $\longrightarrow$ 1 2 3

Remember, to go from a larger to a smaller unit we always move the decimal point to the right.

To change a physical quantity from a smaller to a larger unit, we move the decimal point to the left.

<u>Example</u>:   2000 millimeters = 2 meters
2000.0 millimeters becomes 2.0.0.0.0 meters

3 2 1 ⟵ left

Note that "milli-" means "a thousandth"; thus, the meter is a larger unit than the millimeter.

DRILL EXERCISES

I. <u>EXERCISES</u> <u>IN</u> <u>METRIC</u> <u>UNITS</u>

The purpose of these exercises is to help you become very skillful in converting from a quantity to one of its multiples or sub-multiples. Every conversion in these exercises involves nothing more than moving the decimal point. Fill in the blanks of the second column with a number that, together with its unit, is equivalent to the quantity in the first column. Another purpose of the exercises is to learn the accepted abbreviations of the various SI units (consult the table in the text for these).

	Quantity	**Equivalent**	
Examples:	0.001 m	<u>1</u>	mm
	1000 m	<u>1.000</u>	km
	1654 g	<u>1.654</u>	kg

1.	0.01 m	_____ cm	11.	0.527 mg	_____ μg
2.	0.10 m	_____ cm	12.	0.120 ml	_____ μL
3.	2985 m	_____ km	13.	15 mm	_____ cm
4.	1564 mg	_____ g	14.	15 cm	_____ m
5.	156.4 mg	_____ g	15.	0.98 m	_____ cm
6.	15.64 mg	_____ g	16.	1.620 g	_____ mg
7.	1640 mL	_____ liter	17.	0.101 liter	_____ mL
8.	0.002 liter	_____ mL	18.	0.067 g	_____ mg

9. 454 g   _____ kg   19. 250 mL   _____ liter

10. 150 mL   _____ liter   20. 0.0001 mL   _____ μL

II.   EXERCISES IN SCIENTIFIC NOTATION

The purpose of these problems is to help you become skilled in converting back and forth between numbers expressed in the usual way and their equivalents in scientific notation.

SET A. Change each number to the form it has in scientific notation.

1. 1,062,457   _____

2. 0.00543   _____

3. 111.6   _____

4. 0.00000521   _____

5. 5.025   _____

SET B. Change each number given in scientific notation to its equivalent stated in the usual way.

1. 6.150×10^3 _____

2. 5.362×10^2 _____

3. 2.35×10^{-2} _____

4. 8.79×10^{-5} _____

5. 6,542,001   _____

III. EXERCISES IN SIGNIFICANT FIGURES

Rule 1.   Zeros are never counted as significant if they are used to set off the decimal point on their left.

Rule 2.   Zeros are always significant if nonzero numbers occur on both their right and their left.

Rule 3.   Trailing zeros are always counted as significant when they occur in numbers written in scientific notation. (In the text and in this study guide we'll also consider trailing zeros to be significant regardless of how the number in which they occur is expressed, unless some contrary statement is specifically given.)

Rule 4.   Exact numbers are treated as having an infinite number of significant figures.

Examples: 4.0054   5   (Rule 2)
            4540   4   (Rule 3 as further noted in the parentheses)
            0.000454   3   (Rule 1)
            4.5400   5   (Rule 4)

1. 60,500   _____   2. 6.0500   _____

3. 0.605   _____   4. 0.6050   _____

5. 0.000605   _____   6. 0.0006050   _____

7. 1.98×10^{10} _____   8. 1.9800×10^{10} _____

9. 123.50   _____   10. 12,350   _____

IV. EXERCISES IN THE FACTOR-LABEL METHOD FOR CONVERTING FROM ONE SYSTEM OF UNITS TO ANOTHER

The key idea: Get the final unit you want fixed in your mind and work toward your answer being in that unit.

The key operation: Multiply and divide (cancel) units like numbers.

The key strategy: Set up multiplications or divisions to get all unwanted units to cancel. If they don't, you know you have set up the solution to the problem incorrectly.

We shall study and learn to do this by means of examples.

Sample Problem 1:   Change 10.0 inches into centimeters.

A table of conversion factors (e.g., Table 1.2 in the text) tells us that there are 2.54 cm in 1 inch.

We may write this basic relation in either of two equivalent ways. Both state the same relation between the inch and the centimeter. One is simply the inverse, or the reciprocal, of the other.

$$\frac{2.54 \text{ cm}}{1 \text{ in.}} \quad \text{or} \quad \frac{1 \text{ in.}}{2.54 \text{ cm}}$$

The divisor line is read as "per." The first ratio says "2.54 cm per 1 in." The second says "1 in. per 2.54 cm."

We want our answer to be in centimeters. Therefore, we must use the ratio that will give us that answer. That is the "key idea" for this problem.

Step 1. Write down the given (be sure
to include the unit.)   10.0 in.

Step 2. Multiply the given by the right
conversion factor.

Do we write it as   (A) $10.0 \text{ in.} \times \dfrac{2.54 \text{ cm}}{1 \text{ in.}}$?

or

Do we write it as   (B) $10.0 \text{ in.} \times \dfrac{1 \text{ in.}}{2.54 \text{ cm}}$?

In other words, which way do we use the relation between inches and centimeters? In (B), units will not cancel to leave the answer solely in "cm." In (A), units cancel correctly.

(A) $10.0 \text{ in.} \times \dfrac{2.54 \ \text{cm}}{1 \ \text{in.}}$

Step 3. Do the arithmetic. Do whatever multiplications or divisions of numbers that there are left to do.

$$10.0 \times 2.54 \text{ cm} = 25.4 \text{ cm}$$

SET A. Before continuing, work enough of the following until you are sure you understand and can set up conversion factors in either of two ways.
1. There are 5280 feet in a mile.
2. There are 2.20 pounds in a kilogram.
3. There are 454 grams in a pound.
4. There are 36.4 inches in a yard.
5. There are 1000 milligrams in a gram.
6. There are 1000 meters in a kilometer.
7. There are 29.6 milliliters in a fluid ounce.
8. There are 7000 grains in 454 grams.
9. There are 1000 kilograms in a metric ton.

10. There are 1000 grams in a kilogram.

<u>Sample Problem 2</u>:   How many inches are there in 127 centimeters?

Step 1.   Write down the given   127 cm
Step 2.   Multiply the given by   Do we write it
          the right conversion   2.54 cm
          factor.   (C) 127 cm x ───────── ?
                                                           1 in.
                                       Or do we write it
                                                           1 in.
                                       (D) 127 cm x ───────── ?
                                                          2.54 cm

          Only (D) lets us cancel
          units in a way that leaves
          the answer in the unit
          called for in the problem.   (cm)(cm)   cm^2
          The units for (C) would be   ───────── or ───── !
                                           in.   in.

                                        1 in.   127
Step 3.   Do the arithmetic.   127 x ─────── = ───── in. = 50.0 in.
                                        2.54   2.54

<u>SET B</u>. Work the following problems using the conversion factors stated in Set A.
1. How many feet are there in 0.5000 mile?
2. How many miles are there in 15,000 feet?
3. How many yards are there in 100 inches?
4. How many milligrams are there in 0.100 gram?
5. How many grams are there in 5.0 grains?

These problems were actually quite simple. The real power of the factor-label method comes when you don't have a single conversion factor to use to work a problem and you have to improvise from two or more factors. Using the conversion factors given in Set A, let us see how this works by studying examples.

<u>Sample Problem 3</u>:   How many metric tons are there in 1000 pounds?

We don't have a factor to convert pounds to metric tons directly. We have to improvise. We do have a factor relating metric tons to kilograms (No. 9 in Set A).

$$\frac{1000 \text{ kilograms}}{1 \text{ metric ton}} \quad \text{or} \quad \frac{1 \text{ metric ton}}{1000 \text{ kilograms}}$$

We also have a factor relating kilograms to pounds (No. 2 in Set A).

$$\frac{2.20 \text{ pounds}}{1 \text{ kilogram}} \quad \text{or} \quad \frac{1 \text{ kilogram}}{2.20 \text{ pounds}}$$

Following our steps, we have:

Step 1. Write down the given.   1000 pounds

Step 2.

Pick a factor that allows you to cancel "pounds."

$$1000 \text{ pounds} \times \frac{1 \boxed{\text{kilogram}}}{2.20 \text{ pounds}}$$

Multiply all this by another factor that allows you to cancel kilograms.

$$1000 \text{ pounds} \times \frac{1 \text{ kilogram}}{2.20 \text{ pounds}} \times \frac{1 \boxed{\text{metric ton}}}{1000 \text{ kilograms}}$$

(In principle, you would keep doing this, picking additional conversion factors that let you cancel unwanted units, until only the desired unit(s) remains. As you can see, we are through using conversion factors in this sample problem.)

Step 3. Do the arithmetic.

$$1000 \times \frac{1}{2.20} \times \frac{1}{1000} \text{ metric tons} = 0.455 \text{ metric ton}$$

The advantage of leaving all the arithmetic to the end is that one can often cancel some of the numbers too.

Sample Problem 4:   How many milligrams are there in 1.00 grain?

(The needed units are given in Set A.)   Here is how the final solution will look before the arithmetic is done.

$$1.00 \text{ grain} \times \frac{454 \text{ grams}}{7000 \text{ grains}} \times \frac{1000 \text{ milligram}}{1 \text{ gram}}$$

Final answer: 64.9 milligrams per grain

SET C. Practice what you have learned from Sample Problems 3 and 4 by working the following.
1. How many grains are there in 1.00 kilogram?
2. How many pounds are there in 2.0 metric tons?
3. How many milligrams are there in 0.5 kilogram?

SELF-TESTING QUESTIONS

COMPLETION. Fill in the blanks with the words or phrases that best complete each statement or answer the question.

1. The operation by which we compare an unknown physical quantity with a known quantity such as a reference standard is called _____.

2. Properties that we can measure without changing a substance into some other substance are called _____ properties.

3. The inertia of an object is its ability to _____ _____.

4. The quantitative measure of an object's inertia is called its _____.

5. Which of these are base units and which are derived?

   (a) volume _____   (c) area   _____
   (b) mass   _____   (d) length _____

6. The successor to the metric system is the _____.

7. The physical embodiment of a base unit is called a reference _____.

8. The base unit of length in the metric system is the _____.

9. The base unit of length in the successor to the metric system is the _____.

10. The reference standard for the measurement of length in the metric

system is _____.

11. Is this also the reference standard for the measurement of length in the successor to the metric system? _____

12. If not, how do the two differ in length? _____

13. The base unit of mass in the metric system is the _____.

14. The base unit of mass in the successor to the metric system is _____.

15. Scientists would like to have a reference standard for mass that is not an artifact (not a manufactured object) because _____ _____.

16. The base unit of time in the SI is the _____.

17. The base unit for the temperature degree in the SI is the _____ _____.

18. The degree _____ was formerly called the degree Centigrade.

19. There are _____ degree divisions between the freezing point and the boiling point of water on the Fahrenheit scale.

20. The meter is a little longer than _____ feet.

21. Half an inch would be _____ than 1 centimeter.
                          (shorter or longer)

22. In 1.0 milliliter of water there are about 16 drops. One drop is therefore _____ microliters.

23. One kilogram of water occupies one _____ of volume (in SI terms), and this volume is roughly one _____ in the "English" system.

24. To change 0.00056 to scientific notation we move the decimal point 4 places to the _____ and use an exponent of _____
                    (left or right)

for the 10 part of the expression; e.g., $5.6 \times 10^{\underline{\quad}}$.

25. Express these numbers in scientific notation in which the first part of the number is between 1 and 10.

    (a) 156 _____

    (b) 4,360,890,000,000 _____

    (c) 0.00000043 _____

    (d) 0.10004 _____

26. Complete the following conversions of the first numbers given to alternative expressions in scientific notation in which the exponents are numbers divisible by 3.

    (a) $94,500,000 = 94.5 \times 10^{\underline{\quad}}$

    (b) $0.000896 = 896 \times 10^{\underline{\quad}}$ or $0.896 \times 10^{\underline{\quad}}$

27. Write the accepted SI abbreviations for each unit.

    (a) milligram _____   (c) deciliter _____
    (b) microliter _____   (d) milliliter _____

28. Reexpress these quantities using the SI prefix supplied.

    (a) 1500 g = _____ kg

    (b) 0.0000080 liter = _____ μL

    (c) 0.0045 mL = _____ μL

    (d) 4502 mg = _____ g

    (e) 0.015 kg = _____ g

29. If the correct value of the mass of an object is 14.5068 g but someone reported it as 24.5068 g, we'd say that the reported measurement isn't very _____.
    (precise or accurate)

30. If someone reported that the volume of a liquid is 12 mL and it is known to have a volume of 12.478 mL, the report can be described as not very _____.
    (precise or accurate)

31. State how many significant figures are in each physical quantity.

    (a) 1.00050 gram   _____

    (b) 1.000 x 10^4 liter   _____

    (c) 0.0105 mL   _____

32. The inch is legally defined as 2.54 cm, exactly. In doing calculations, how many significant figures can be assumed are in 2.54 cm when it occurs in the conversion factor: $\dfrac{2.54 \text{ cm}}{1 \text{ in.}}$? _____

33. When we subtract 24.567 mL from 482.4 mL, how do we round off the difference and how is the difference correctly expressed? _____ (The calculator answer is 457.833.)

34. If we subtract 0.458 m from 362 m, how is the difference correctly expressed? _____ (The calculator answer is 361.542.)

35. In multiplying 2.3 in. times $\dfrac{2.54 \text{ cm}}{1 \text{ in.}}$, how is the product correctly expressed? _____ (The calculator answer is 5.842.)

36. If we multiply 3.678 cm by the conversion factor $\dfrac{1 \text{ in.}}{2.54 \text{ cm}}$, the answer is correctly expressed as _____. (The calculator answer is 1.4480314.)

37. If we multiply 13.97 cm by the conversion factor $\dfrac{1 \text{ in.}}{2.54 \text{ cm}}$, the answer is correctly expressed as _____. (The calculator answer is 5.5.)

38. The relationship 1 liter = 1.057 quart can be expressed by means of what two ratios that we can use as conversion factors?

_____   _____

39. If 1 gram = 15.43 grain and 1 dram = 60 grains (exactly), then how many drams are in 425 gram? _____

40. What is 75 $^{\circ}$F in $^{\circ}$C? _____   In K? _____

41. The mass of a sample of water is an example of an _____
                                                              (extensive or
    _____ property.
    intensive)

42. The weight of an object divided by its volume equals its _____
    _____.

43. An organic liquid that cannot dissolve in water has a specific
    gravity of 1.5.

    (a) Will this liquid sink or float in water? _____

    (b) Why isn't the value of 1.5 given any units? _____

    _____

44. The density of mercury at room temperature is (rounded) 13.5 g/mL.
    The density of water at room temperature (again rounded) is 1.0
    g/mL. What is the specific gravity of mercury at room tempera-
    ture? _____

45. At the same temperature, object A has a density of 1.6 g/mL and
    object B has a density of 1.1 g/mL. Suppose a sample of A and a
    sample of B weigh the same. Which would occupy the larger volume?

    _____

46. Suppose someone invented a new temperature scale - call it the X
    scale - and arbitrarily (a) set 200-degree intervals between the
    freezing point and the boiling point of water, and (b) set the
    freezing point of water at +10° X on the new scale. What would be
    the equation for converting from degrees X to degrees Celsius?
    (Check your equation by doing the conversions at known values for
    the freezing point and the boiling point of water in degrees
    Celsius). _____

MULTIPLE-CHOICE. For each of the following select all of the correct
answers found among the choices (in some questions two or more choices
will correctly answer the question).

  1. An example of a chemical property is
     (a) the melting of ice
     (b) the digestion of a slice of bread

(c) the golden luster of polished brass
(d) the evaporation of water

2. When properly expressed in scientific notation, the quantity 0.01205 liter becomes
   (a) 12.05 mL
   (b) 12.05 x 10^3 liter
   (c) 1.205 x 10^3 liter
   (d) 1.205 x 10^{-2} liter

3. A distance of 100 cm is the same as
   (a) 1 meter   (b) 10 cc   (c) 1000 mm   (d) 0.001 kilometer

4. A temperature of 60 °C is the same as
   (a) 15.6 °F   (b) 213 K   (c) 108 °F   (d) 140 °F

5. A length of 25.4 mm is the same as
   (a) 1 in.   (b) 10 in.   (c) 2.54 cm   (d) 0.0254 meter

6. A measurement reported as 1.5050 x 10^{-3} g has how many significant figures?
   (a) 4   (b) 5   (c) 7   (d) 8

7. A substance that is denser than water might have
   (a) a density of 1000 grams per liter
   (b) a specific gravity of 0.5
   (c) a density of 18 micrograms per microliter
   (d) a specific gravity of 1

8. One milliliter is the same as
   (a) 0.001 deciliter
   (b) 1000 microliters
   (c) 1 mm
   (d) 0.001 liter

9. Which unit is closest in size to 500 grams?
   (a) ounce   (b) 500 mL   (c) pound   (d) kilogram

10. A temperature of 27 °C is the same as
    (a) 246 K   (b) −27 K   (c) 300 K   (d) 27 K

11. A volume of 145 mL is the same as
    (a) 0.145 liter   (b) 0.145 kL   (c) 1450 μL   (d) 1.45 liter

12. A mass of 0.203 g is the same as
    (a) 2.03 kg   (b) 0.0203 kg   (c) 20.3 mg   (d) 203 mg

13. A mass of 5,000,000 µg is the same as
    (a) 5000 mg   (b) 5 kg   (c) 5 g   (d) 0.5 g

14. How many meters are in 1.00 bolt of cloth, if
        1 bolt = 120 feet (exactly), and
        1 foot = 0.3048 meter (exactly)?
    (a) 36.6 m   (b) 394 m   (c) 0.00254 m   (d) 40.0 m

15. How many furlongs are in 5.00 mile if
        1 furlong = 220 yards (exactly)
        1 yard   =   3 feet (exactly)   and
        1 mile   = 5280 feet (exactly) ?
    (a) 360 furlong   (c) 1.29×10^{-4} furlong
    (b) 1.94×10^{6} furlong   (d) 40.0 furlong

16. The density of liquid water at 20 °C may be accurately given in
    any one of these ways. Which has the greatest precision?
    (a) 1 g/mL   (b) 1.0 g/mL   (c) 0.999 g/mL   (d) 0.99862 g/mL

17. A sample of a mineral has a volume of 4.50 mL and a mass of 13.5
    grams. What is its density?
    (a) 3.00 g/mL   (b) 0.33 mL/g   (c) 0.333 g/mL   (d) 3.00 mL/g

18. The density of gem quality diamond is 3.513 g/mL at 20 °C. To
    three significant figures what is its specific gravity? (See
    question 16 for the density of water at 20 °C.)
    (a) 3.5165   (b) 3.52   (c) 3.52 g/mL   (d) 3.52 mL/g

19. Benzene boils at 80 °C. What is its boiling point in degrees F?
    (a) 27 °F   (b) 176 °F   (c) 353 °F   (d) 160 °F

ANSWERS

ANSWERS TO DRILL EXERCISES

   I.   Exercises in Metric Units

         1. 1 cm   11. 527 µg
         2. 10 cm   12. 120 µL

3. 2.985 km
4. 1.564 g
5. 0.1564 g
6. 0.01564 g
7. 1.640 liter
8. 2 mL
9. 0.454 kg
10. 0.150 liter

13. 1.5 cm
14. 0.15 m
15. 98 cm
16. 1620 mg
17. 101 mL
18. 67 mg
19. 0.250 liter
20. 0.1 μL

II. Exercises in Exponential Notation

Set A.
1. 1.062457×10^6
2. 5.43×10^{-3}
3. 1.116×10^2
4. 5.21×10^{-6}
5. 5.025 (not 5.025×10^0; this is not done.)

Set B.
1. 6150
2. 536.2
3. 0.0235
4. 0.0000879
5. 6.542001×10^6

III. Exercises in Significant Figures

1. 5 (Rule 2, Rule 3, as modified)
2. 5 (Rules 2 and 3)
3. 3 (Rules 1 and 2)
4. 4 (Rules 1, 2, and 3)
5. 3 (Rules 1 and 3)
6. 4 (Rules 1, 2, and 3)
7. 3 (Rule 3)
8. 5 (Rule 3)
9. 5 (Rule 3 as modified)
10. 5 (Rule 3 as modified)

IV. Exercises in the Cancel-Unit Method for Converting from One System of Units to Another

Set A
1. $\dfrac{5280 \text{ feet}}{1 \text{ mile}}$ or $\dfrac{1 \text{ mile}}{5280 \text{ feet}}$

2. $\dfrac{2.20 \text{ pounds}}{1 \text{ kilogram}}$ or $\dfrac{1 \text{ kilogram}}{2.20 \text{ pounds}}$

3. $\dfrac{454 \text{ grams}}{1 \text{ pound}}$ or $\dfrac{1 \text{ pound}}{454 \text{ grams}}$

4. $\dfrac{36.4 \text{ inches}}{1 \text{ yard}}$ or $\dfrac{1 \text{ yard}}{36.4 \text{ inches}}$

5. $\dfrac{1000 \text{ mg}}{1 \text{ g}}$ or $\dfrac{1 \text{ g}}{1000 \text{ mg}}$

6. $\dfrac{1000 \text{ m}}{1 \text{ km}}$ or $\dfrac{1 \text{ km}}{1000 \text{ m}}$

7. $\dfrac{29.6 \text{ mL}}{1 \text{ fl oz}}$ or $\dfrac{1 \text{ fl oz}}{29.6 \text{ mL}}$

8. $\dfrac{7000 \text{ grains}}{454 \text{ grams}}$ or $\dfrac{454 \text{ grams}}{7000 \text{ grains}}$

9. $\dfrac{1000 \text{ kg}}{1 \text{ ton}}$ or $\dfrac{1 \text{ ton}}{1000 \text{ kg}}$

10. $\dfrac{1000 \text{ g}}{1 \text{ kg}}$ or $\dfrac{1 \text{ kg}}{1000 \text{ g}}$

Set B   1. $0.5000 \text{ mile} \times \dfrac{5280 \text{ feet}}{1 \text{ mile}} = 2640 \text{ feet}$

2. $15000 \text{ feet} \times \dfrac{1 \text{ mile}}{5280 \text{ feet}} = 2.841 \text{ miles}$

3. $100 \text{ inches} \times \dfrac{1 \text{ yard}}{36.4 \text{ inches}} = 2.75 \text{ yards}$

4. $0.1 \text{ gram} \times \dfrac{1000 \text{ mg}}{1 \text{ g}} = 100 \text{ mg}$

5. $5.0 \text{ grains} \times \dfrac{454 \text{ grams}}{7000 \text{ grains}} = 0.32 \text{ gram}$

Set C   1. $1.00 \text{ kg} \times \dfrac{1000 \text{ grams}}{1 \text{ kg}} \times \dfrac{7000 \text{ grains}}{454 \text{ grams}} = 1.54 \times 10^4 \text{ grains}$
(The pocket calculator result, 15,418.5022, has to be rounded to 3 significant figures.)

2. $2.0 \text{ metric ton} \times \dfrac{1000 \text{ kilogram}}{1 \text{ metric ton}} \times \dfrac{2.2 \text{ pounds}}{1 \text{ kilogram}} = 4.4 \times 10^3 \text{ pounds}$

3. $0.5 \; \cancel{kg} \; x \; \dfrac{1000 \; \cancel{grams}}{1 \; \cancel{kg}} \; x \; \dfrac{1000 \; milligrams}{1 \; \cancel{gram}} = 5 \; x \; 10^5 \; milligrams$

ANSWERS TO SELF-TESTING QUESTIONS

Completion

1. measurement
2. physical
3. resist a change in motion
4. mass
5. (a) derived   (b) base   (c) derived   (d) base
6. SI (International System of Units)
7. standard
8. meter
9. meter
10. the distance between two scratches on a bar of platinum-iridium alloy kept at Sèvres, France
11. No
12. They do not differ at all in length.
13. kilogram mass (at Sèvres, France)
14. the same kilogram mass
15. manufactured artifacts are subject to corrosion and possibly to theft or loss through natural disaster
16. second
17. Kelvin
18. Celsius
19. 180
20. 3
21. longer
22. 63 microliters. Using the factor-label method it works out like this: $\dfrac{1.0 \; \cancel{mL}}{16 \; drops} \; x \; \dfrac{1000 \; microliters}{1 \; \cancel{mL}} = 63 \; \dfrac{microliters}{drop}$
23. liter, quart (liquid, U.S.)
24. right; -4; $5.6 \; x \; 10^{-4}$
25. (a) $1.56 \; x \; 10^2$ (b) $4.36089 \; x \; 10^{12}$
    (c) $4.3 \; x \; 10^{-7}$ (d) $1.0004 \; x \; 10^{-1}$
26. (a) 6   (b) -6; -3
27. (a) mg   (b) μL   (c) dL   (d) mL
28. (a) 1.500 kg   (b) 8.0 μL   (c) 4.5 μL   (d) 4.502 g   (e) 15 g
29. accurate

30. precise
31. (a) 6   (b) 4   (c) 3
32. an infinite number
33. 457.8 mL
34. 362 m
35. 5.8 cm
36. 1.448 in.
37. 5.500 in.
38. $\dfrac{1 \text{ liter}}{1.057 \text{ quart}}$ or $\dfrac{1.057 \text{ quart}}{1 \text{ liter}}$
39. 109 dram
40. 24 °C; 297 K
41. extensive
42. density
43. (a) Sink   (b) The units cancel. For example 1.5 g/mL divided by 1.0 g/mL gives an answer of 1.5.
44. 13.5
45. Object B. Being less dense, it needs more room to have the same mass as the more dense A.
46. $2C = X - 10$ or $C = \dfrac{1}{2}X - 5$

Multiple-Choice

1. b	2. d	3. a, c, and d
4. d	5. a, c, and d	6. b
7. c	8. b and d	9. c
10. c	11. a	12. d
13. a and c	14. a	15. d
16. d	17. a	18. b
19. b		

2 Matter and Energy

OBJECTIVES

The kinds of matter and the states of matter are described in this chapter in only general terms. More about the nature of matter will be presented later, but first we encounter the laws of chemical combination here.

Because matter changes in important ways when energy is either given to it or taken from it, the broad topic of energy dominates Chapter Two. Energy can cause changes in physical state, and it can cause chemical reactions. Energy can also be produced by changes of state and by chemical reactions. No matter where energy goes or how it is transformed, however, it is always conserved.

When we add or remove energy, we need units for describing quantities involved. The units we use are the calorie and the kilocalorie. They are particularly important in chemistry as applied to the life sciences.

Our physical well-being depends upon our keeping a steady body temperature in the face of internal chemical changes that constantly release heat. Because water, the main body fluid, helps us to do just that, we take a thorough look at the thermal properties of water.

These particular aspects of Chapter Two and the important terms should be given your best attention. In addition after you have studied this chapter and have worked the exercises in it, you should be able to do the following. (Other objectives may also be assigned.)

1. Define "matter," give its three states, and name and define its three kinds.
2. Be able to tell from a chemical equation what are the symbols of the reactants and the products.
3. Write the names and symbols of common elements listed in Table 2 of the text.
4. State the key difference between a compound and a mixture.
5. Define "energy" and give the names for six forms of energy.
6. State the following laws of nature: the law of conservation of energy, the law of conservation of mass, the law of definite proportions, and the law of multiple proportions.
7. Give the five main postulates of Dalton's atomic theory and explain how they fit the three laws of chemical combination.
8. Define "calorie," "kilocalorie," "large calorie (Calorie)," and "joule."
9. Do calculations involving specific heats, heats of fusion, and heats of vaporization.
10. Explain how the specific heat and the latent heats of water make it especially fit as the body fluid.
11. Name five ways by which the body becomes hypothermic (five ways the body loses heat) and explain how each works.
12. Define "basal metabolism" and name other ways by which the body is heated.
13. Explain what is meant by exothermic and endothermic.
14. Define all of the terms listed in the Glossary.

GLOSSARY

Alloy. An intimate mixture of one or more metals in each other made by mixing the metals in their molten forms and then cooling the mixture.

Atom. The smallest particle of a given element that bears the chemical properties of the element.

Basal Activities. The minimum activities of the body needed to maintain muscle tone, control body temperature, circulate the blood, handle wastes, breathe, and carry out other essential activities.

Basal Metabolic Rate. The rate at which energy is expended to maintain basal activities.

Basal Metabolism. The total of all of the chemical reactions that support basal activites.

Boiling Point. The temperature at which a liquid boils.

Calorie. The amount of heat that raises the temperature of 1 g of water by 1 degree Celsius from 14.5 ^{O}C to 15.5 ^{O}C.

Chemical Energy. The potential energy that substances have because their arrangements of electrons and atomic nuclei are not as stable as are alternative arrangements that become possible in chemical reactions.

Compound. A substance made from the atoms of two or more elements that are present in a definite proportion by mass and by atoms.

Condensation. The physical change of a substance from its gaseous state to its liquid state.

Conduction. In the science of heat energy, the transfer of heat from a region of higher temperature to a region of lower temperature by means of the transfers of kinetic energies of atoms, ions, or molecules to their neighbors. In the science of electrical energy, the movement of electrons in a conductor.

Convection. In the science of heat energy, the transfer of heat by the circulation of a warmer fluid throughout the remainder of the fluid.

Dalton's Atomic Theory. A theory that accounts for the laws of chemical combination by postulating that matter consists of indestructible atoms; that all atoms of the same element are identical in mass and other properties; that the atoms of different elements are different in mass and other properties; and that in the formation of a compound, atoms join together in definite, whole-number ratios.

Element. A substance that cannot be broken down into anything that is both stable and more simple.

Endothermic. Describing a change that needs a constant supply of energy to happen.

Energy. A capacity to cause a change that can, in principle, be harnessed for useful work.

Equation, Balanced. A chemical equation in which all atoms that appear among the reactants are found among the products.

Equation, Chemical. A shorthand representation of a chemical reaction that uses formulas instead of names for reactants and products; that separates reactant formulas from product formulas by an arrow; that separates formulas on either side of the arrow by plus signs; and that expresses the proportions of the chemicals (in terms of formula units) by simple numbers (coefficients) placed before the formulas.

Evaporation. The conversion of a substance from its liquid to its vapor state.

Exothermic. Describing a change by which heat energy is released from the system.

Formula. A symbol for a chemical substance that uses atomic symbols and subscripts to describe the ratio of atoms in one formula unit of the substance.

Formula Unit. A single particle that has the composition shown by the chemical formula of the substance.

Heat. The form of energy that transfers between two objects in contact that have initially different temperatures.

Heat Capacity. The quantity of heat that a given object can absorb (or release) per degree Celsius change in temperature.
heat capacity = heat/Δt where Δt is the change in temperature.

Heat of Fusion. The quantity of heat that one gram of a substance absorbs when it changes from its solid to its liquid state at its melting point.

Heat of Vaporization. The quantity of heat that one gram of a substance absorbs when it changes from its liquid to its gaseous state.

Hyperthermia. An elevated body temperature.

Hypothermia. A low body temperature.

Kilocalorie (**kcal**). The quantity of heat equal to 1000 calories.

Kinetic Energy. The energy of an object by virtue of its motion.

$$\text{Kinetic Energy} = \frac{1}{2}(\text{mass})(\text{velocity})^2$$

Law of Conservation of Energy. Energy can be neither created nor destroyed but only transformed from one form to another.

Law of Conservation of Mass. Matter is neither created nor destroyed in chemical reactions; the masses of all products equal the masses of all reactants.

Law of Definite Proportions. The elements in a compound occur in definite proportions by mass.

Law of Multiple Proportions. When two elements can combine to form more than one compound, the different masses of the first that can combine with the same mass of the second are in the ratio of small whole numbers.

Matter. Anything that occupies space and has mass.

Melting Point. The temperature at which a solid changes into its liquid form.

Metal. Any element that is shiny, conducts electricity well, and (if a solid) can be hammered into sheets and drawn into wires.

Mixture. One of the three kinds of matter (together with elements and compounds); any substance made up of two or more elements or compounds combined physically in no particular proportion by mass and separable into its component parts by physical means.

Nonmetal. Any element that is not a metal. (See **Metal**.)

Perspiration, Insensible. The loss of water from the body with no visible sweating; evaporative losses from the skin and the lungs.

Perspiration, Sensible. Visible perspiration released by the sweat glands.

Potential Energy.   Stored or inactive energy.

Product.   A substance that forms in a chemical reaction.

Radiation, Heat.   A process whereby light or heat is emitted; also the emitted light or heat.

Radioactivity. The property of unstable atomic nuclei whereby they emit alpha, beta, or gamma rays.

Reactant.   One of the substances that reacts in a chemical reaction.

Reaction, Chemical. Any event in which substances change into different chemical substances.

Sensible Perspiration. Visible perspiration released by the sweat glands.

Specific Heat. The amount of heat that one gram of a substance can absorb per degree Celsius increase in temperature.

Heat capacity $= \dfrac{\text{heat}}{\text{g } \Delta t}$ where Δt = the change in temperature.

(The unit of heat is usually the calorie, but kilocalorie, joule, or kilojoule can be used.)

States of Matter. The three possible physical conditions of aggregation of matter—solid, liquid, and gas.

Substances, Chemical. The materials of which matter consists.

Vaporization. The change of a liquid into its vapor.

SELF-TESTING QUESTIONS

COMPLETION

1. The general name we give to anything that occupies space and has

mass is _____.

2. The three states of matter are _____, _____
   and _____.

3. The three kinds of matter are _____, _____
   and _____.

4. Two classes of matter that are organized around such physical
   properties as their abilities to conduct electricity or to be
   hammered into sheets are _____ and _____.

5. In the chemical equation: Mg + S → MgS   the symbol(s) of the
   reactant(s) is (are) _____ and of the product(s):
   _____.

6. Referring to the equation of question 5, which is the more
   elementary substance, Mg or MgS?   _____

7. Write the symbols of the following elements.

   (a) calcium   _____   (b) oxygen   _____

   (c) sodium   _____   (d) phosphorus _____

   (e) potassium _____   (f) iron   _____

   (g) chlorine   _____   (h) nitrogen   _____

   (i) magnesium _____   (j) manganese   _____

   (k) lead   _____   (l) copper   _____

   (m) lithium   _____   (n) sulfur   _____

8. Write the names of the following elements.

   (a) Ag _____   (b) H   _____

   (c) I   _____   (d) Hg _____

   (e) Ba _____   (f) Br _____

   (g) Zn _____   (h) Na _____

   (i) K   _____   (j) Mg _____

9. Which two kinds of matter, by definition, obey the law of definite

proportions? _____ and _____

10. Examples that illustrate the law of multiple proportions are limited to which kind of matter? _____

11. Element A and element B, both colorless, odorless gases, are mixed to give a colorless, odorless gas that, when cooled sufficiently, changes to a liquid identified as the liquid form of A with a gas above it identified as B. The mixing of A and B is an example of a _____ change.

12. Element X and element Y, both colorless gases that can be kept in glass containers are mixed. A great deal of heat is generated, and, after cooling, the mixture, still a colorless gas, slowly dissolves the glass container. The mixing of X and Y is an example of a _____ change.

13. If the separate masses of elements X and Y (question 12) are 2 grams each, then according to the law of _____ _____ the mass of the result of mixing X and Y would be _____.

14. The five main postulates of Dalton's Atomic Theory are

   (a) _____

   (b) _____

   (c) _____

   (d) _____

   (e) _____

15. If atoms are indestructable, then the ratios of the atoms present in a compound can be expressed only in _____ numbers.

16. If atoms are present in a compound in a definite ratio by atoms, and if atoms are indestructable, then the atoms must also be present in a definite ratio by _____.

17. In the formula Na_2SO_4, the 4 is called a _____, and the ratio of sulfur atoms to sodium atoms is 1 to _____.

18. A car in motion, by virtue of this motion, possesses _____

_____ energy.

19. If a bar of metal warmed to 40 °C is placed in physical contact with a solid object and the temperature of the metal falls, then the solid object either experiences an increase in its temperature or some of it (or all of it) _____.

20. If a bar of metal warmed to 40 °C is placed in physical contact with a solid object and the temperature of the metal remains the same, then the solid object has the same _____ _____ as the metal.

21. The three ways by which an object, such as a bar of metal, can lose some of its heat to its surroundings are _____, _____, and _____.

22. Which of these would be ruled out if the metal were in a vacuum (a complete absence of air) but still in contact with a colder object? _____.

23. Which of these means of heat loss could still take place even if the bar of metal were in a vacuum and suspended in such a way as to be out of contact with any other object (except for a very thin string holding it suspended that could absorb no heat)? _____.

24. In winter, the uncovered human head loses heat principally by _____ and _____ when little physical activity is in progress.

25. When a substance changes from its liquid state to its vapor state, we call this in general, a _____ change; the specific name for this change is _____.

26. To change a specified quantity of a substance from its liquid to its vapor state requires a certain amount of heat; the amount of heat required per gram is called the _____ of the liquid.

27. If one gram of water at 100 °C is changed to its vapor at 100 °C, _____ calories are absorbed by the water and are present in the water vapor.

28. What happens to the heat in water vapor when the vapor changes back into a liquid at the same temperature? _____ _____

29. Besides radiation, conduction, and convection, the human body loses heat energy by _____.

30. The set of chemical changes which occur in the body even when it is between meals and at complete rest is called the body's _____.

31. If insensible perspiration is insufficient to help cool the body, then _____ occurs.

32. Loss of heat via insensible perspiration occurs at the skin and in the _____.

33. The amount of heat that one gram of an object must absorb in order to change its temperature by one degree Celsius is called the _____ of that object, and it has common units of _____.

34. The amount of heat absorbed by an object in changing only its physical state but not its temperature is called the object's _____ of fusion or _____ of vaporization, and it usually has units of _____.

35. Ice at 0 $^{\circ}$C is better for making a cold pack than water at 0 $^{\circ}$C because the ice can absorb heat by the physical change of _____, which requires _____ calories per gram, whereas the liquid water can absorb heat only by the physical change of _____, which requires only _____ calories per gram per degree Celsius.

36. The amount of heat that a specific object or a specific mass of a substance can absorb while changing temperature by one degree Celsius is called that substance's _____.

37. The temperature at which a substance boils is called its _____ _____.

38. The capacity of an object to transfer heat to its surroundings or take heat from its surroundings without changing its own

temperature is called that object's _____ of fusion
or vaporization.

39. The burning of coal is a _____ change that,
thermally, would be described as _____.

MULTIPLE-CHOICE

1. One of the states of matter is
   (a) a mixture   (b) liquid   (c) element   (d) the atom

2. One of the kinds of matter is
   (a) a mixture   (b) liquid   (c) solid   (d) gas

3. One of the "building blocks" of matter is
   (a) a mixture   (b) liquid   (c) calorie   (d) the atom

4. The correct symbols for calcium, carbon, copper, and chlorine, in
   this order are:
   (a) C, Cl, Ca, Cu   (c) Ca, Cb, Cr, Ch
   (b) Cl, Cu, Ca, C   (d) Ca, C, Cu, Cl

5. The names that correspond to the following symbols for elements N,
   Na, K, Ag are, in the order given:
   (a) silver, sulfur, nitrogen, potassium
   (b) nitrogen, sodium, potassium, silver
   (c) sodium, nitrogen, mercury, aluminum
   (d) neon, iron, phosphorus, argon

6. Water from any source anywhere in the world always has hydrogen
   and oxygen combined in the ratio of 11.1 g of hydrogen to 89.9 g
   of oxygen. This fact illustrates the law of
   (a) definite proportions   (c) conservation of mass
   (b) multiple proportions   (d) conservation of energy

7. Substances that obey the law of definite proportions include all
   (a) elements   (b) compounds   (c) mixtures   (d) both a and b

8. Most of the elements are
   (a) metals   (b) nonmetals   (c) liquids   (d) gases

9. One formula unit of a compound with the formula Al_2O_3 is made of
   (a) 2 aluminum atoms and 6 oxygen atoms

(b) 3 aluminum atoms and 2 oxygen atoms

(c) 6 atoms of all kinds

(d) 2 aluminum atoms and 3 oxygen atoms

10. One of the forms of energy is
    (a) temperature   (b) hypothermia   (c) heat   (d) convection

11. When liquid water at 0 °C changes to ice at 0 °C, the water will give up to the surroundings its
    (a) specific heat   (c) heat of fusion
    (b) heat of vaporization   (d) specific gravity

12. When liquid water at 0 °C changes to ice at 0 °C, the water gives up to the surroundings
    (a) 80 cal/g   (b) 80 kcal/g   (c) 540 cal/g   (d) 540 kcal/g

13. When ethyl chloride is sprayed on the skin to cool it, the heat drawn from the skin goes mostly to the ethyl chloride to
    (a) warm it   (b) evaporate it   (c) sublime it   (d) condense it

14. Ice is a better coolant than water (even when the water is at 0 °C) because ice has a relatively large
    (a) specific heat   (c) heat of vaporization
    (b) heat capacity   (d) heat of fusion

15. When the body allows the evaporation of water to carry away body heat, it takes advantage of water's relatively large
    (a) specific heat   (c) heat of vaporization
    (b) heat capacity   (d) heat of fusion

16. A breeze has a cooling effect because it aids in the loss of body heat by
    (a) convection   (b) conduction   (c) radiation   (d) condensation

17. Metabolism is described best by which (one) word:
    (a) basal   (b) exothermic   (c) endergonic   (d) endothermic

18. The ability of a given mass of a substance to absorb heat while undergoing a change in temperature is called its
    (a) latent heat   (c) exothermic capacity
    (b) heat capacity   (d) kinetic energy

19. If atoms of A and of B chemically combine in a ratio of 1:1 by

atoms but a ratio of 2:1 by weight, then compared to the atoms of B those of A are
(a) half as heavy
(b) equal in weight
(c) twice as heavy
(d) three times as heavy

20. If 50 g of water underwent a temperature change from 15 $^{\circ}$C to 25 $^{\circ}$C, the water
(a) absorbed 500 calories
(b) absorbed 500 kcal
(c) released 500 calories
(d) released 500 kcal

21. Dalton's theory proposed that matter is made of particles called
(a) elements   (b) atoms   (c) mixtures   (d) compounds

22. Hot steam at 100 $^{\circ}$C is more dangerous to exposed skin than hot water at 100 $^{\circ}$C because the steam contains the water's
(a) heat of vaporization
(b) specific heat
(c) kinetic energy
(d) sensible heat

23. To change one gram of steam at 100 $^{\circ}$C to one gram of ice at 0 $^{\circ}$C, one would have to remove how many calories from the water?
(a) 640 cal   (b) 180 cal   (c) 720 cal   (d) 100 cal

24. How much heat does it take to warm a 500-g piece of granite from 15.5 $^{\circ}$C to 37.0 $^{\circ}$C? The specific heat of granite is

$0.192 \dfrac{cal}{g \; ^{\circ}C}$.

(a) 21.5 cal   (b) 2.06 x 10^3 cal   (c) 4.66 cal   (d) 1.08 x 10^4 cal

25. When 3.0 g of benzene freezes, how much heat is released to the surroundings? The heat of fusion of benzene is 30 cal/g.
(a) 10 cal   (b) 0.10 cal   (c) 3.0 cal   (d) 90 cal

ANSWERS

ANSWERS TO SELF-TESTING QUESTIONS

Completion

1. matter
2. solid, liquid, and gas
3. elements, compounds, and mixtures
4. metals and nonmetals

5. Mg and S; MgS
6. Mg
7. (a) Ca   (b) O   (c) Na
   (d) P   (e) K   (f) Fe
   (g) Cl   (h) N   (i) Mg
   (j) Mn   (k) Pb   (1) Cu
   (m) Li   (n) S
8. (a) silver   (b) hydrogen
   (c) iodine   (d) mercury
   (e) barium   (f) bromine
   (g) zinc   (h) sodium
   (i) potassium   (j) magnesium
9. elements and compounds
10. compounds
11. physical
12. chemical
13. conservation of mass; 4 grams
14. (a) Matter consists of atoms.
    (b) Atoms are indestructible. (Even in chemical change they only
        rearrange.)
    (c) Atoms of the same element are identical, particularly in mass.
    (d) Atoms of different elements differ in mass.
    (e) In compounds, atoms occur in definite ratios or proportions.
15. whole
16. mass
17. subscript; 2
18. kinetic
19. melts
20. temperature
21. radiation, conduction, and convection
22. convection
23. radiation
24. radiation and convection
25. physical; evaporation
26. heat of vaporization
27. 540 (rounded from 539.6)
28. The heat is released to the surrounding air.
29. evaporation (or perspiration)
30. basal metabolism
31. sensible perspiration
32. lungs
33. specific heat; calories per gram per degree (or calories per
    degree per gram)

34. heat; heat; calories per gram
35. melting; 80 (rounded from 79.6); an increase in temperature; 1.0
36. heat capacity
37. boiling point
38. heat
39. chemical; exothermic

Multiple-Choice

1. b		2. a		3. d	
4. d		5. b		6. a	
7. d		8. a		9. d	
10. c		11. c		12. a	
13. b (Note 1)		14. d (Note 2)		15. c	
16. a		17. b		18. b	
19. c		20. a		21. b	
22. a (Note 3)		23. c (Note 4)		24. b (Note 5)	
25. d (Note 6)					

NOTES

1. Some warming (choice a) also occurs, but heats of vaporization are generally much greater than specific heats. Only the specific heat of ethyl chloride is involved in warming; the heat of vaporization is involved in evaporation.

2. Ice can cool an object in contact with it simply by drawing heat from the object in order to melt, and ice's heat of fusion is much larger than liquid water's specific heat.

3. Both forms, of course, are very dangerous. However, cooler skin in contact with the steam would make some of the steam condense, releasing the heat of vaporization which is relatively large. This extra heat could then cause more injury.

4.

heat of vaporization	540 cal/g
cooling liquid water by 100 degrees at 1 cal/deg/g	100 cal/g
heat of fusion	80 cal/g
Total	720 cal/g

5. $\Delta t = (37.0 - 15.5)\ ^\circ C = 21.5\ ^\circ C$. Then, $0.192\ \frac{cal}{g\cdot{}^\circ C} \times 500\ g \times 21.5\ ^\circ C = 2064$ cal, which has to be rounded and expressed to show 3 significant figures.

6. $30\ \frac{cal}{g} \times 3\ g = 90$ cal

3 Atomic Theory and the Periodic System of the Elements

OBJECTIVES

While this chapter may seem remote from our theme — the molecular basis of life — mastering its contents will be of vital importance in understanding the information presented in the rest of the book.

The specific objectives of this chapter can be classified under three main objectives.

1. Continue to accumulate the basic vocabulary of chemistry. Most of the terms in the glossary will be used again and again throughout the course, particularly in the next chapter. They must be mastered.
2. Learn how to write the electron configuration for an atom of any element with atomic number from 1 to 20. We need a thorough knowledge of electron configurations in order to be able to understand how these elements form various kinds of chemical bonds.
3. Learn how the elements can be sorted into families having somewhat similar properties. The Periodic Table will be a helpful aid in correlating the properties of the elements as we proceed to later chapters.

After studying this chapter and working all of the assigned exercises, you should be able to do the following.

1. Name and give the relative masses and electric charges for the three subatomic particles of interest to chemists.
2. Give Rutherford's contribution to atomic theory.
3. State Bohr's two fundamental postulates and describe his atomic model.
4. Define and distinguish between: (a) atom and element; (b) orbit and orbital; (c) period and group; (d) atomic number and atomic weight; (e) the Bohr model and the orbital model; and (f) an atom and its isotope.
5. Use atomic numbers and mass numbers to find the charge on the nucleus, the numbers of electrons, protons, and neutrons in one atom of any given isotope, and to write the correct symbol for the isotope.
6. Use the atomic number and atomic weight to construct an electron configuration, including the composition of the nucleus, for an atom of any element from 1 through 20, showing the distribution among the orbitals.
7. Repeat objective six for an atom that would reasonably be an isotope of any given atom haveing an atomic number of 1 through 20.
8. Illustrate the application of Pauli's Principle and Hund's Rule.
9. Tell what is meant by an atomic weight and explain why atomic weights are almost never whole numbers like mass numbers.
10. Explain what a representative element is.
11. Given only the atomic number of an element (among the first 20), write the electron configuration of the element occurring next to it, on either side, in the Periodic Table.
12. Define all of the terms in the Glossary.

GLOSSARY

Alkali Metals. The elements of Group IA of the Periodic Table – lithium, sodium, potassium, rubidium, cesium, and francium.

Alkaline Earth Metals. The elements of Group IIA of the Periodic Table – beryllium, magnesium, calcium, strontium, barium, and radium.

Atom. A small particle with one nucleus and zero charge; the smallest particle of a given element that bears the chemical properties of the element.

Atomic Mass Number. (See Mass Number.)

Atomic Mass Unit (amu). 1.66606×10^{-24} g. A mass very close to that of a proton or a neutron.

Atomic Number. The positive charge on an atom's nucleus; the number of protons in an atom's nucleus.

Atomic Orbital. A region in space close to an atom's nucleus in which one or two electrons can reside.

Atomic Weight. The average mass, in amu, of the atoms of the isotopes of a given element as they occur naturally.

Aufbau Rules. The rules for constructing electron configurations.

Binary Compound. A compound made from just two elements.

Bohr Model of the Atom. The solar system model of the structure of an atom, proposed by Niels Bohr, that pictures the electrons circling the nucleus in discrete energy states called orbits.

Carbon Family. The Group IVA elements in the Periodic Table – carbon, silicon, germanium, tin, and lead.

Electron. A subatomic particle that bears one unit of negative charge and has a mass that is 1/1836 the mass of a proton.

Electron Cloud. A mental model that views the one or two rapidly moving electrons of an orbital as creating a cloud-like distribution of negative charge.

Electron Configuration. The most stable arrangement (that is, the arrangement of lowest energy) of the electrons of an atom, ion, or molecule.

Electron Shell. An alternative name for principal energy level.

Element. A substance that cannot be broken down into anything that is both stable and more simple; a substance in which all of the atoms have the same atomic number and the same electron configuration; one of the three broad kinds of matter, the others being compounds and mixtures.

Group. A vertical column in the Periodic Table; a family of elements.

Halogens. The elements of Group VIIA of the Periodic Table — fluorine, chlorine, bromine, iodine, and astatine.

Heisenberg Uncertainty Principle. It is impossible simultaneously to determine with precision and accuracy both the position and the velocity of an electron.

Hund's Rule. Electrons become distributed among different orbitals of the same general energy level insofar as there is room.

Inner Transition Elements.   The elements of the lanthanide and actinide series of the Periodic Table.

Isotope. A substance in which all of the atoms are identical in atomic number, mass number, and electron configuration.

Mass Number. The sum of the numbers of protons and neutrons in one atom of an isotope.

Metalloids. Elements that have some metallic and some nonmetallic properties.

Model, Scientific. A mental construction, often involving pictures or diagrams, that is used to explain a number of facts.

Neutron. An electrically neutral subatomic particle with a mass of 1 amu.

Nitrogen Family. The elements of Group VA of the Periodic Table — nitrogen, phosphorus, arsenic, antimony, and bismuth.

Noble Gases. The elements of Group 0 of the Periodic Table — helium, neon, argon, krypton, xenon, and radon.

Nucleus. In chemistry and physics, the subatomic particle that serves as the core of an atom and that is made up of protons and neutrons.

Orbital. (See Atomic Orbital.)

Outside Level. In an atom the highest principal energy level that holds at least one electron.

Oxygen Family. The elements in Group VIA of the Periodic Table —
oxygen, sulfur, selenium, tellurium, and polonium.

Pauli Exclusion Principle. No more than two electrons can occupy the
same orbital at the same time, and two can be present only if they
have opposite spin.

Period. A horizontal row in the Periodic Table.

Periodic Law. Many properties of the elements are periodic functions
of their atomic numbers.

Periodic Table. A display of the elements that emphasizes the family
relationships.

Photon. A package of energy released when an electron in an atom moves
from a higher to a lower energy state; a unit of light energy.

Principal Energy Level. A space near an atomic nucleus where there
are one or more sublevels and orbitals in which electrons can
reside; an electron shell.

Proton. A subatomic particle that bears one unit of positive charge
and has a mass of 1 amu.

Quantum. A quantity of energy possessed by a photon.

Representative Element. Any element in any A-group of the Periodic
Table; any element in Groups IA — VIIA and those in Group 0.

Subatomic Particle. An electron, a proton, or a neutron; the atomic
nucleus as a whole is also a subatomic particle.

Sublevel. A region that makes up part (sometimes all) of a principal
energy level and that can itself be subdivided into individual
orbitals.

Transition Elements. The elements between those of Group IIA and Group
IIIA in the long periods of the Periodic Table; a metallic element
other than one in Group IA or IIA or in the actinide or lanthanide
families.

SELF-TESTING QUESTIONS

COMPLETION.   Some of the sentences in these exercises may well be completed by various answers. If you have picked an answer not given at the end of this chapter, remember that the sentence as completed by the answer given is meant to give you a new slant on the concept. Let it work that way for you. The object is not to get a score of 100. The object is to understand and then to know.

1. The names of the three subatomic particles are _____, _____, and _____.

2. A larger subatomic particle present in all atoms and discovered by Rutherford is the _____.

3. A substance whose atomic nuclei all have the same amount of electric charge is called _____.

4. In units of amu, the mass of the proton is _____, of the neutron is _____, and of the electron is _____.

5. The mass number equals the sum of the _____ and _____.

6. The difference between the mass number and the atomic number of an atom corresponds to _____.

7. If X and Y are isotopes and the symbol of one is $^{14}_{7}X$, then the symbol of the other might reasonably be what? _____

8. Which number can be used to determine how many electrons must be arranged in writing an electron configuration, the atomic number or the mass number? _____

9. The two fundamental postulates made by Niels Bohr concerning electron configurations and atomic structure were:

   (a) _____

   (b) _____

10. Of the two terms, orbit and orbital, which describes a region of

space in the vicinity of a nucleus? _____ Which
describes a particular path around the nucleus? _____

11. When one electron jumps from a higher energy level to a lower
energy level it emits one _____

12. According to the Heisenberg uncertainty principle, it is not
possible to determine with precision and accuracy both the
_____ and the _____ of an electron
simultaneously.

13. Complete the following table by writing in the numbers of the
different kinds of sublevels and the total number of atomic
orbitals at each given principal or main level.

Main Level	Number of Different Kinds of Sublevels	Total Number of Atomic Orbitals
1	_____	_____
2	_____	_____
3	_____	_____

14. A p-type sublevel consists of how many orbitals? _____

15. According to the Pauli exclusion principle, an orbital can hold a
maximum of how many electrons? _____ What must be true
about the electrons corresponding to this maximum if they actually
are in the same orbital? _____

16. If the electron configuration of an atom is $1s^2 2s^2 2p_x^2 2p_y^2 2p_z^1$ then
its atomic number is _____.

17. The total number of electrons in the outside level of the atom
whose electron configuration was given in question 16 is _____.

18. Write the electron configuration of an element of atomic number

(a) 14 _____

(b) 19 _____

19. Write the electron configuration of the element with the lowest
atomic number that illustrates Hund's rule. _____

20. As a first (and very close) approximation, when the mass numbers of all of the isotopes present in an element are averaged, due consideration given to the relative abundances of the isotopes, the average mass is called the element's _____.

21. The scientist given the most credit for recognizing the periodic relations among the elements was _____.

22. Metallic elements generally have _____, _____ or _____ electrons in their outside levels.

23. Elements whose outside levels all have seven electrons are in Group _____.

24. As one crosses from left to right in the first main period of the Periodic Table, energy level number _____ is filling.

25. The periodic law states that _____
_____.

26. The representative elements in Group IA are called _____; those in VIIA are _____; those in IIA are the _____.

27. Referring to the Periodic Table, the symbols of the elements in the carbon family are, in order _____.

28. The atoms of the oxygen family all have _____ electrons in their highest occupied main energy levels.

29. A very large group of elements occurring roughly in the center of the Periodic Table are called _____ elements.

30. Referring to the Periodic Table, the element of atomic number 79 is likely a metal or a nonmetal? _____

31. The formula of calcium oxide is CaO. Considering the Periodic Table, the formula of magnesium oxide is likely _____.

MULTIPLE-CHOICE

1. Lord Rutherford's contribution to our understanding of atomic structure was that atoms

(a) are like solar systems   (c) have dense inner cores

(b) contain electrons   (d) have neutrons

2. Bohr postulated that electrons occur in atoms
   (a) in discrete orbitals
   (b) in discrete energy states
   (c) in a condition of constant shifting between states
   (d) in hard, dense, inner cores

3. An atom with 20 protons would have
   (a) a nuclear charge of 20+
   (b) a nuclear charge of 20−
   (c) 20 electrons
   (d) 40 neutrons

4. An element whose atoms have 10 protons, 10 neutrons, and 10 electrons would have an atomic number of
   (a) 10   (b) 20   (c) 30   (d) 40

5. All of the atoms of any naturally occurring element will have identical numbers of
   (a) protons   (b) nuclei   (c) neutrons   (d) electrons

Problems 6 through 10 should be answered without reference to tables or charts.

6. An element of atomic number 19 consists of atoms having how many electrons in their outside levels?
   (a) eight   (b) nine   (c) one   (d) 19

7. An element of atomic number 14 is probably
   (a) a metal   (c) a transition element
   (b) a nonmetal   (d) a noble gas

8. Standing immediately to the left of the element of atomic number 17 in the Periodic Table is an element with
   (a) a charge of 16+ on its nuclei
   (b) a charge of 18+ on its nuclei
   (c) a charge of 17+ on its nuclei
   (d) a charge of 9+ on its nuclei

9. Standing immediately below the element of atomic number 17 in the Periodic Table is an element with

    (a) 17 electrons in its outside level
    (b) 7 electrons in its outside level
    (c) 17 protons in its nuclei
    (d) 7 protons in its nuclei

10. If element X is a gas, then the element immediately above it in the Periodic Table is most likely a
    (a) metal   (b) nonmetal   (c) gas   (d) transition element

11. An element whose atoms have one, two, or three electrons in their outside levels is probably
    (a) a nonmetal   (b) a gas   (c) a liquid   (d) a metal

12. An element whose atoms have 7 or 8 electrons in their outside levels is probably
    (a) a nonmetal (b) a solid (c) a transition element (d) a metal

13. If an element of atomic number 20 has two isotopes and one isotope had 20 neutrons while the other had 22 neutrons, and if these isotopes were present in a 50:50 ratio, what would be the atomic weight of the element as determined chemically?
    (a) 20.5   (b) 21   (c) 40.5   (d) 41

14. The 3s orbital of the element of atomic number 20 has how many electrons?
    (a) 0   (b) 1   (c) 2   (d) 10

ANSWERS

ANSWERS TO SELF-TESTING QUESTIONS

Completion

1. electrons, protons, and neutrons
2. nucleus
3. an element
4. 1 amu; 1 amu; 1/1836 amu
5. neutrons and protons
6. the number of neutrons

7. $^{15}_{7}Y$ or $^{13}_{7}Y$. (What is important is that the left subscript is 7 and that the left superscript is not 14 but is near 14.)

8. atomic number
9. (a) Electrons may be in an atom only in certain allowed energy
      states.
   (b) An atom neither absorbs nor radiates energy so long as its
       electrons remain in their basic energy states.
10. orbital, orbit
11. quantum of energy
12. position, velocity
13. 1   1   1
      2   2   4
      3   3   9
14. three
15. two; they must have opposite spin.
16. 9
17. 7
18. (a) $1s^2 2s^2 2p_x^2 2p_y^2 2p_z^2 3s^2 3p_x^1 3p_y^1$ (b) $1s^2 2s^2 2p^6 3s^2 3p^6 4s^1$
19. $1s^2 2s^2 2p_x^1 2p_y^1$
20. atomic weight
21. Dimitri Mendeleev
22. one, two, three
23. VIIA
24. Two
25. The properties of the elements are a periodic function of their
    atomic numbers.
26. the alkali metals; the halogens; the alkaline earth metals
27. C, Si, Ge, Sn, Pb
28. 6
29. transition
30. metal
31. MgO

Multiple-Choice

1. c	2. b	3. a and c
4. a	5. a, b, and d	6. c (Note)
7. b (Note)	8. a	9. b
10. b and c	11. d	12. a
13. d	14. c	

NOTE. Problems 6 and 7 illustrate problems that are solved by first
writing an electronic configuration (main levels only) and then
applying the rules that correlate the number of electrons in the
outside level with particular properties.

4 Chemical Compounds and Chemical Bonds

OBJECTIVES

Here we will complete our basic introduction to the structure of matter. In the last chapter, we found out what holds subatomic particles together in atoms. In this chapter, we will answer the following questions: What holds atoms together in compounds? What determines the ratios in which atoms of different elements join together? Why do only certain specific ratios occur and not others?

Broadly speaking, the main topics in this chapter are chemical bonds (ionic and covalent) and types of compounds (ionic and molecular). The ideas and the vocabulary introduced are basic. Terms such as ion, molecule, polar molecule, molecular compound, and ionic compound will be used frequently in succeeding chapters. We cannot speak of a molecular basis of life without knowing what a molecule is.

After you have studied this chapter and worked the exercises in it, you should be able to do the following.

1. Name two kinds of compounds and describe the kinds of particles that make up each type.
2. Define and explain the difference between (a) an atom, an ion, and a molecule and (b) an ionic bond and a covalent bond.

3. Give the electrical charges of all the ions in Tables 4.1 and 4.3.
4. Among the representative elements that can form ions, use their location in the Periodic Table to tell what charges their ions have.
5. Write formulas from the names of ionic compounds and names from their formulas, limiting ourselves to information in Tables 4.1 and 4.3.
6. Use covalence numbers to write structures of molecular compounds if given only their molecular formulas.
7. Use the octet rule to predict the likely oxidation number or covalence for any element of atomic number one through twenty from its electronic configuration or its location in the Periodic Table.
8. Explain a covalent bond in terms of the formation of a molecular orbital.
9. Explain the difference between a covalent bond and a coordinate covalent bond.
10. Define "oxidation" and "reduction" and give an illustration of each.
11. Define "electronegativity" and use its values, when supplied, to predict if given bonds will be polar or nonpolar.
12. Explain how molecules can sometimes stick to each other even though they are electrically neutral.
13. Define all of the terms in the Glossary.

GLOSSARY

Acid. Any substance that makes H^+ available in water.

Base. A compound that provides the hydroxide ion.

Bond, Chemical. A net electrical force of attraction that holds atomic nuclei near each other within compounds.

Coordinate Covalent Bond. A covalent bond in which both of the electrons of the shared pair originated from one of the atoms involved in the bond.

Covalence Number. The number of electron-pair (covalent) bonds that an atom can have in a molecule.

Covalent Bond. The net force of attraction that arises as two atomic nuclei share a pair of electrons. One pair is shared in a single bond, two pair in a double bond, and three electron pairs are shared in a triple bond.

Diatomic Molecule. A molecule made of two atoms.

Dipole, Electrical. A pair of equal but opposite (and usually partial) electrical charges separated by a small distance in a molecule.

Electrolyte. Any substance whose solution in water conducts electricity; or the solution itself of such a substance.

Electronegativity. The ability of an atom joined to another by a covalent bond to attract the electrons of the bond toward itself.

Electron Sharing. The joint attraction of two atomic nuceli toward a pair of electrons situated between the nuclei and between which, therefore, a covalent bond exists.

Formula, Empirical. A chemical symbol for a compound that gives just the ratios of the atoms and not necessarily the composition of a complete molecule.

Formula, Molecular. A chemical symbol for a substance that gives the composition of a complete molecule.

Formula, Structural. A chemical symbol for a substance that uses atomic symbols and lines to describe the pattern in which the atoms are joined together in a molecule.

Ion. An electrically charged, atomic or molecular-sized particle; a particle that has one or a few atomic nuclei and either one or two (seldom, three) too many or too few electrons to render the particle electrically neutral.

Ionic Bond. The force of attraction between oppositely charged ions in an ionic compound.

Ionic Compound. A compound that consists of an orderly aggregation of oppositely charged ions that assemble in whatever ratio ensures overall electrical neutrality.

Molecular Compound. A compound whose smallest representative particle is a molecule; a covalent compound.

Molecular Orbital. A region in the space that envelopes two (or sometimes more) atomic nuclei where a shared pair of electrons of a covalent bond resides.

Molecule. An electrically neutral (but often polar) particle made up of the nuclei and electrons of two or more atoms and held together by covalent bonds; the smallest representative sample of a molecular compound.

Neutralization, Acid-Base. A reaction between an acid and a base.

Octet, Outer. A condition of an atom or ion in which its highest occupied energy level has eight electrons – a condition of stability.

Octet Rule. The atoms of a reactive element tend to undergo those chemical reactions that most directly give them the electron configuration of the noble gas that stands nearest the element in the Periodic Table (all but one of which have outer octets).

Orbital Overlap. The interpenetration of one atomic orbital by another from an adjacent atom to form a molecular orbital.

Oxidation. The loss of one or more electrons from an atom, molecule, or ion.

Oxidation Number. For simple monoatomic ions, the quantity and sign of the electrical charge on the ion.

Oxidizing Agent. A substance that can cause an oxidation.

Polar Bond. A bond at which we can write $\delta+$ at one end and $\delta-$ at the other end, the end that has the more electronegative atom.

Polar Molecule. A molecule that has sites of partial positive and partial negative charge and therefore a permanent electrical dipole.

Polyatomic Ion. Any ion made from two or more atoms, such as OH^-, SO_4^{2-}, and CO_3^{2-}.

Redox Reaction. Abbreviation of reduction-oxidation reaction; a reac-
   in which electrons transfer between reactants.

Reducing Agent. A substance that can cause another to be reduced.

Reduction. The gain of one or more electrons by an atom, ion, or
   molecule.

Structure. Synonym for structural formula.

Subscripts. Numbers placed to the right and a half space below the
   atomic symbols in a chemical formula.

SELF-TESTING QUESTIONS

COMPLETION

1. The fundamental physical force that is responsible for all kinds
   of chemical bonds is an _____ force.

2. A small particle with one or a few atomic nuclei but bearing a net
   negative or positive charge is called _____.

3. A compound consisting of an orderly aggregation of oppositely
   charged ions is classified as _____.

4. The formula unit of a molecular compound is called _____.

5. When the nuclei and electrons of atoms are reorganized to form
   particles of opposite electrical charge, the product is classified
   as _____ compound.

6. The net force of attraction that operates in an ionic compound to
   keep the ions from flying apart has the special name of
   _____.

7. The reaction of lithium with chlorine, $2Li + Cl_2 \rightarrow 2LiCl$, changes

   the lithium atom (Li) to the _____, which has the

   symbol _____. It changes the chlorine atom (in Cl_2) into

_____, for which the symbol is _____. In this oxidation-reduction reaction, the lithium atom is _____

(reduced or

_____ and the chlorine atom is _____.

oxidized)

The oxidizing agent is _____ and its symbol is

_____. The reducing agent is _____ and its symbol is

(name)

_____.

8. The formula unit of an ionic compound consists of a pair or a small cluster of _____.

9. The ions Al^{3+} and SO_4^{2-} would aggregate in a ratio of _____ ions of Al^{3+} to _____ ions of SO_4^{2-}.

10. They must aggregate in that ratio because chemical compounds are, in general, electrically _____.

11. Tests how well you have learned the names and formulas for the ions in Tables 4.1 and 4.3 in the text by completing the following. Give the formula for each ion; for example,

sulfite ion $\underline{SO_3^{2-}}$

(You must include the electric charge, for instance, SO_3 by itself is the symbol of sulfur trioxide, an air pollutant.)

(a) chloride ion _____

(b) sodium ion _____

(c) magnesium ion _____

(d) hydroxide ion _____

(e) sulfate ion _____

(f) potassium ion _____

(g) iodide ion _____

(h) calcium ion _____

(i) hydronium ion _____

(j) nitrate ion _____

(k) hydrogen sulfate ion _____

(1) permanganate ion _____

(m) bromide ion _____

(n) nitrite ion _____

(o) ammonium ion _____

(p) dihydrogen phosphate ion _____

12. Using your knowledge of the names and formulas of ions, write formulas for each of the following ionic compounds.

(a) sodium iodide _____

(b) sodium sulfate _____

(c) sodium phosphate _____

(d) sodium nitrate _____

(e) magnesium sulfate _____

(f) calcium sulfate _____

(g) potassium sulfate _____

(h) aluminum sulfate _____

(i) ammonium nitrate _____

(j) ammonium sulfate _____

(k) calcium phosphate _____

(1) sodium monohydrogen phosphate _____

(m) potassium nitrite _____

(n) silver nitrate _____

(o) potassium permanganate _____

(p) iron(III) oxide _____

13. Write the name of each of the following compounds.

(a) NaBr _____

(b) Na_2SO_4 _____

(c) $NaNO_3$ _____

(d) Na_2CO_3 _____

(e) $NaMnO_4$ _____

(f) Na_3PO_4 _____

(g) $NaHSO_4$ _____

(h) KH_2PO_4 _____

(i) KCl _____

(j) $KHCO_3$ _____

(k) KNO_2 _____

(l) $MgCl_2$ _____

(m) $MgHPO_4$ _____

(n) $Ca(NO_3)_2$ _____

(o) $CaSO_4$ _____

(p) $(NH_4)_2SO_4$ _____

(q) NH_4Cl _____

(r) NH_4NO_3 _____

(s) $FeCl_3$ _____

(t) $HgCl_2$ _____

(u) $AgCl$ _____

(v) $FeCl_2$ _____

(w) $HgCl$ _____

(x) K_2CO_3 _____

(y) $CaCO_3$ _____

(z) $MgCO_3$ _____

(aa) $NaHCO_3$ _____

(bb) $LiHCO_3$ _____

(cc) $Ca(HCO_3)_2$ _____

(dd) $Al_2(SO_4)_3$ _____

14. Being guided by the locations of the following elements in the

Periodic Table, write the correct symbols for the most reasonable ions these elements could form. Be sure to show the electrical charge. If an ionic existence is highly unlikely, write "no ion." The numbers in parentheses are the atomic numbers of the elements.

(a) Cs (55) _____   (c) Sr (38) _____

(b) I (53) _____   (d) Xe (54) _____

15. Each of the following formulas represents an ionic compound, and you should know the electrical charge on one of the ions involved. Using this knowledge, figure out the electrical charge on the ion specified. Write the correct symbol of the ion.

(a) Rb_2SO_4 Rb ion   _____

(b) $Pb(NO_3)_2$ Pb ion   _____

(c) $Cd_3(PO_4)_2$ Cd ion   _____

(d) $NiCO_3$ Ni ion   _____

16. The particle that results when two or more atoms have their nuclei and electrons reorganized so that some electrons are shared between the nuclei is called _____.

17. A small particle having two or more nuclei that is capable of isolated existence and that is electrically neutral is called _____.

18. A small particle that is the smallest representative sample of a molecular compound is called _____.

19. The net electrical force of attraction that operates in molecules to keep their nuclei from flying apart has the special name of _____.

20. If two electrically neutral molecules can still attract each other, they must be _____.

21. When two atomic orbitals overlap the result is a _____ _____; this will make possible a _____ bond provided it is occupied by two _____.

22. Using their atomic symbols, arrange these elements in the order of

their relative electronegativities: carbon, oxygen, nitrogen, and hydrogen.

⟨   ⟨   ⟨

----

    least   most
electronegative   electronegative

23. If element X has a relative electronegativity of 3.5 and that of Y is 2.5, then in the molecule X–Y, the $\delta+$ will be close to _____ and the $\delta-$ will be close to _____.
   (X or Y)   (X or Y)

24. The covalence of oxygen is two, not three or four, because the outside level of an oxygen atom has _____ electrons, just _____ electrons short an outer octet.

25. Complete the following table according to the pattern provided by the example. Write the electron configuration (main levels only) of the atom whose atomic number is given. Deduce if the atom can form an ion; if so, write the formula of the ion using just the general symbol X for the atom in each case. Deduce next if the atom can participate in covalent bond formation; if so, write the most reasonable covalence number it will have on the basis of the octet theory. (Many elements have multiple covalences, but we will not study this here.)

	Atomic Number	Electron Configuration (Main Levels)[a]	Symbol of Ion (if any)	Covalence Number (if any)
Example	16	2 8 6	X^{2-}	2
(a)	9	_____	_____	_____
(b)	20	_____	_____	_____
(c)	5	_____	_____	_____
(d)	17	_____	_____	_____
(e)	8	_____	_____	_____
(f)	13	_____	_____	_____

[a]In order, left to right, of level 1, level 2, etc.

26. Because selenium, Se, is in the same family of the Periodic Table as oxygen, one reasonable covalence of selenium must be _____ because members of the same family have the identical number of electrons in their _____.

27. Using your knowledge of covalence numbers, write molecular formulas of the compounds of each of the following elements combined with hydrogen; for example.

    oxygen H_2O

    (a) sulfur   _____

    (b) chlorine   _____

    (c) nitrogen   _____

    (d) iodine   _____

    (e) arsenic, As   _____
        (arsenic is in the same family as nitrogen)

    (f) phosphorus, P   _____
        (phosphorus is in the same family as nitrogen)

    (g) silicon, Si   _____
        (silicon is in the same family as carbon)

MULTIPLE-CHOICE

1. If an atom, X, can accept two electrons from another atom in the process of changing into a relatively stable ion, then an atom of X
   (a) has two protons in its nuclei
   (b) is a metal
   (c) will be in Group IIA in the Periodic Table
   (d) is also capable of forming a molecule of the formula H_2X

2. Atoms of elements with atomic numbers 20 and 17 are known to combine. If X is the symbol of element 20 and Y is the symbol of element 17, the correct formula for the most stable compound between these two elements is
   (a) X_2Y (b) XY_2 (c) YX_2 (d) XY

3. Pure substances classified as molecular are characterized as being
   (a) mixtures   (b) made of oppositely charged ions
   (c) aggregations of molecules   (c) metallic

4. The type of bonding found in sodium bromide is
   (a) ionic   (b) covalent   (c) coordinate   (d) nonpolar

5. A particle having eleven protons and ten electrons would bear a charge of
   (a) −1   (b) 0   (c) +1   (d) +11

6. If elements X and Y are in the same family and X forms the compound Na_2X, then one of the compounds of Y would be
   (a) K_2Y (b) CaY   (c) Na_2Y (d) Al_2Y_3

7. If X just precedes Y in a horizontal row of the Periodic Table (i.e., the atomic number of X is one less than that of Y), and if X forms the compound HX, then a compound of Y would be
   (a) HY   (b) H_2Y (c) H_3Y (d) Y is a noble gas; therefore it will form no compound.

8. Each formula unit of $Mg(H_2PO_4)_2$ contains a total of how many atomic nuclei?
   (a) sixteen   (b) fifteen   (c) eight   (d) fourteen

9. In the equation, $2Na + Cl_2 \rightarrow 2NaCl$, the underlined number is called a
   (a) subscript   (c) multiplier
   (b) superscript   (d) coefficient

10. The symbol Fe is the symbol for
    (a) an iron atom   (c) an iron molecule
    (b) an iron ion   (d) the element iron

11. The symbol $3Cl_2$ is the symbol for
    (a) three molecules of chlorine   (c) three grams of chlorine
    (b) three atoms of chlorine   (d) six molecules of chlorine

12. In the equation, $Zn + 2AgNO_3 \rightarrow Zn(NO_3)_2 + 2Ag$
    (a) the equation is balanced
    (b) the equation is not balanced
    (c) Zn is oxidized
    (d) Ag^+ is reduced

13. In the equation of problem 12,
    (a) Zn is an oxidizing agent
    (b) Zn is a reducing agent

(c) Ag^+ is an oxidizing agent

(d) Ag^+ is a reducing agent

14. The substance calcium chloride, $CaCl_2$, is an aggregation of which particles?
    (a) calcium atoms and chlorine molecules (Cl_2)
    (b) calcium atoms and chlorine atoms
    (c) calcium ions and chlorine molecules
    (d) calcium ions and chloride ions

15. On the basis of electronegativities, the bonds in the molecule NH_3 are in which state of polarity?

$$
\begin{array}{llll}
\overset{\delta+}{} \quad \overset{\delta-}{} & \overset{+}{} \quad \overset{-}{} & \overset{\delta-}{} \quad \overset{\delta+}{} & \overset{-}{} \quad \overset{+}{} \\
\text{(a) } N-H & \text{(b) } N-H & \text{(c) } N-H & \text{(d) } N-H
\end{array}
$$

16. If element A has a relative electronegativity of 3.7 and that of B is 2.7, and the molecule whose formula is A_2B is nonpolar, then the structure of A_2B is most likely
    (a) $A-A-B$ (b) $A-A$ (c) B (d) $A-B-A$
                              $\backslash$ $/\,\backslash$
                              B A A

17. A substance, Z, consists of molecules. Z is a gas at room temperature and remains a gas even at $-140\ ^\circ C$. The molecules of Z, therefore, are most likely
    (a) very polar   (c) relatively nonpolar
    (b) made of oppositely charged ions   (d) negative ions

ANSWERS

ANSWERS TO SELF-TESTING QUESTIONS

Completion

1. electrical
2. an ion
3. an ionic compound
4. a molecule
5. an ionic compound
6. the ionic bond
7. lithium ion, Li^+; chloride ions, Cl^-; oxidized, reduced; chlorine, Cl_2; lithium, Li

8. oppositely charged ions
9. two, three
10. neutral
11. (a) Cl^- (b) Na^+ (c) Mg^{2+} (d) OH^-
    (e) SO_4^{2-} (f) K^+ (g) I^- (h) Ca^{2+}
    (i) H_3O^+ (j) NO_3^- (k) HSO_4^- (l) MnO_4^-
    (m) Br^- (n) NO_2^- (o) NH_4^+ (p) $H_2PO_4^-$

12. (a) NaI (b) Na_2SO_4 (c) Na_3PO_4
    (d) $NaNO_3$ (e) $MgSO_4$ (f) $CaSO_4$
    (g) K_2SO_4 (h) $Al_2(SO_4)_3$ (i) NH_4NO_3
    (j) $(NH_4)_2SO_4$ (k) $Ca_3(PO_4)_2$ (l) Na_2HPO_4
    (m) KNO_2 (n) $AgNO_3$ (o) $KMnO_4$
    (p) Fe_2O_3

13. (a) sodium bromide
    (b) sodium sulfate
    (c) sodium nitrate
    (d) sodium carbonate
    (e) sodium permanganate
    (f) sodium phosphate
    (g) sodium hydrogen sulfate
    (h) potassium dihydrogen phosphate
    (i) potassium chloride
    (j) potassium bicarbonate
    (k) potassium nitrite
    (l) magnesium chloride
    (m) magnesium monohydrogen phosphate
    (n) calcium nitrate
    (o) calcium sulfate
    (p) ammonium sulfate
    (q) ammonium chloride
    (r) ammonium nitrate
    (s) iron(III) chloride [ferric chloride]
    (t) mercury(II) chloride [mercuric chloride]
    (u) silver chloride
    (v) iron(II) chloride [ferrous chloride]
    (w) mercury(I) chloride [mercurous chloride]
    (x) potassium carbonate
    (y) calcium carbonate
    (z) magnesium carbonate
    (aa) sodium bicarbonate
    (bb) lithium bicarbonate

(cc) calcium bicarbonate
(dd) aluminum sulfate

14. (a) Cs^+ (b) I^- (c) Sr^{2+} (d) no ion
15. (a) Rb^+ (b) Pb^{2+} (d) Cd^{2+} (d) Ni^{2+}
16. a molecule
17. a molecule
18. a molecule
19. covalent bond
20. polar
21. molecular orbital; covalent, electrons
22. H < C < N < O
23. Y, X (since X is more electronegative than Y)
24. six, two
25.

	Atomic Number	Electronic Configuration	Symbol of Ion	Covalence Number
(a)	9	2 7	X^-	1
(b)	20	2 8 8 2	X^{2+}	none
(c)	5	2 3	X^{3+}	none
(d)	17	2 8 7	X^-	1
(e)	8	2 6	X^{2-}	2
(f)	13	2 8 3	X^{3+}	none

26. 2, outside levels
27. (a) H_2S (b) HCl
    (c) NH_3 (H_3N is all right, too)   (d) HI
    (e) AsH_3 (f) PH_3
    (g) SiH_4

Multiple-Choice

1. d (X has six electrons in its outer level and therefore can also form two covalent bonds.)
2. b
3. c
4. a
5. c [(11+) + (10−) = 1+]
6. a, b, c, and d (an ion of X must have a charge of 2−; therefore, the charge on an ion of Y must be 2−, also.)
7. d
8. b
9. a
10. a and d
11. a
12. a, c, and d

13. b and c
14. d
15. c
16. d (Choices a and b could not exist because they show element A
    with two covalences. If the compound were choice c the substance
    would be polar. Only in choice d, the linear molecule, will there
    be bond polarities which cancel each other out and make the
    molecule nonpolar. In choice d, the center of density of the
    positive charge and the center of density of the negative charge
    coincide.)
17. c

5 Quantitative Relationships in Chemical Reactions

OBJECTIVES

After you have studied this chapter and worked the assigned exercises, you should be able to do the following.

1. Examine any chemical equation and tell if it is balanced.
2. For those equations that can be balanced by the procedures developed in this chapter, balance such equations, given the formulas of the reactants and products.
3. Explain what the coefficients in a chemical equation represent and contrast this meaning with the information given by the subscripts within formulas.
4. Give the name and value of the number that chemists use as a standard number of formula units.
5. Calculate how many grams a sample must weigh, given its chemical formula, if the sample is known to contain Avogadro's number of formula units.
6. Calculate how many formula units are in a sample of a substance when you know the mass of the sample and the formula weight.
7. Calculate a formula weight from a chemical formula.
8. Give the SI definition of the "mole," and relate it to the definition of Avogadro's number.
9. Set up the two possible conversion factors that the molar mass of a sample provides.
10. Calculate how many moles of a substance are in a given mass of a sample.

11. Calculate how many grams of a substance are in a given number of moles.

12. Given a balanced chemical equation, calculate how many moles of one substance in the reaction are required if a certain number of moles of any other substance in the reaction is given.

13. Given a balanced equation, calculate how many grams of one substance must be involved if a certain number of grams of any other substance is involved.

14. Describe the relationships among the terms: solution, solute, and solvent.

15. Know how and when to use the qualitative descriptions of concentration such a dilute, concentrated, unsaturated, saturated, and supersaturated.

16. Describe what occurs when we say that something precipitates from a solution.

17. Define "molar concentration" and attach specific units to the numerical value of the molarity of a solution.

18. Set up the two possible conversion factors that the molarity of a solution provides.

19. Calculate the grams of a solute that must be measured to prepare a solution that has a given volume and molarity.

20. Calculate how many milliliters of a solution of known molarity must be measured to obtain a given number of moles of the solute.

21. Do stoichiometric calculations that involve solutions of known molarity whose solutes participate in a given reaction.

22. Do the calculations needed to prepare a dilute solution of known molarity from a more concentrated solution.

GLOSSARY

Avogadro's Number. 6.023×10^{23}. The number of formula units in one mole of any element or compound.

Coefficients. Numbers placed before formulas in chemical equations to indicate the mole proportions of reactants and products.

Concentration. The quantity of some component of a mixture in a unit of volume or a unit of mass of the mixture; the ratio of quantity of solute to quantity of solvent or quantity of solution.

Equation, Balanced. A chemical equation in which all of the atoms represented in the formulas of the reactants are present in identical numbers among the products, and in which any net electrical charge provided by the reactants equals the same charge indicated by the products. (See also Chemical Equation.)

Formula Weight. The sum of the atomic weights of the atoms represented in a chemical formula.

Molar Concentration (M). A solution's concentration in units of moles of solute per liter of solution; molarity.

Molarity. (See Molar Concentration.)

Mole (mol). A mass of a compound or of an element that equals its formula weight in grams; Avogadro's number of a substance's formula units.

Molar Mass. The number of grams per mole of a substance.

Molecular Weight. The formula weight of a substance.

Precipitate. A solid that separates from a solution as the result of a chemical reaction.

Precipitation. The formation and separation of a precipitate.

Solute. The component of a solution that is understood to be dissolved in or dispersed in a continuous solvent.

Solution. A homogeneous mixture of two or more substances that are at the smallest levels of their states of subdivision——at the ion, atom, or molecule level.

Solution, Aqueous. A solution in which water is the solvent.

Solution, Concentrated. A solution with a high ratio of quantity of solute to that of solvent.

Solution, Dilute. A solution with a low ratio of solute to solvent.

Solution, Saturated. A solution into which no more solute can be dissolved at the given temperature.

Solution, Supersaturated. An unstable solution that has a higher concentration of solute than that of the saturated solution.

Solution, Unsaturated. A solution into which more solute could be dissolved without changing the temperature.

Solvent. That component of a solution into which the solutes are considered to have dissolved; the component that is present as a continuous phase.

DRILL EXERCISES

I. EXERCISES IN CALCULATING FORMULA WEIGHTS

Calculate the formula weights of the following substances. Round atomic weights to three significant figures before using them in these calculations, and round the answers to the first decimal place. Then write the two conversion factors made possible for each substance by its formula weight. The answers will be used in later drill exercises.

1. NH_3 _____ Conversion factors:

2. H_2O _____ Conversion factors:

3. SO_3 _____ Conversion factors:

4. NO_2 _____ Conversion factors:

5. $MgCl_2$ _____ Conversion factors:

6. I_2 _____ Conversion factors:

7. NaOH   _____ Conversion factors:

8. H_2SO_4 _____ Conversion factors:

9. KNO_3 _____ Conversion factors:

10. $C_{12}H_{22}O_{11}$ _____ Conversion factors:

II. EXERCISES IN CALCULATING GRAMS FROM MOLES

Calculate the number of grams in the following samples. The conversion factors were prepared in the previous exercise. Show the full setup for the solution and make the correct cancel lines. (Notice that the mole quantities that are specified limit each answer to three significant figures.)

1. 3.00 mol of NH_3 _____ g NH_3

2. 9.00 mol of H_2O _____ g H_2O

3. 0.200 mol of SO_3 _____ g SO_3

4. 0.0100 mol of NO_2 _____ g NO_2

5. 6.00 mol of $MgCl_2$ _____ g $MgCl_2$

6. 0.300 mol of I_2 _____ g I_2

7. 0.0500 mol of NaOH   _____ g NaOH

8. 0.300 mol of H_2SO_4 _____ g H_2SO_4

9. 1.50 mol of KNO_3 _____ g KNO_3

10. 0.100 mol of $C_{12}H_{22}O_{11}$ _____ g $C_{12}H_{22}O_{11}$

III. EXERCISES IN CALCULATING MOLES FROM GRAMS

Calculate the number of moles in the following samples. The conversion factors were prepared in Drill Exercise I. Show the full setup for the solution and make the correct cancel lines. Include the right unit with the answer, and leave only the correct number of significant figures.

1. 34.0 g of NH_3 _____

2. 54.0 g of H_2O _____

3. 400 g of SO_3 _____

4. 4.60 g of NO_2 _____

5. 0.950 g of $MgCl_2$ _____

6. 12.7 g of I_2 _____

7. 8.00 g of NaOH   _____

8. 5.65 g of H_2SO_4 _____

9. 32.0 g of KNO_3 _____

10. 28.4 g of $C_{12}H_{22}O_{11}$ ————————————————

IV. EXERCISES IN BALANCING CHEMICAL EQUATIONS

Convert the following sentences into balanced chemical equations. Always remember that once you've set down the correct formula for a substance you are not allowed to alter the formula to balance the equation. All you can alter are the coefficients. Also remember that many nonmetals occur as diatomic molecules, for example, O_2, H_2, N_2, Cl_2, etc.

1. Magnesium reacts with oxygen to form magnesium oxide.

2. Calcium reacts with oxygen to form calcium oxide.

3. Sulfur reacts with oxygen to form sulfur dioxide (SO_2).

4. Sulfur reacts with oxygen to form sulfur trioxide (SO_3).

5. Hydrogen reacts with chlorine to form hydrogen chloride.

6. Hydrogen reacts with oxygen to form water.

7. Carbon reacts with oxygen to form carbon dioxide (CO_2).

V. EXERCISES IN WEIGHT-RELATION PROBLEMS INVOLVING BALANCED EQUATIONS

We can summarize the basic steps for working these problems as follows.

Step 1.   Be sure you are working with a balanced equation.

Step 2.   Calculate the formula weights and set them off to one side for reference.

Step 3.   Beneath the formulas of the balanced equation, draw two rows of blank lines, the upper row representing the "mole level" and the lower the "gram level." For example:

$$2H_2 \quad + \quad O_2 \longrightarrow 2H_2O \qquad \left(\begin{array}{ll} H_2 & 2.02 \\ O_2 & 32.0 \\ H_2O & 18.0 \end{array}\right.$$

moles   _____   _____   _____

grams   _____   _____   _____

Step 4.   Write in the "given" on the proper line. For example, let the question be as follows: If you start with 64.0 grams of oxygen, how much hydrogen can be consumed and how much water will be produced (both in grams)? If the given is in a weight, convert it at once to moles and write in the answer on the proper line.

To convert the given into moles:

$$64.0 \ \cancel{g \ O_2} \times \frac{1}{32.0} \ \frac{\boxed{\text{mole } O_2}}{\cancel{g \ O_2}} = 2.00 \text{ moles } O_2$$

Here is how the setup should now look:

$$2H_2 \quad + \quad O_2 \longrightarrow 2H_2O \qquad \left(\begin{array}{ll} H_2 & 2.02 \\ O_2 & 32.0 \\ H_2O & 18.0 \end{array}\right.$$

(calculated)

moles   _____   2.00 mol O_2 _____

grams   _____   64.0 g O_2 _____

(the "given")

This step illustrates the most important rule in weight

relation problems: SOLVE THE PROBLEM AT THE MOLE LEVEL;
THEN GO BACK TO THE GRAM LEVEL AS NEEDED.

Step 5. Use the coefficients of the equation to construct conversion factors that are used in calculating the numbers of
moles of anything else requested. This step fills in the
blanks at the mole-level of the answer. For example, to
find out how many moles of hydrogen are needed for a reaction with 2.00 mol of oxygen:

$$2.00 \text{ mol } O_2 \times \frac{2 \text{ mol } H_2}{1 \text{ mol } O_2} = 4.00 \text{ mol } H_2$$

conversion factor that uses the
coefficients of O_2 and H_2

To find the number of moles of water made by the reaction
of 2.00 mol of O_2:

$$2.00 \text{ mol } O_2 \times \frac{2 \text{ mol } H_2O}{1 \text{ mol } O_2} = 4.00 \text{ mol } H_2O$$

conversion factor that uses the
coefficients of O_2 and H_2O

The setup should now look like this:

	$2H_2$	$+$	$O_2 \longrightarrow$	$2H_2O$	H_2 2.02 O_2 32.0 H_2O 18.0 (all
moles	4.00 mol H_2		2.00 mol O_2	4.00 mol H_2O	calculated)
grams			64.0 g O_2		

(the given)

Step 6.   Convert moles into grams for the final answers to the original question.

For hydrogen: $4.00 \text{ mol } H_2 \times \dfrac{2.01 \text{ g } H_2}{1 \text{ mol } H_2} = 8.04 \text{ g } H_2$

For water: $4.00 \text{ mol } H_2O \times \dfrac{18.0 \text{ g } H_2O}{1 \text{ mol } H_2O}$ $72.0 \text{ g } H_2O$

When completed, the setup should look like this:

$$2H_2 \quad + \quad O_2 \ \text{------------>} \ 2H_2O \qquad \left(\begin{array}{ll} H_2 & 2.02 \\ O_2 & 32.0 \\ H_2O & 18.0 \end{array} \right.$$

moles $\dfrac{4.00 \text{ mol } H_2}{\rule{2cm}{0.4pt}} \quad \dfrac{2.00 \text{ mol } O_2}{\rule{2cm}{0.4pt}} \quad \dfrac{4.00 \text{ mol } H_2O}{\rule{2cm}{0.4pt}}$

grams $\dfrac{8.04 \text{ g } H_2}{\rule{2cm}{0.4pt}} \quad \dfrac{64.0 \text{ g } O_2}{\rule{2cm}{0.4pt}} \quad \dfrac{72.0 \text{ g } H_2O}{\rule{2cm}{0.4pt}}$

As a final check, add up the weights of the chemicals used as reactants. Then add up the weights of the chemicals made as products. These two sums must be identical. Remember: matter is neither created nor destroyed by a chemical reaction.

$8.04 \text{ g} + 64.0 \text{ g} = 77.0 \text{ g}$ (when correctly rounded)

Work the following problems using the steps that we have just covered. Check your answers at the end of the chapter.

PROBLEM A

The combustion of methane is shown by the following equation:

$$CH_4 \quad + \quad 2O_2 \quad \rightarrow \quad CO_2 \quad + \quad 2H_2O$$

methane   oxygen   carbon dioxide   water

If 48.0 grams methane are burned in this way, how much oxygen is needed and how much carbon dioxide and water are produced? State your answer both in moles and in grams.

PROBLEM B

The combustion of ethane is shown by the following equation:

$$2C_2H_6 + 7O_2 \rightarrow 4CO_2 + 6H_2O$$
ethane

If 15.0 grams of ethane are burned this way, how much oxygen is needed and how much carbon dioxide and water are produced? State your answer both in moles and in grams.

SELF-TESTING QUESTIONS

COMPLETION

1. An amount of carbon containing Avogadro's number of carbon atoms has a mass of _____ grams.

2. The formula weight of carbon dioxide (CO_2) is 44.0. Therefore, one mole of carbon dioxide has a mass of _____ grams, and 0.500 mole has a mass of _____.

3. A sample of water containing 6.02×10^{23} molecules has a mass of _____ (include the unit); this amount of water is one standard reacting unit of water or, to use the scientific term, one _____.

4. Sodium chloride has the formula NaCl, and it is an ionic compound. Its formula weight is 58.5 One formula unit of sodium chloride consists of one _____, whose chemical symbol is _____, and one _____, whose chemical symbol is _____.

5. A sample of sodium chloride containing 6.02×10^{23} of these formula units has a mass of _____ (include the unit); this amount is one standard reacting unit of sodium chloride, or one _____.

6. The smallest representative sample of a covalent substance such as water is called a _____; a sample of a covalent compound that contains 6.02×10^{23} of these tiny particles makes up one _____ of that substance.

7. If you have a relatively large sample of any pure substance – an element or a compound – that consists of 6.02×10^{23} particles of its own particular smallest representative sample, then you would have one _____ of it.

8. Compounds X, Y, and Z have the following formula weights: X = 50.0; Y = 100; and Z = 150. We shall assume they are all covalent compounds and therefore consist of molecules.

   (a) Suppose you had 50.0 g of X, 100 g of Y, and 150 g of Z in separate containers. What would each of these samples have in common? _____

   _____

   (b) Suppose you had 100 g of Y and 100 g of X in separate containers. Which container would have the greater number of molecules? _____

   (c) Suppose that X and Y react to produce Z and that you elect to use 50.0 g of X in a particular experiment. How much of Y (in grams) would you have to use (assuming you want a "clean" reaction, one in which nothing of X or Y was left over)? _____. The equation or the reaction is

   X + Y → Z .

   (d) As individual molecule of Z has a mass that is _____ times as much as the mass of a molecule of X; one billion molecules of Z would have a mass that is _____ times as much as one billion molecules of X.

   (e) If you wanted to prepare 300 g of Z, how many moles of X and Y would you need? _____ moles X, _____ moles Y. How many grams? _____ g X, _____ g Y.

9. To prepare 250 mL of 0.150 M H_2SO_4 requires that we dissolved _____ mol of H_2SO_4 or _____ g of H_2SO_4 in enough water to make the final volume _____ mL.

10. If one took 175 mL of 0.250 M NaOH, the sample would contain _____ mol of NaOH and _____ g of NaOH.

11. If 80.0 mL of a 0.500 M solution are diluted with water to a final volume of 400 mL, the final concentration will be _____ .

12. How many milliliters of 0.250 M H_2SO_4 are needed to react with 5.46 g of $NaHCO_3$ according to the following equation? _____

$2NaHCO_3(s) + H_2SO_4(aq) \rightarrow Na_2SO_4(aq) + 2CO_2(g) + H_2O (\ell)$

MULTIPLE-CHOICE

1. When the following equation is balanced, which term(s) will appear?

   ____Na_3PO_4 + ____$CaCl_2$ → ____$Ca_3(PO_4)_2$ + ____$NaCl$

   (a) $2Na_3PO_4$ (c) $Ca_3(PO_4)_2$

   (b) $2CaCl_2$ (d) $6NaCl$

2. In the reaction, $CaCl_2 + 2AgNO_3 \rightarrow Ca(NO_3)_2 + 2AgCl$, if one mole of $AgNO_3$ is to be consumed, then
   (a) one mole of AgCl will form
   (b) one-half mole of $CaCl_2$ will be needed
   (c) one mole of $Ca(NO_3)_2$ will also be produced
   (d) one-half mole of calcium will also be produced

3. Suppose that the formula weight of X is 30 and the formula weight of Y is 60. Which of the following statements are correct:
   (a) X weighs 30 g and Y weighs 60 g
   (b) 30 g of X will have the same number of molecules as 60 g of Y
   (c) 60 g of X will have the same number of molecules as 120 g of Y
   (d) Molecules of Y are twice as heavy as molecules of X
   (e) Molecules of Y weigh five times as much as atoms of carbon-12

4. If a 40-g sample of substance X is known to contain the same number of molecules as a 120-g sample of substance Y, then the formula weight of X must be related to the formula weight of Y in which way(s)? The formula weight of X is
   (a) equal to the formula weight of Y
   (b) one-third the formula weight of Y
   (c) three times the formula weight of Y
   (d) 4.8 times the formula weight of Y

5. The atoms of element X are one-third as heavy as the atoms of carbon-12. The formula weight of X is
   (a) 36   (c) 4
   (b) 3   (d) 12

6. An important unit of concentration in chemistry is the
   (a) molar unit   (c) molecule unit
   (b) mole unit   (d) molecular weight unit

7. Compound X, whose solubility in water at 50 °C is 60g per 100 mL, is available as a 6% solution. We would call this solution
   (a) concentrated   (c) dilute
   (b) supersaturated   (d) isotonic

8. A 0.50 M solution of compound Y is known to contain 50 g Y per liter. The formula weight of Y is
   (a) 50   (c) 0.50
   (b) 25   (d) 100

9. To prepare 250 mL of 0.0400 M glucose ($C_6H_{12}O_6$) we would have to obtain a sample of glucose with a mass of how many grams?
   (a) 0.180 g glucose   (c) 18.0 g glucose
   (b) 1.80 g glucose   (d) 10.0 g glucose

10. If we obtained 125 mL of 0.0450 M NaCl, the sample would contain how much solute?
    (a) 0.329 g NaCl   (c) 0.0450 g NaCl
    (b) 0.00231 mol NaCl   (d) 0.0112 g NaCl

11. How many milliliters of 1.25 M HCl would it take to react completely with 2.56 g of magnesium hydroxide according to the following equation?

$$Mg(OH)_2(s) + 2HCl(aq) \rightarrow MgCl_2(aq) + 2H_2O$$

(a) 35.3 mL   (b) 18,200 mL   (c) 1.25 mL   (d) 70.3 mL

12. If 10.5 mL of 0.150 M HCl were diluted to a final volume of 100 mL, the concentration of the resulting solution would be
    (a) 1.43 M HCl   (c) 0.00150 M HCl
    (b) 0.0158 M HCl   (d) 10.5 M HCl

ANSWERS

I. Exercises in Calculating Formula Weights

1. $\dfrac{17.0 \text{ grams } NH_3}{1 \text{ mole } NH_3}$ or $\dfrac{1 \text{ mole } NH_3}{17.0 \text{ grams } NH_3}$

2. $\dfrac{18.0 \text{ grams } H_2O}{1 \text{ mole } H_2O}$ or $\dfrac{1 \text{ mole } H_2O}{18.0 \text{ grams } H_2O}$

3. $\dfrac{80.1 \text{ grams } SO_3}{1 \text{ mole } SO_3}$ or $\dfrac{1 \text{ mole } SO_3}{80.1 \text{ grams } SO_3}$

4. $\dfrac{46.0 \text{ grams } NO_2}{1 \text{ mole } NO_2}$ or $\dfrac{1 \text{ mole } NO_2}{46.0 \text{ grams } NO_2}$

5. $\dfrac{95.3 \text{ grams } MgCl_2}{1 \text{ mole } MgCl_2}$ or $\dfrac{1 \text{ mole } MgCl_2}{95.3 \text{ grams } MgCL_2}$

6. $\dfrac{254 \text{ grams } I_2}{1 \text{ mole } I_2}$ or $\dfrac{1 \text{ mole } I_2}{254 \text{ grams } I_2}$

7. $\dfrac{40.0 \text{ grams } NaOH}{1 \text{ mole } NaOH}$ or $\dfrac{1 \text{ mole } NaOH}{40.0 \text{ grams } NaOH}$

8. $\dfrac{98.1 \text{ grams } H_2SO_4}{1 \text{ mole } H_2SO_4}$ or $\dfrac{1 \text{ mole } H_2SO_4}{98.1 \text{ grams } H_2SO_4}$

9. $\dfrac{101.1 \text{ grams } KNO_3}{1 \text{ mole } KNO_3}$ or $\dfrac{1 \text{ mole } KNO_3}{101.1 \text{ grams } KNO_3}$

10. $\dfrac{342.2 \text{ grams } C_{12}H_{22}O_{11}}{1 \text{ mole } C_{12}H_{22}O_{11}}$ or $\dfrac{1 \text{ mole } C_{12}H_{22}O_{11}}{342.2 \text{ grams } C_{12}H_{22}O_{11}}$

II. Exercises in Calculating Grams from Moles

1. 51.0 g NH_3 (3.00 ~~mol NH$_3$~~ x $\dfrac{17.0 \text{ g NH}_3}{1 \text{ mol NH}_3}$)

2. 162 g H_2O 3. 16.0 g SO_3

4. 0.460 g NO_2 5. 572 g $MgCl_2$

6. 76.2 g I_2 7. 2.00 g NaOH

8. 29.4 H_2SO_4 9. 152 g KNO_3

10. 34.2 g $C_{12}H_{22}O_{11}$

III. Exercises in Calculating Moles from Grams

1. 2.00 mol NH_3 (34.0 ~~g NH$_3$~~ x $\dfrac{1 \text{ mol NH}_3}{17.0 \text{ g NH}_3}$)

2. 3.00 mol H_2O 3. 4.99 mol SO_3

4. 0.100 mol NO_2 5. 0.00997 mol $MgCl_2$

6. 0.0500 mol I_2 7. 0.200 mol NaOH

8. 0.0576 mol H_2SO_4 9. 0.317 mol KNO_3

10. 0.0830 mol $C_{12}H_{22}O_{11}$

IV. Exercises in Balancing Chemical Equations

1. $2Mg + O_2 \rightarrow 2MgO$

2. $2Ca + O_2 \rightarrow 2CaO$

3. $S + O_2 \rightarrow SO_2$

4. $2S + 3O_2 \rightarrow 2SO_3$

5. $H_2 + CL_2 \rightarrow 2HCl$

6. $2H_2 + O_2 \rightarrow 2H_2O$

7. $C + O_2 \rightarrow CO_2$

V. Exercises in Weight-Relation Problems Involving Balanced Equations

Problem A. When completed the setup should look like this.

Form Wts.

CH_4 16.0

O_2 32.0

CO_2 44.0

H_2O 18.0

$$CH_4 \quad + \quad 2O_2 \longrightarrow CO_2 + 2H_2O$$

moles	3.00 mol CH_4	6.00 mol O_2	3.00 mol CO_2	6.00 mol H_2O
grams	48.0 g CH_4 (given)	192 g O_2	132 g CO_2	108 g H_2O

Check   48.0 + 192 = 240 = 132 + 108

masses of   masses of
reactants   products

Problem B. When completed the setup should look like this.

Form Wts.

C_2H_6 30.0

O_2 32.0

CO_2 44.0

H_2O 18.0

$$2C_2H_6 \quad + \quad 7O_2 \longrightarrow 4CO_2 + 6H_2O$$

moles	0.500 mol C_2H_6	1.75 mol O_2	1.00 mol CO_2	1.50 mol H_2O
grams	15.0 g C_2H_6 (given)	56.0 g O_2	44.0 g CO_2	27.0 g H_2O

Check   15 + 56 = 71 = 44 + 27

ANSWERS TO SELF-TESTING QUESTIONS

Completion

1. 12.0
2. 44.0 g; 22.0 g
3. 18.0 g; mole
4. sodium ion, Na^+, chloride ion, Cl^-
5. 58.5 g; mole
6. molecule; mole
7. mole
8. (a) Each would consist of 1.00 mole of that substance (or each would contain Avogadro's number — 6.02×10^{23} — of molecules).
   (b) X (Since each molecule of X is lighter by half than each molecule of Y, you get twice as many X molecules as Y in the same mass of samples.)
   (c) 100 g
   (d) three; three
   (e) 2.00, 2.00; 100, 200
9. 0.0375 mol H_2SO_4; 3.68 g H_2SO_4; 250 mL

10. 0.0438 mol NaOH; 1.75 g NaOH

11. 0.100 M.   (Using Equation 6.5: $conc_f = \dfrac{80.0 \text{ mL} \times 0.500 \text{ M}}{400 \text{ mL}}$
12. 130 mL

Multiple Choice

1. a, c, and d
2. a and b
3. b, c, d, and e
4. b
5. c
6. a
7. c
8. d
9. b
10. a
11. d
12. b

6 States of Matter, Kinetic Theory, and Equilibria

OBJECTIVES

After you have studied this chapter and worked the exercises, you should be able to do the following.

1. Name, define, and give the normally used units for four properties of a gas that are used to describe its physical state.
2. Use the pressure-volume law (Boyle's law) to calculate final values of pressure and volume.
3. Use the law of partial pressures (Dalton's law) to convert between total pressure and partial pressures.
4. Use the temperature-volume law (Charles' law) to calculate final volume or temperature.
5. State the volume-mole relationship for gases compared at the same pressure and temperature.
6. Give the values of STP.
7. Use the universal gas law to calculate gas properties.
8. State the pressure-temperature law for gases and use it in calculations.
9. Define an "ideal gas" and contrast it with a "real gas."
10. Give the postulates of the kinetic theory of gases.
11. Relate the temperature of a gas to kinetic theory in general terms.
12. Compute the total pressure of gases over water on a dry basis when water vapor is present.
13. Describe (in general terms only) how the kinetic model of a gas

applies (a) to the liquid and the solid states and (b) to the phenomena of melting, boiling, and vapor pressure.

14. Know when to use such terms as "volatile" and "nonvolatile" when describing a liquid.

15. Use such examples as the earth's heat budget, or the concentration of oxygen in the atmosphere, or a change of physical state to explain what is involved (in general terms) in a dynamic equilibrium.

16. Write (using words) equations for the equilibria between a liquid and its vapor, or between a solid and its liquid form, equations that include a heat term, and explain how these equilibria shift when heat is added or removed from the system.

17. Know how and when to use such terms associated with equilibria such as forward and reverse changes, upsetting or shifting an equilibrium, and favored and unfavored reactions.

18. State Le Chatelier's principle and use it to explain the influences of concentration, temperature, and catalysis on rates of reactions.

19. Explain how London forces affect normal boiling points, and how these boiling points change with formula weights among otherwise very similar substances.

20. Explain the relationship between a collision between two reactant particles and the energy of activation for the reaction.

21. Interpret a progress of reaction diagram for a reaction and indicate which features of the diagram represent the energy of activation, the energy of the reaction, the reactants and the products

22. Describe how a catalyst affects the energy of activation of a reaction, the energy of the reaction, and the rate of the reaction.

23. Describe how an increase in the temperature of a mixture of reactants affects exothermic and endothermic reactions.

24. Describe how changes in concentrations of reactants can influence the rates of reactions.

25. Define each of the terms in the Glossary.

GLOSSARY

Atmosphere, Standard (atm). The pressure that supports a column of mercury 760 mm high when the mercury has a temperature of 0 $^{\circ}$C.

Avogadro's Law. Equal volumes of gases contain equal numbers of moles when they are compared at identical temperatures and pressures.

Barometer.   An instrument for measuring atmospheric pressure.

Boiling.   The turbulent behavior in a liquid when its vapor pressure
equals the atmospheric pressure and when the liquid absorbs heat
while experiencing no rise in temperature.

Boiling Point, Normal.   The temperature at which a substance boils
when the atmospheric pressure is 760 mm Hg (1 atm).

Boyle's Law.   (See Pressure-Volume Law.)

Catalysis.   The phenomenon of an increase in the rate of a chemical
reaction brought about by a relatively small amount of a chemical
– the catalyst – that is not permanently changed by the reaction.

Catalyst.   A substance that is able, in relatively low concentrations,
to accelerate the rate of a chemical reaction without itself being
permanently changed. (In living systems, the catalysts are called
enzymes.)

Charles' Law.   (See Temperature-Volume Law.)

Dalton's Law.   (See Law of Partial Pressures.)

Diffusion.   A physical process whereby particles, by random motions,
intermingle and spread out so as to erase concentration gradients.

Energy of Activation.   The minimum energy that must be provided by the
collision between reactant particles to initiate the rearrangement
of electrons relative to nuclei that must happen if the reaction
is to occur.

Enzyme.   A catalyst in a living system.

Equilibrium, Dynamic.   A situation in which two opposing events occur
at identical rates so that no net change happens.

Gas Constant, Universal (R).   The ratio of PV to nT for a gas, where
P = the gas pressure, V = volume, n = number of moles, and T =
the Kelvin temperature. When P is in mm Hg and V is in mL,
$$R = 6.23 \times 10^4 \text{ mm Hg mL/mol K}$$

Heat of Reaction. The net energy difference between the reactants and the products of a reaction.

Ideal Gas. A hypothetical gas that obeys the gas laws exactly.

Kinetics. The field of chemistry that deals with the rates of chemical reactions.

Kinetic Theory of Gases. A set of postulates about the nature of an ideal gas: that it consists of a large number of very small particles in constant, random motion; that in the collisions the particles lose no frictional energy; that between collisions the particles neither attract nor repel each other; and that the motions and collisions of the particles obey all the laws of physics.

Law of Partial Pressures (Dalton's Law). The total pressure of a mixture of gases is the sum of their individual partial pressures.

Le Chatelier's Principle. If a system is in equilibrium and a change is made in its conditions, the system will change in whichever way most directly restores equilibrium.

London Force. A net force between molecules that arises from temporary polarities induced in the molecules by collisions or near-collisions with neighboring molecules.

Manometer. A device for measuring gas pressure.

Melting Point. The temperature at which a solid changes into its liquid form; the temperature at which equilibrium exists between the solid and liquid forms of a substance.

Millimeter of Mercury (mm Hg). A unit of pressure equal to 1/760 atm.

Nonvolatile Liquid. Any liquid with a very low vapor pressure at room temperature and that does not readily evaporate.

Partial Pressure. The pressure contributed by an individual gas in a mixture of gases.

Photosynthesis. The synthesis in plants of complex compounds from carbon dioxide, water, and minerals with the aid of sunlight captured by the plant's green pigment, chlorophyll.

Polarization. Any increase in the polarity of a molecule.

Pressure. Force per unit area.

Pressure-Temperature Law (Gay-Lussac's Law). The pressure of a gas is directly proportional to its Kelvin temperature when the gas volume is constant.

Pressure-Volume Law (Boyle's Law). The volume of a gas is inversely proportional to its pressure when the temperature is constant.

Rate of Reaction. The number of successful (product-forming) collisions that happen each second in each unit of volume of the reacting mixture.

Standard Conditions of Temperature and Pressure (STP). 0 $^{\circ}$C (or 273 K) and 1 atm (or 760 mm Hg).

Temperature-Volume Law (Charles' Law). The volume of a gas is directly proportional to its Kelvin temperature when the pressure is kept constant.

Torr. A unit of pressure; 1 torr = 1 mm Hg; 1 atm = 760 torr.

Universal Gas Law. $PV = nRT$ (See also **Gas Constant, Universal**.)

Vacuum. An enclosed space in which there is no matter.

Vapor Pressure. The pressure exerted by the vapor that is in equilibrium with its liquid state at a given temperature.

Volatile Liquid. A liquid that has a high vapor pressure and readily evaporates at room temperatyure.

HOW TO STUDY THE GAS LAWS

The properties of gases are described by various laws. What is important here are the properties of gases, not the names of the laws. Even after you forget which law is Boyle's and which is Charles' or Gay-Lussac's, you should remember the basic properties of gases.

Start by learning the list of variables used in the gas laws. These variables are the properties of a sample of gas that can be measured and that can be varied or adjusted by the experimenter. They are:

Pressure	P	Temperature	T (in kelvins)
Volume	V	Quantity	n (in moles)

Often the subscripts 1 and 2 are used to designate the two different states of the gas. Usually, a fixed quantity of a gas is involved. This means that n, the number of moles of the gas, is a constant. That leaves just three variables: P, V, and T. Always remember that T must be in kelvins. (kelvins = degrees Celsius + 273).

If a question involves changes in all three variables – P, V, and T – then the universal gas law is generally used:

$$PV = nRT$$

where R is the universal gas constant whose numerical value depends upon the units selected for P and V. Thus, if P is in mm Hg and V is in mL, then $R = 6.23 \times 10^4 \frac{\text{mm Hg mL}}{\text{mole K}}$.

If you forget the value of R or if a reference containing it is not handy, then you can always set up a ratio in which R cancels; this was worked out in Exercise 5.92 to give:

$$\frac{P_1 V_1}{T_1} = \frac{P_2 V_2}{T_2}$$

This equation is perhaps the best single equation to memorize. From it, if needed, Boyle's, Charles', and Gay-Lussac's equations can be readily derived.

Boyle's Law is the pressure-volume law: both n and T are fixed. If $T_1 = T_2$, then they cancel in the basic equation and we have the equation: $P_1 V_1 = P_2 V_2$

Charles' Law is the temperature-volume law; both n and P are fixed. If $P_1 = P_2$, then they cancel in the basic equation and we have:

$$\frac{V_1}{T_1} = \frac{V_2}{T_2}$$

Gay-Lussac's Law is the temperature-pressure law; both n and V are fixed. If $V_1 = V_2$, they cancel in the basic equation, and we have:

$$\frac{P_1}{T_1} = \frac{P_2}{T_2}$$

Since the relations among the variables are more important than the names of the laws, be sure you learn these basic relations.

1. Pressure varies inversely with volume (T and n are fixed).
   Increase the pressure – reduce the volume
   Decrease the pressure – increase the volume
2. Volume varies directly with Kelvin temperature (P and n are fixed).
   Increase the temperature – increase the volume
   Decrease the temperature – decrease the volume
3. Pressure varies directly with Kelvin temperature (V and n are fixed).
   Increase the temperature – increase the pressure
   Decrease the temperature – decrease the pressure

The method of ratios described and illustrated in the text is probably the best single method for working problems dealing with pressure, volume, and temperature. To practice that method, but more importantly, to fix the basic relations firmly in mind, complete the Table "Exercises in T, V, and P Variables for Gases."

We are also concerned with Dalton's Law and Avogadro's Law. Be especially certain to understand and learn Dalton's Law of Partial Pressure.

The kinetic theory of gases is particularly helpful in its explanation of the gas laws; it gives something of a "molecule's-eye view" of gases. It is also a beautiful illustration of the interplay between experimental work and theoretical concepts. The theory tells us that the pressure of a gas is proportional to the mean kinetic energy of the gas particles. Likewise, it tells us that the temperature of a gas is proportional to the mean kinetic energy of the gas particles. If we raise the temperature of a gas, we increase its mean kinetic energy; if we raise the temperature of a confined gas, we also raise its pressure.

DRILL EXERCISES

Exercises in T, V and P Variables for Gases
(fixed quantities)

Exercise	The Change in the Sample of the Gas	The Correct Ratio for Calculating the Final Value	Calculated Final Value
1.	The pressure changes from 640 mm Eg to 740 mm Hg for 1.00 L of gas at a fixed T	_____	V_2 = _____
2.	The pressure changes from 1.0 atm to 0.80 atm for 20 L of gas at a fixed T	_____	V_2 = _____
3.	The temperature changes from 25 °C to 30 °C for a gas initially at a pressure of 760 mm Hg confined in a fixed volume	_____	P_2 = _____
4.	The temperature changes from 100 °C to 0 °C for a gas in a fixed volume initially at a pressure of 760 mm Hg	_____	P_2 = _____
5.	The temperature changes from 20 °C to 36 °C for 750 mL of a gas under conditions in which the pressure is kept the same.	_____	V_2 = _____

The kinetic theory also gives us a "molecule's-eye view" of the motion or vibration of atoms and molecules in liquids and solids. Concentrate on learning the vocabulary in Section 5.6.

The kinetic theory also helps us understand how the rate of a chemical reaction is influenced by the concentration of the reactants, the temperature, and the presence or absence of a catalyst. Be sure to learn what is meant by the energy of activation and what, in general, a catalyst does to it. We shall need these concepts to understand how enzymes work.

SELF-TESTING QUESTIONS

<u>COMPLETION</u>

1. The atmosphere at sea level presses down with a force of _____ lb/in.^{2} and will hold up a column of mercury that is _____ mm in height.

2. Another name for 1 mm Hg is _____.

3. Robert Boyle discovered the relation between the pressure and the _____ of a fixed amount of gas when its _____ is constant.

4. Boyle found that this relation appears to be true for all _____.

5. Using symbols of P_1, P_2, V_1, and V_2, Boyle's law may be written in mathematical form as : _____

6. In a general way, Boyle's law tells us that if greater pressure is placed on a gas, its volume will _____.

7. Alternatively, if the pressure on a gas is reduced, its volume will _____.

8. If the pressure on a gas is reduced by one-half, its volume will _____.

9. If a mixture of gases is made up of three individual gases, A, B, and C, and the total pressure on the gas mixture is 760 mm Hg, what is the partial pressure of C if the partial pressures of A and B are each 350 mm Hg? _____

10. Comparing inhaled air with exhaled air, the partial pressure of _____ is less in exhaled air partly because it has been consumed during chemical reactions in the body.

11. A sample of oxygen was collected a 22 °C over water. The volume of this wet oxygen was 475 mL at a pressure of 742 mm Hg. What volume would the oxygen occupy if completely dry at 742 mm Hg? The vapor pressure of water at 22 °C is 19.8 mm Hg.

_____

12. The standard conditions of temperature and pressure are _____ K and _____ atm.

13. At STP, one mole of oxygen occupies _____ liters, one mole of hydrogen occupies _____ liters, and one mole of nitrogen occupies _____ liters.

14. Hydrogen and oxygen combine to form water: $2H_2 + O_2 \rightarrow 2H_2O$. For a complete reaction, one liter of oxygen at STP would require _____ liters of hydrogen also at STP, because according to Avogadro's law, "when measured at the same T and P, equal volumes of gases contain equal numbers of _____."

15. The basic postulates of the kinetic theory of gases that describe an ideal gas are:

   (a) _____

   (b) _____

   (c) _____

   (d) _____

16. If somehow the average kinetic energy of all the particles in a sample of helium gas contained in a tight flask is made to increase, then both the _____ and the _____ of the helium will increase.

17. What important physical constant of a liquid will likely have a relatively low value if the liquid is considered to be very volatile?   _____

18. When the number of molecules of a liquid that escape from it exactly equals the number that return to it, a condition of _____ exists.

19. The minimum collision energy of two reacting particles above which all collisions of the proper orientation will produce products is called the _____.

20. If a reaction has a high energy of activation, it probably will proceed _____.
    (slowly or rapidly)

21. We could speed up a slow reaction by one or more of these operations.

    (a) Increase the _____ of the reactants.

    (b) Raise the _____ of the reacting mixture.

    (c) Introduce a special _____.

22. Which of the three factors in question 21 best explains why substances burn more vigorously in pure oxygen than in air?

    _____

23. Which of the three factors in question 21 has the greatest influence on how rapidly reactions proceed in living things?

    _____

24. Increasing the concentration of the reactants increases the rates of most reactions because higher concentrations mean higher frequencies of _____ between the reactants.

25. Raising the temperature of a reacting mixture increases the rate of the reaction primarily by increasing the _____ of the collisions.

26. Introducing a catalyst into a reacting mixture increases the rate of the reaction largely by lowering the reaction's _____
    _____.

27. The special catalysts found in living things are called _____.

28. The combustion of coal in air is an example of an energy-releasing or _____ event.

29. A reaction normally requiring a high temperature in order to take place is an energy-consuming or _____ event.

30. Consider the following physical equilibrium
        liquid ⇌ solid + heat
    (a) As this expression is written, which is the forward change?
        (Write the equation for just this change.)

    _____

    (b) What could be done to the temperature to shift this
        equilibrium to the left? _____

31. Consider the following chemical equilibrium.

        $H_2 + I_2 \rightleftharpoons 2HI$

    (a) An increase in the initial concentration of H_2 would favor a
        shift in which direction for this equilibrium? _____
    (b) If the forward reaction is endothermic, in which direction
        will this equilibrium shift if the temperature is lowered?

        _____
    (c) If a catalyst were found for this equilibrium system, what
        would the catalyst do to the position of equilibrium?

        _____

MULTIPLE-CHOICE

1. The two gases important in respiration are
   (a) nitrogen   (b) carbon dioxide   (c) oxygen   (d) water vapor

2. Pressure is the same as
   (a) force   (c) force per unit distance
   (b) force per unit volume   (d) force per unit area

3. One atmosphere of pressure may be expressed as
   (a) 1 atm   (b) 760 mm Hg   (c) 760 torr   (d) 20 lb/in^2

4. If 1.00 L of a gas at 20 °C is made to change its pressure from 1
   atm to 2 atm, the new volume will be
   (a) 2.00 L   (b) 1.00 L   (c) 0.500 L   (d) 0.100 L

5. If 1.00 L of a gas at 50 °C is made to change its temperature to
   100 °C under conditions where it is allowed to freely expand in
   order to maintain a constant pressure, the new volume of the gas
   will be
   (a) 2.00 L   (b) 0.500 L   (c) 4.00 L   (d) 1.15 L

6. If a gas mixture consists of 1 mol of helium and 1 mol of argon at STP, the partial pressure of helium will be
   (a) 1 atm   (b) 0.5 atm   (c) 2 atm   (d) 1.15 atm

7. A sample of an unknown gas occupies 22.4 L at STP and the sample has a mass of 32.0 grams. The formula weight of the gas is, therefore,
   (a) 22.4   (b) 32.0   (c) 11.2   (d) 16.0

8. The postulate of the kinetic theory stating that the molecules of a gas move in straight lines, not curved lines, between collisions is another way of postulating that between the molecules
   (a) no forces of attraction exist
   (b) no forces of repulsion exist
   (c) balanced forces of attraction and repulsion exist
   (d) collisions are perfectly elastic

9. If the molecules of a gas confined within a sealed container are somehow made to have a lower average velocity but still remain in the gaseous state,
   (a) the density of the gas will decrease
   (b) the pressure of the gas will decrease
   (c) the temperature of the gas will decrease
   (d) the volume of the gas will decrease

10. The temperature of a substance is a measure of
    (a) the average kinetic energy of its formula units
    (b) the number of its formula units present
    (c) the gram-formula weight
    (d) the chemical energy of the formula units

11. The opposite of condensation is
    (a) melting   (b) boiling   (c) freezing   (d) evaporation

12. A chemical reaction that is extremely rapid at room temperature
    (a) is probably highly endothermic
    (b) probably has a very low energy of activation
    (c) very likely has an unusually high energy of activation
    (d) may have the benefit of a catalyst

13. The following equilibrium would be shifted in which direction by increasing the temperature?

$$C_2H_4 \; + \; H_2 \rightleftharpoons C_2H_6 \; + \; heat$$

(a) to the left   (c) both directions equally
(b) to the right   (d) in neither direction at all

ANSWERS

ANSWERS TO DRILL EXERCISES

Exercises in T, V, and P for Gases of Fixed Quantities

	Correct Ratio	Calculated Final Value		
1.	$\dfrac{640}{740}$	$V_2 = \dfrac{640}{740} \times 1.00$ L	$= 0.865$ L	(Note 1)
2.	$\dfrac{1.0}{0.80}$	$V_2 = \dfrac{1.0}{0.80} \times 20$ L	$= 25$ L	(Note 2)
3.	$\dfrac{303}{298}$	$P_2 = \dfrac{303}{298} \times 760$ mm Hg	$= 773$ mm Hg	(Note 3)
4.	$\dfrac{273}{373}$	$P_2 = \dfrac{273}{373} \times 760$ mm Hg	$= 556$ mm Hg	(Note 3)
5.	$\dfrac{309}{293}$	$V_2 = \dfrac{309}{293} \times 750$ mL	$= 791$ mL	(Note 4)

Notes

1. Since the pressure is increasing, the volume must decrease and the fraction must be less than one.
2. The pressure decreases; therefore, the volume must increase. The fraction must be greater than one.
3. 25 $^{\circ}$C + 273 = 303 K; 30 + 273 = 303 K. Temperature must always be in kelvins. Since an increase in the temperature must increase the pressure, we need a ratio of temperatures greater than one. Likewise, in exercise four, since the temperature decreases so will the pressure; this time we need a ratio of temperatures less than one.
4. Since an increase in the temperature must increase the volume, we need a temperature ratio greater than one.

ANSWERS TO SELF-TESTING QUESTIONS

Completion

1. 14.7; 760
2. torr
3. volume, temperature
4. gases
5. $$P_1V_1 = P_2V_2$$
6. decrease
7. increase
8. double
9. 60 mm Hg (350 + 350 + 60 = 760; Dalton's law)
10. oxygen
11. 462 mL
12. 273; 1
13. 22.4, 22.4, 22.4
14. two, moles
15. (a) Gases consist of large numbers of small particles in random motion.
    (b) These particles are very hard and collide in frictionless collisions.
    (c) These particles neither attract nor repel each other.
    (d) The motions of these particles obey the laws of physics.
16. pressure, temperature
17. boiling point
18. dynamic equilibrium
19. energy of activation
20. slowly
21. (a) concentrations
    (b) temperature
    (c) catalyst
22. a (increased concentration)
23. c (special catalysts)
24. collisions
25. energies (or average kinetic energies)
26. energy of activation
27. enzymes
28. exothermic
29. endothermic
30. (a) liquid $\rightarrow$ solid + heat
    (b) raise the temperature
31. (a) to the right

(b) to the left
(c) nothing

Multiple-Choice

1. b and c
3. a, b, and c
5. d
7. b
9. b and c
11. d
13. a

2. d
4. c
6. b
8. a and b
10. a
12. b and d

7 Water, Solutions, and Colloids

OBJECTIVES

The topics in this chapter that are most important for an understanding of the molecular basis of life are:

1. The hydrogen bond

    Because it occurs in all carbohydrates, proteins, and genes.

2. How water can dissolve some substances well and others poorly

    Because water is the basic fluid of cells and blood.

3. How Le Châtelier's principle applies to certain equilibria

    Because living systems must maintain some kind of equilibrium in the face of forces that unbalance equilibria.

4. Dialysis

    Because that is how many substances migrate into and out of cells.

5. Ways to express concentrations other than in units of moles per liter

    Because these are often used to describe a number of fluids found in living systems as well as fluids that are administered to patients.

After you have studied this chapter and worked the exercises in it, you should be able to do the following.

1. Use the valence shell electron pair repulsion theory to explain how molecules get their bond angles and polarities.
2. Explain how the high polarity of water helps it to function as a solvent.
3. Draw an illustration of the hydrogen bond and define it citing specific evidence for its existence in the liquid state.
4. Explain how surface tension arises and how it is responsible for the beading of water on a waxy surface and the forming of a meniscus on glass.
5. Describe the effect of a surfactant and cite two examples in which surfactants are vital to our lives.
6. Compare and contrast true solutions, colloidal dispersions, and suspensions with respect to particle sizes and four physical properties.
7. Explain why hydrates are called compounds, not mixtures, and give formulas for plaster of paris and gypsum.
8. Use Le Châtelier's principle to explain how temperature affects solubility.
9. Describe the dependence of the solubility of a gas in water on (a) the temperature and (b) the pressure of the gas over the solution.
10. Use the pressure-solubility law (Henry's law) to calculate the solubility of a gas in water.
11. Explain how chemical reactions enable carbon dioxide, sulfur dioxide, and ammonia to be much more soluble in water than gases such as oxygen or nitrogen.
12. Describe how changes in the oxygen tension and the carbon dioxide tension of blood and cell fluids help deliver these gases in the body.
13. Do the calculations required for the preparation and use of solutions of known percent concentrations (a) from pure solutes and solvents and (b) from more concentrated solutions.
14. Give the circumstances necessary for osmosis or dialysis.
15. Describe how dialysis is necessary for life.
16. Predict the direction in which water will flow in osmosis or dialysis, given the concentrations of solutes expressed as molarities or osmolarities.
17. Define each of the terms in the Glossary.

GLOSSARY

Active Transport. The movement of a substance through a biological membrane against a concentration gradient and caused by energy-consuming chemical changes that involve parts of the membrane.

Anhydrous Form. A substance without any water of hydration.

Brownian Movement. The random, chaotic movements of particles in a colloidal dispersion that can be seen with a microscope.

Colligative Property. A property of a solution that depends only on the concentrations of the solute and the solvent and not on their chemical identities (e.g., osmotic pressure).

Colloidal Dispersion. A relatively stable, uniform distribution in some dispersing medium of colloidal particles – those with at least one dimension between 1 and 1000 mm.

Colloidal Osmotic Pressure. The contribution made to the osmotic pressure of a solution by substances colloidally dispersed in it.

Deliquescence. The ability of a substance to attract water vapor to itself to form a concentrated solution.

Desiccant. A substance that combines with water vapor to form a hydrate and thereby reduces the concentration of water vapor in the air space around the substance.

Dialysis. The passage through a dialyzing membrane of water and particles in solution, but not of particles that have colloidal size.

Emulsion. A colloidal dispersion of tiny microdroplets of one liquid in another liquid.

Gas Tension. The partial pressure of a gas over its solution in some liquid when the system is in equilibrium.

Gel. A colloidal dispersion of a solid in a liquid that has adopted a semisolid form.

Hemolysis. The bursting of a red blood cell.

Henry's Law. (See Pressure-Solubility Law.)

Homogeneous Mixture. A mixture in which the composition and properties are uniform throughout.

Hydrate. A compound in which intact molecules of water are held in a definite molar proportion to the other components.

Hydration. The association of water molecules with dissolved ions or polar molecules.

Hydrogen Bond. The force of attraction between a $\delta+$ on a hydrogen held by a covalent bond to oxygen or nitrogen (or fluorine) and a $\delta-$ charge on a nearby atom of oxygen or nitrogen (or fluorine).

Hygroscopic. Describing a substance that can reduce the concentration of water vapor in the surrounding air by forming a hydrate.

Hypertonic. Having an osmotic pressure greater than some reference; having a total concentration of all solute particles higher than that of some reference.

Hypotonic. Having an osmotic pressure less than some reference; having a total concentration of dissolved solute particles less than that of some reference.

Isotonic. Having an osmotic pressure identical to that of a refer- ence; having a concentration equivalent to the reference with respect to the ability to undergo osmosis.

Macromolecule. Any molecule with a very high formula weight – gener- ally several thousand or more.

Osmolarity. The molar concentration of all osmotically active solute particles in a solution.

Osmosis. The passage of water only, without any solute, from a less concentrated solution (or pure water) to a more concentrated solution when the two solutions are separated by a semipermeable membrane.

Osmotic Membrane. A semipermeable membrane that permits only osmosis, not dialysis.

Osmotic Pressure. The pressure that would have to be applied to a solution to prevent osmosis if the solution were separated from water by an osmotic membrane.

Parts per Billion (ppb). The number of parts in a billion parts. (Two drops of water in a railway tank car that holds 34,000 gallons of water correspond roughly to 1 ppb.)

Parts per Million (ppm). The number of parts in a million parts. (Two drops of water in a large, 32-gallon trash can correspond roughly to 1 ppm.)

Percent (%). A measure of concentration.

   Vol/vol percent: The number of volumes of solute in 100 volumes of solution.
   Wt/wt percent: The number of grams of solute in 100 g of the solution.
   Wt/vol percent: The number of grams of solute in 100 mL of the solution.
   Milligram percent: The number of milligrams of the solute in 100 mL of the solution.

Physiological Saline Solution. A solution of sodium chloride with an osmotic pressure equal to that of blood.

Pressure-Solubility Law (Henry's Law). The concentration of a gas in a liquid at any given temperature is directly proportional to the partial pressure of the gas on the solution.

Reagent. Any mixture of chemicals, usually a solution, that is used to carry out a chemical test.

Saturated Solution. A solution into which no more solute can be dissolved at the given temperature; a solution in which dynamic equilibrium exists between the dissolved and the undissolved solute.

Semipermeable. Descriptive of a membrane that permits only certain kinds of molecules to pass through and not others.

Solution. A homogeneous mixture of two or more substances that are at the smallest levels of their states of subdivision — at the ion, atom, or molecule level.

Surface-Active Agent. (See Surfactant.)

Surface Tension. The quality of a liquid's surface by which it be-
haves as if it were a thin, invisible, elastic membrane.

Surfactant. A substance, such as a detergent, that reduces the surface tension of water.

Suspension. A homogeneous mixture in which the particles of at least
one component have average diameters greater than 1000 nm.

Tetrahedral. Descriptive of the geometry of bonds at a central atom
in which the bonds project to the corners of a regular
tetrahedron.

Tyndall Effect. The scattering of light by colloidal sized particles
in a colloidal dispersion.

Valence-Shell Electron-Pair Repulsion Theory (VSEPR). Bond angles at
a central atom are caused by the repulsions of the electron clouds
of valence-shell electron pairs.

Water of Hydration. Water molecules held in a hydrate in some definite mole ratio to the rest of the compound.

SELF-TESTING QUESTIONS

COMPLETION

1. Write the structure of the water molecule correctly showing its
bond angle. _____ The theory that
explains this bond angle is called _____

2. In terms of the net electrical charge for the whole molecule, the
water molecule is _____.

3. However, the water molecule is _____, because
the center of density of all its _____ charges
(contributed by its atomic nuclei) does not coincide with the
center of density of all its _____ charges

(contributed by its _____).

4. Write the structure of the water molecule again and place the symbols δ+ and δ— by the appropriate parts of the structure to show the locations of sites of relative electron deficiency and electron richness.

_____

5. Write the structure of two water molecules, orienting them in a manner we would expect because they can attract each other.

_____

6. This force of attraction in water is called the _____. It extends from a site of δ+ associated with the _____ end
                                                            (H or O)
of one molecule to a site of δ— associated with the _____
                                                            (H or O)
end of the other molecule.

7. If water forms tight beads on the surface of a plastic bottle, the molecules in the plastic must be _____.
                                              (polar or nonpolar)

8. The surface tension of water would be enough to collapse the lungs by collapsing the _____ in the lungs if the water in the lungs did not contain some dissolved _____.

9. Bile contains a _____ agent that aids in the digestion of _____.

10. If water contains a dispersion of particles larger than simple ions or molecules but not heavy enough to be forced by gravity to settle or otherwise coagulate, this mixture is called a _____.

11. If you direct a light beam through this mixture (question 10), you will observe the _____ effect.

12. A colloidal dispersion of one liquid in another is called an _____.

13. The motions of colloidal dispersed particles in a fluid are called
    the _____.

14. Potassium chloride is similar in structure to sodium chloride. At
    20 °C the solubility of KCl in water is 24.7 g/100 mL solution.

    (a) What particles make up the formula unit of KCl? (Give both
        their names and their symbols.)

        _____ whose symbol is _____,   and

        _____ whose symbol is _____.

    (b) What particles separate from the crystal of KCl as it goes
        into solution in water? (Formulas only)

        _____ and _____

    (c) What happens to them as they enter the solution (something
        involving the water molecules)? _____

    (d) In the crystal of KCl, the potassium ions were generally sur-
        rounded by what electron-rich particles? (Name)

        _____

    (e) By going into solution, have the potassium ions given up being
        surrounded by an electron-rich envelope? _____

        What electron-rich substance surrounds each K^+ ion in the
        solution? _____

    (f) As for the chloride ions in KCl, they are surrounded by what
        relatively electron-poor particles?   (Give name)

        _____ (Formula) _____

    (g) When the chloride ions go into solution in water, what is it
        that is relatively electron-poor that crowds around them? (Be
        specific; it is not just the particle's name we seek but also
        a particular part of it.)

        _____

15. Le Chatelier's principle enables us to predict that the following
    equilibrium will shift to the _____ if it is cooled.
                                    (left or right)

$$\text{heat} + CuSO_4(s) \underset{\text{water}}{\rightleftharpoons} Cu^{2+}(aq) + SO_4^{2-}(aq)$$

The symbol (s) stands for the solid substance and (aq) for the dissolved particles.

16. Le Chatelier's principle helps us to predict that the gas tension of carbon dioxide over water will _____ if the
    (increase or decrease)
    following equilibrium is heated.

$$CO_2(aq) + \text{heat} \rightleftharpoons CO_2(g)$$

The symbol (g) stands for the gaseous state, and (aq) means the aqueous solution.

17. We would use the _____ law to predict the solubility of oxygen in water at 640 mm Hg if we knew its solubility at 760 mm Hg (at the same temperature).

18. The solubility of oxygen in water at 20 $^\circ$C and 760 mm Hg is 4.30 mg/dL. What is its solubility at a pressure of 270 mm Hg?

19. Suppose you have a solution of KCl at 20 $^\circ$C with a concentration of 34.7 g/100 g (which is also the solubility of KCl in water at 20 $^\circ$C).

    (a) What qualitative expression describes this solution most accurately? _____
    (b) What is the percent of concentration (w/w)? _____
    (c) What would you see if you added some crystals of KCl to this solution? _____
    (d) If you separated this solution from pure water by a semi-permeable membrane, _____ would occur and the volume of this solution would _____ and its
       (increase or decrease)
       concentration would _____.
       (increase or decrease)

20. To prepare 500 mL of a 10% (vol/vol) solution of ethanol in water, you have to mix _____ mL of ethanol with _____ mL of water (assuming that these volumes, when mixed actually give 500 mL of solution).

21. In 100 mL of 5% (wt/vol) of salt in water there is _____ g of salt.

22. Three colligative properties of solutions containing nonvolatile solutes are

    (a) _____

    (b) _____

    (c) _____

23. Which solution will have the lower freezing point, a solution with 1 mole of KCl in 1 kg of water or a solution with 1 mole of $MgCl_2$ in 1 kg of water? _____

    Explain. _____

    _____

    _____

24. Which of the two solutions in the previous question would have the higher osmotic pressure? _____

25. If pure water were fed intravenously, a patient's red blood cells would tend to _____; technically, this is called
    (shrink or burst)

    _____.

26. A compound with the formula $Na_3PO_4 \cdot 12H_2O$ is an example of a

    _____.

27. If this compound (in question 26) were heated intensely, it would lose its water of _____ and be changed into its _____ form.

28. Calcium chloride, $CaCl_2$, will take water vapor from humid air and change to $CaCl_2 \cdot 6H_2O$. What would this product be called?

    _____

29. Because anhydrous calcium chloride will do this, it is called a

    _____.

MULTIPLE CHOICE

1. In the water molecule there are four pairs of electrons in the outside level of the oxygen atom. According to the VSEPR theory, what should be the bond angle?
   (a) 90° (b) 109.5^O (c) 120° (d) 180°

2. In the boron trifluoride molecule, BF_3, there are only <u>three</u> pairs of the electrons in the valence shell of boron. Each pair is part of a covalent bond to a fluorine atom. The VSEPR theory would enable one to predict that the three fluorine atoms are
   (a) at three of the four corners of a tetrahedron and make bond angles of 109.5^O
   (b) at the three corners of a regular triangle and make bond angles of 120°
   (c) make bond angles of 90°
   (d) make bond angles of 60°

3. The curved arrow points to what kind of bond?
   (a) covalent
   (b) ionic
   (c) hydrogen bond
   (d) nonpolar

   ```
                                          O              H
                                         / \            /
                                        H   H  ...  O
                                                     \
                                                      H
   ```

4. Hydrogen bonds between molecules will be the strongest for which substance?
   (a) NH_3 (b) CH_4 (c) H_2O (d) $H - H$

5. If water tends to spread out and form a very thin film on a surface, then the molecules in that surface are very likely
   (a) nonpolar (b) polar   (c) dense   (d) mobile

6. Substance A is a molecular compound that dissolves in gasoline but not in water. The molecules of A are very likely
   (a) polar   (b) nonpolar   (c) solvated   (d) crenated

7. When water molecules in liquid water are attracted to and crowd around solute ions (or molecules), the phenomenon is called (one choice)
   (a) hydrolysis   (b) solvation   (c) hydration   (d) hydrogenation

8. The property of the surface of water that can be likened to its having a thin, elastic invisible membrane is called

(a) surface action   (c) surface pressure
(b) surface polarity   (d) surface tension

9. One property a colloidal dispersion has that a solution does not is
   (a) homogeneity   (c) the Tyndall effect
   (b) filterability   (d) osmotic pressure

10. The dispersion of oil droplets in water is called
    (a) a smoke   (b) an emulsion   (c) a sol   (d) a gel

11. In a dilute solution of sugar in water, water is the
    (a) solute   (b) solvent   (c) emulsifying agent   (d) desiccant

12. An aqueous solution whose concentration effectively matches the concentration of blood is said to be
    (a) isotopic   (b) isobaric   (c) isotonic   (d) isoelectronic

13. To prepare 50.0 g of 1% (w/w) $AgNO_3$, you would need to weigh out how many grams of $AgNO_3$?
    (a) 0.0500 g   (b) 0.500 g   (c) 5.00 g   (d) 50.0 g

14. To prepare 500 g of 2.00% (w/w/) NaCl from 5.00% (w/w) NaCl, how many grams of the more concentrated solution would you have to weigh out and dilute?
    (a) 2.00 g   (b) 20.0 g   (c) 200 g   (d) 50.0 g

15. In 500 mL of 2.00% (mg/vol) of glucose in water there is how much glucose?
    (a) 10.0 mg   (b) 1.00 mg   (c) 10.0 g   (d) 1.00 g

16. To make 300 mL of a 4.00% (wt/vol) solution of sugar we need to weigh out
    (a) 4.00 g of sugar
    (b) 40.0 g of sugar
    (c) 12.0 g of sugar
    (d) we need to know the formula weight of sugar before making an answer

17. If a bag made of a dialyzing material and containing a mixture of water, dissolved salt, and colloidally dispersed protein were placed in a beaker of pure water, the water would eventually contain

(a) dissolved salt

(b) protein

(c) both salt and protein

(d) no additional material

18. The weak force of attraction acting between water molecules is called

(a) a hydrogen to oxygen covalent bond

(b) a hydrogen bond

(c) an ionic bond

(d) a hydration bond

19. In the substance whose formula is $Na_2CO_3 \cdot 10H_2O$, for every five formula units of Na_2CO_3 there are how many water molecules?

(a) 50   (b) 2   (c) 10   (d) 100

ANSWERS

<u>ANSWERS</u> <u>TO</u> <u>SELF-TESTING</u> <u>QUESTIONS</u>

<u>Completion</u>

1. H   H;   valence shell electron pair repulsion theory
     \ /
      O
2. electrically neutral
3. polar, positive, negative, electrons
4. δ+   δ+
     H   H
      \ /
       O
       δ-
5.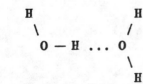
6. hydrogen bond; H, O
7. nonpolar
8. alveoli, surfactant (or surface-active agent)
9. surface-active, fats and oils (or lipids) as well as other foods coated with fats and oils
10. colloidal dispersion
11. Tyndall
12. emulsion
13. Brownian movement

14. (a) the potassium ion, K^+, the chloride ion, Cl^-
    (b) K^+ and Cl^-
    (c) They become hydrated.
    (d) chloride ions
    (e) No, they have not; water molecules
    (f) potassium ions, K^+
    (g) the hydrogen ends of water molecules
15. left
16. increase
17. the pressure-solubility law (Henry's law)
18. 1.53 mg O_2/dL H_2O
19. (a) saturated
    (b) 34.7% (w/w)
    (c) The crystals would sink to the bottom and not dissolve.
    (d) osmosis, increase, decrease
20. 50 mL of ethanol and 450 mL of water
21. 5 g
22. (a) lowering of the freezing point of the solvent
    (b) raising of the boiling point of the solvent
    (c) osmotic pressure of the solution
23. The $MgCl_2$ solution; 1 mole $MgCl_2$ will provide 3 moles of ions (1 of Mg^{2+} and 2 of Cl^-) whereas 1 mole KCl will provide only 2 moles of ions (1 of K^+ and 1 of Cl^-) when the solutes are fully dissolved.
24. The $MgCl_2$ solution; it has the higher concentration of solute particles.
25. burst; hemolysis
26. hydrate
27. hydration, anhydrous
28. a hydrate (calcium chloride hexahydrate)
29. desiccant

Multiple-Choice

1. b	2. b	3. a	4. c
5. b	6. b	7. c	8. d
9. c	10. b	11. b	12. c
13. b	14. c	15. a	16. c
17. a	18. b	19. a	

8 Acids, Bases, and Salts

OBJECTIVES

In our study of the molecular basis of life, we need to be familiar with the major topics of this chapter for several reasons.

The acid and base theories and acid-base equilibria	Because acids and bases (particularly as viewed in the Brønsted Theory) are involved in most reactions of metabolism.
Net ionic equations	Because from now on we shall use them almost exclusively to describe reactions.
The reactions of ions	Because we need to know how to tell which ions can remain unaffected in the presence of others and which cannot, since ions occur in all body fluids.

Before you attempt any problems or exercises, you should commit to memory the names and formulas for the five strong acids plus phosphoric acid listed in Table 8.1 and the strong bases listed in Table 8.2. These are short lists, and we make the important but simplified assumption that any acids or bases not on the lists of the strong acids or bases will be weak. Also commit to memory the solubility rules for salts. Then you will have automatically learned what the insoluble salts are – any salts not mentioned in the solubility rules.

Next learn the properties of acids and bases in water and the reactions of strong acids with active metals and their hydroxides, carbonates, and bicarbonates. Practice writing both full and net ionic equations. Acquiring skill now in these matters will make later studying that much easier. The summary of the circmstances in which the various ions react with one another given in section 7.4 is very important.

If only one goal were to be stated for this chapter, it would be to learn the chief chemical properties of ions and ion-producing substances – acids, bases, and salts.

After studying this chapter and working the exercises in it, you should be able to do the following.

1. Give the main points in the Arrhenius Theory and the Brønsted Theory.
2. Define and give examples of strong and weak electrolytes.
3. Write the names and formulas of five strong acids and two strong, highly soluble bases and explain why they are classified as "strong."
4. Given the formula of an acid or base, tell whether it is strong or weak. Do the same from K_a or K_b values as well.
5. Write the formulas of the hydronium ion, the ammonium ion, the carbonate ion, and the bicarbonate ion as well as the structures of water, ammonia, and carbonic acid.
6. Write the equation for the dynamic equilibrium involved in the self-ionization of water.
7. Write equations, full and net ionic, for the reactions of strong aqueous acids with active metals and their hydroxides, bicarbonates, and carbonates.
8. Describe what it means when a metal is high in the activity series.
9. Write net ionic equations involving proton transfers between (a) the ammonia molecule and the hydronium ion and (b) the ammonium ion and the hydroxide ion.
10. Recognize and write formulas of conjugate acid-base pairs.
11. Recognize from a formula if a compound is a salt and if it is, write equations for making that salt in four different ways using

a strong acid.

12. Use the solubility rules to predict double decompositions.
13. Summarize the three circumstances under which ions usually react, giving the equations of two specific examples for each.
14. Explain how the common ion effect is an example of Le Chatelier's Principle, giving a specific example.
15. Define (and, where substances are involved, illustrate) all of the terms listed in the Glossary.

GLOSSARY

Acid. Arrhenius Theory: Any substance that makes H^+ available in water.

Brønsted Theory: Any substance that can donate a proton (H^+).

Acid-Base Neutralization. The reaction of an acid with a base.

Acid Ionization Constant (K_a) A modified equilibrium constant for the following equilibrium: $HA + H_2O \rightleftharpoons H_3O^+ + A^-$

$$K_a = \frac{[H_3O^+][A^-]}{[HA]}$$

Acidic Solution. A solution in which the molar concentration of hydronium ions is greater than that of hydroxide ions.

Activity Series. A list of elements (or other substances) in the order of the ease with which they release electrons under standard conditions and become oxidized.

Alkali. A strongly basic substance such as sodium hydroxide or potassium hydroxide.

Anion. A negatively charged ion.

Anode. The positive electrode to which negatively charged ions – anions – are attracted during electrolysis.

Arrhenius Theory. Acids in water release hydrogen ions. Bases in water release hydroxide ions.

Base. Arrhenius Theory: A compound that provides the hydroxide ion.

Brønsted Theory: A proton-acceptor; a compound that neutralizes hydrogen ions.

Base Ionization Constant (K_b)   For the equilibrium (where $\underline{B}$ is some base) $\underline{B} + H_2O \rightleftharpoons \underline{B}H^+ + OH^-$

$$\underline{K_b} = \frac{[\underline{B}H^+][OH^-]}{[\underline{B}]}$$

Basic Solution.   A solution in which the molar concentration of hydroxide ions is greater than that of hydronium ions.

Brønsted Theory.   An acid is a proton-donor and a base is a proton-acceptor.

Cathode.   The negative electrode to which positively charged ions – cations – are attracted during electrolysis.

Cation.   A positively charged ion.

Common Ion Effect.   The reduction in the solubility of a salt in some solution by the addition of another solute that furnishes one of the ions of this salt.

Conjugate Acid-Base Pair.   Two particles whose formulas differ by only one H^+, such as NH_4^+ and NH_3, or HCl and Cl^-.

Diprotic Acid.   An acid with two protons available per molecule to neutralize a base, e.g., H_2SO_4.

Double Decomposition.   A reaction in which a compound is made by the exchange of partner-ions between two salts.

Electrical Balance.   The condition of a net ionic equation wherein the algebraic sum of the positive and negative charges of the reactants equals that of the products.

Electrode.   A metal object, usually a wire, suspended in an electrically conducting medium through which electricity passes to or from an external circuit.

Electrolysis.   A procedure in which an electrical current is passed through a solution that contains ions, or through a molten salt, for the purpose of bringing about a chemical change.

Electrolyte. Any substance whose solution in water conducts electricity; or the solution itself of such a substance.

Equilibrium Constant. The value that the mass action expression has
when a chemical system is at equilibrium. (See Law of Mass
Action.)

Hydronium Ion. H_3O^+

Hydroxide Ion. OH^-

Inorganic Compound. Any compound that is not an organic compound.

Ionic Equation. A chemical equation that explicitly shows all of the
particles – ions, atoms, or molecules – that are involved in a
reaction even if some are only spectator particles. (See also Net
Ionic Equation; Equation, Balanced.)

Ionization. A change, usually involving solvent molecules, whereby
molecules or ionic crystals break apart into ions.

K_a (See Acid Dissociation Constant.)

K_b (See Base Dissociation Constant.)

Law of Mass Action (Law of Guldberg and Waage). The molar proportions
of the interacting substances in a chemical equilibrium are
related by the following equation (in which the ratio on the left
is called the mass action expression for the system).

$$\frac{[C]^c[D]^d}{[A]^a[B]^b} = K_{eq}$$ The symbols refer to the following general-

ized equilibrium: $aA + bB \rightleftharpoons cC + dD$ and the brackets,
[ ], denote molar concentrations. (When an equilibrium involves
additional substances, the equation for the equilibrium constant
is adjusted accordingly.)

Material Balance. The condition of a chemical equation in which all
of the atoms present among the reactants are also found in the
products.

Molecular Equation. An equation that shows the complete formulas of
all of the substances present in a mixture undergoing a reaction.

(See also Net Ionic Equation; Equation, Balanced.)

Monoprotic Acid. An acid with one proton per molecule that can neutralize a base.

Net Ionic Equation. A chemical equation in which all spectator particles are omitted so that only the particles that participate directly are represented.

Neutral Solution. A solution in which the molar concentration of hydronium ions exactly equals the molar concentration of hydroxide ions.

Nonelectrolyte. Any substance that cannot furnish ions when dissolved in water or when melted.

Organic Compounds. Compounds of carbon other than those related to carbonic acid and its salts, or to the oxides of carbon, or to the cyanides.

Salt. Any crystalline compound that consists of oppositely charged ions (other than H^+ or OH^-).

Simple Salt. A salt that consists of just one kind of cation and one kind of anion.

Strong Acid. An acid with a high percentage ionization or a high value of acid ionization constant, K_a.

Strong Base. A metal hydroxide with a high percentage ionization in solution.

Strong Brønsted Acid. Any species, molecule or ion, that has a strong tendency to donate a proton to some acceptor.

Strong Brønsted Base. Any species, molecule or ion, that binds an accepted proton strongly.

Strong Electrolyte. Any substance that has a high percentage ionization in solution.

Triprotic Acid. An acid that can supply three protons per molecule.

Weak <u>Acid</u>. An acid with a low percentage ionization in solution and with a low value of acid ionization constant, $\underline{K}_a$.

Weak <u>Base</u>. A metal hydroxide with a low percentage ionization in solution.

Weak <u>Brønsted</u> <u>Acid</u>. Any species, molecule or ion, that has a weak tendency to donate a proton and poorly serves as a proton donor.

Weak <u>Brønsted</u> <u>Base</u>. Any species, molecule or ion, that weakly holds an accepted proton and poorly serves as a proton-acceptor.

Weak <u>Electrolyte</u>. Any electrolyte that has a low percentage ionization in solution.

SELF-TESTING QUESTIONS

<u>COMPLETION</u>

1. If you see the noun "sodium" in a sentence, unless it is immediately followed by the word _____, it means either the element sodium or an _____ of sodium metal.

2. The self-ionization of water can be written by what equilibrium equation? _____

3. The three broad classes of substances that can liberate ions in water are _____, _____ and _____.

4. Negative ions are called _____ and positive ions are called _____.

5. When water is the solvent and its molecules react with molecules of hydrogen chloride to form hydronium ions and chloride ions, water molecules are acting as Brønsted _____ and hydrogen chloride molecules are acting as Brønsted _____.

6. The hydronium ion has the formula _____; the hydrogen ion has the formula _____.

7. After dissolving 100 g of the covalent compound HX in a liter of water, an average of 10% of the HX molecules were always in their intact molecular form and an average of 90% were split up into the ions X^- nd H^+ (actually, H_3O^+). HX may be classified as a

_____ .

   (strong or weak)   (acid, base, or salt)

What kind of an electrolyte is it? _____

8. Give the chemical formulas for:
   (a) sulfuric acid   _____

   (b) carbonic acid   _____

   (c) nitric acid   _____

   (d) phosphoric acid   _____

   (e) hydrochloric acid   _____

   (f) ammonia   _____

9. Write the names and formulas of the strong acids and bases.

Strong Acids

Name	Formula
_____	_____
_____	_____
_____	_____
_____	_____
_____	_____

Strong Bases

Name	Formula
_____	_____
_____	_____
_____	_____
_____	_____
_____	_____

10. If HCl is a strong Brønsted acid, this must mean that the ion Cl^- is a _____ binder of a proton and that Cl^- is therefore
    (good or poor)
    a _____ (weak or strong) Brønsted base.

11. We can generalize from this. The conjugate base obtainable from any strong acid will be a relatively _____ Brønsted base. The conjugate base obtainable from any weak acid will be a relatively _____ Brønsted base.

12. The conjugate base of the water molecule is named the _____ _____ and has the formula _____. It is a _____ Brønsted base.
    (strong or weak)

13. Suppose we contrive to make a neutral particle, B, take a proton to form BH. What will be the net electrical charge on BH, if any?
    _____

14. If B holds the proton very weakly, would we classify HB^+, the conjugate acid, as weak or strong? _____

15. Which is the stronger Brønsted base, PO_4^{3-} or $H_2PO_4^-$_____, NH_3 or NH_4^+ _____?

16. When metals are arranged in an order corresponding to their relative tendencies to become ionic, we call this series the _____.

17. The most reactive metals are in Group _____ of the Periodic Table.

18. Write the net ionic equation for the reaction (if any) of
    (a) calcium with nitric acid
    _____

    (b) potassium hydroxide with hydrochloric acid
    _____

    (c) sodium chloride with sulfuric acid
    _____

    (d) sodium bicarbonate with hydrochloric acid
    _____

19. Test your knowledge of the solubility rules for the salts given in the text by predicting whether each of the following would be soluble in water. (Use a plus sign if soluble, a minus sign if insoluble.)

(a) sodium carbonate   _____   (b) calcium bromide   _____

(c) potassium nitrate   _____   (d) lead nitrate   _____

(e) magnesium nitrate   _____   (f) nickel sulfide   _____

(g) copper(II) arsenate   _____   (h) ammonium sulfate   _____

(i) lead chloride   _____   (j) sodium sulfide   _____

20. Test your knowledge of the solubility rules, of the typical reactions of acids and bases, and of which acids and bases are strong by predicting the chemical reactions (if any) that would occur if you mixed together aqueous solutions of the following pairs. Write net ionic equations for reactions you predict. (If there is no reaction, write "none.")

(a) $LiCl$ and $AgNO_3$ _____

(b) $NaNO_3$ and $CaCl_2$ _____

(c) KOH and H_2SO_4 _____

(d) $Pb(NO_3)_2$ and $NaCl$ _____

(e) K_2S and $CuSO_4$ _____

(f) Na_2CO_3 and HBr _____

21. The K_a of acid HX is 1×10^{-3} and the K_a of acid HY is 1×10^{-5}. The formula of the stronger acid is _____. Both acids would be described as _____ relative to, say,
(weak or strong)
nitric acid.

22. Amino acids typically have K_a values in the order of 1×10^{-10} to 1×10^{-11}. Are amino acids weak or strong acids? _____

MULTIPLE-CHOICE

1. When nitric acid reacts with sodium bicarbonate, which of the following is not formed?
(a) $NaNO_3$ (b) CO_2 (c) Na   (d) H_2O

2. Which equation represents the neutralization of hydrochloric acid?

(a) $HCl_{aq} + AgNO_3 \rightarrow AgCl + HNO_3$

(b) $HCl_{aq} + NaBr \rightarrow NaCl + HBr$

(c) $2HCl_{aq} + Ca(NO_3)_2 \rightarrow CaCl_2 + 2HNO_3$

(d) $HCl_{aq} + NaHCO_3 \rightarrow NaCl + H_2O + CO_2$

3. The net ionic equation for $Mg + 2HBr \rightarrow MgBr_2 + H_2$ (assuming that $MgBr_2$ is a water-soluble salt) is

(a) $Mg^{2+} + 2H^+ \rightarrow Mg + H_2$

(b) $Mg + 2H^+ \rightarrow Mg^{2+} + H_2$

(c) $Mg + 2HBr \rightarrow MgBr_2 + 2H^+$

(d) $Mg + 2HBr \rightarrow Mg^{2+} + Br_2 + H_2$

4. After a solution of potassium hydroxide has been carefully neutralized by hydrochloric acid, the solution will contain (besides water)

(a) K^+ and Cl^- (c) K^+ and OH^-

(b) K and Cl   (d) K^-, H^+, and Cl_2

5. The covalent substance HA was soluble enough in water to make a very concentrated solution; however, whether concentrated or dilute, the percentage of ionization of HA into A^- and H^+ (actually H_3O^+) was less than 5%. Another covalent substance, HB, was virtually insoluble in water, but what did dissolve ionized by about 95% into B^- and H^+. Which would be classified as the stronger acid?

(a) the concentrated solution of HA

(b) the dilute solution of HA

(c) simply HA

(d) HB

6. The conjugate acid of PO_4^{3-} is

(a) HPO_4^{2-} (b) $H_2PO_4^-$ (c) H_3PO_4 (d) $H_4PO_4^+$

7. The strongest base among the following is

(a) NO_3^- (b) HSO_4^- (c) CO_3^{2-} (d) I^-

8. Which of the following substances could be added to aqueous hydrobromic acid to reduce the concentration of hydronium ions?

(a) silver nitrate   (c) ammonium chloride
(b) ammonia   (d) potassium nitrate

9. An example of a weak acid is
   (a) HBr   (b) HNO_3 (c) H_2SO_4 (d) HNO_2

10. If you mixed a solution of sodium bicarbonate with a solution of
    sulfuric acid, which particles would interact to form new
    products?

    (a) SO_4^{2-} and Na^+ (c) Na^+ and H_3O^+

    (b) HCO_3^- and SO_4^{2-} (d) H_3O^+ and HCO_3^-

11. If you mixed a solution of hydrochloric acid with ammonia water,
    which particles would be involved in the primary chemical change?

    (a) NH_3 and Cl^- (c) NH_3 and H_2O

    (b) NH_3 and H_3O^+ (d) NH_3 and HCl (molecules)

12. The hydronium ion would readily transfer a proton to which of the
    following?

    (a) PO_4^{3-} (b) CO_3^{2-} (c) OH^- (d) NO_3^-

13. The reaction of hydrochloric acid with sodium hydroxide is called
    (a) hydrolysis   (c) neutralization
    (b) crenation   (d) proton transfer

14. Reactions involving acids, bases, or salts can be predicted to
    take place if there is the possibility of forming
    (a) an insoluble precipitate
    (b) a strong acid
    (c) a gas
    (d) an un-ionized (but soluble) substance

15. Consider the equilibrium between solid potassium chloride and its
    ions in the dissolved state in a saturated solution:

$$\text{Heat} + KCl_{(solid)} \rightleftharpoons K^+_{aq} + Cl^-_{aq}$$

    If somehow the equilibrium "shifts to the left," then more solid
    KCl will
    (a) dissolve   (b) form   (c) vaporize   (d) melt

16. The equilibrium (of question 15) could be shifted to the right by
adding
    (a) solid KCl   (b) HCl_{aq} (c) $NaBr_{(solid)}$ (d) heat

17. The equilibrium (of question 15) would be shifted to the left by
adding
    (a) solid KCl   (b) HCl_{aq} (c) $NaBr_{(solid)}$ (d) heat

ANSWERS

ANSWERS TO SELF-TESTING QUESTIONS

Completion

1. ion, atom
2. $2H_2O \rightleftharpoons H_3O^+(aq) + OH^-(aq)$
3. acids, bases, salts (any order)
4. anions; cations
5. bases, acids
6. H_3O^+, H^+
7. strong acid; strong electrolyte
8. (a) H_2SO_4 (b) H_2CO_3 (c) HNO_3

   (d) H_3PO_4 (e) HCl   (f) NH_3

9.
Strong Acids		Strong Bases	
hydrochloric acid	HCl	sodium hydroxide	NaOH
hydrobromic acid	HBr	potassium hydroxide	KOH
hydriodic acid	HI	magnesium hydroxide	$Mg(OH)_2$
nitric acid	HNO_3	calcium hydroxide	$Ca(OH)_2$
sulfuric acid	H_2SO_4		

10. poor, weak
11. weak; strong
12. hydroxide ion, OH^-; strong
13. +1 (as in BH^+)
14. strong   15. PO_4^{3-}; NH_3
16. activity series   17. IA
18.
(a) $Ca + 2H^+ \rightarrow Ca^{2+} + H_2$ (or $Ca + 2H_3O^+ \rightarrow Ca^{2+} + H_2 + 2H_2O$)

(b) $OH^- + H^+ \rightarrow H_2O$ (or $OH^- + H_3O^+ \rightarrow 2H_2O$)

(c) none

(d) $HCO_3^- + H^+ \rightarrow H_2O + CO_2$ (or $HCO_3^- + H_3O^+ \rightarrow 2H_2O + CO_2$)

19. (a) +   (b) +   (c) +   (d) +   (e) +
    (f) −   (g) −   (h) +   (i) −   (j) +
20. (a) $Cl^- + Ag^+ \rightarrow AgCl$
    (b) none
    (c) $OH^- + H^+ \rightarrow H_2O$
    (d) $Pb^{2+} + 2Cl^- \rightarrow PbCl_2$
    (e) $S^{2-} + Cu^{2+} \rightarrow CuS$
    (f) $CO_3^{2-} + 2H^+ \rightarrow H_2O + CO_2$
21. HX; weak
22. weak

Multiple-Choice

1. c
2. d   (in the other three choices the concentration of hydronium ions
   does not change. Therefore, regardless of the fate of the
   chloride ion the acid has not been neutralized.)
3. b   4. a   5. d
6. a   7. c   8. b
9. d   10. d   11. b
12. a, b, and c   13. c and d   14. a, c, and d
15. b   16. d   17. b

9 Acidity Detection, Control, Measurement

OBJECTIVES

Broadly speaking, the most important topics in this chapter are:

The pH concept

Because we use pH to describe the relative acidities of body fluids.

Substances that affect the pH of a solution of ions other than H_3O^+ and OH^-

Because anything that can affect the pH of a solution must be taken seriously when dealing with life at the molecular level.

Dissolved substances that can stabilize or buffer the pH of a solution

Because buffers in the blood are vital to health and because some tend to wear out during certain diseases.

Under the general heading of acid—base titration, there are several topics particularly important in clinical laboratories and in the work of medical technologists. You may never perform such work yourself, but you may send samples to labs for analysis; therefore, an appreciation of the difficulties of analysis is professionally valuable. Moreover, the concept of an equivalent weight does surface from time to time in the labels of fluids administered to patients and in descriptions of patients' clinical conditions. Thus, you should be familiar with the meaning of the term "equivalent" and "milli-equivalent."

After you have studied this chapter and have worked the Practice Exercises and Review Exercises, you should be able to do the following.

1. Write the equation of K_w and use it to calculate a value of $[H^+]$ from a value of $[OH^-]$ or vice versa.
2. Relate $[H^+]$ to pH.
3. State the range of pH values that corresponds to an acidic solution and the range that corresponds to a basic solution.
4. Describe the purpose of an indicator.
5. Give the color of litmus in an acid and in a base. (Do the same for any other indicators that may be assigned.)
6. Use the facts of which acids and bases are strong to predict whether an aqueous solution of a given salt will test acidic, basic, or neutral.
7. State what is meant by the expression, "this aqueous solution is buffered."
8. Write the ionic equations that show how a carbonate buffer works; do the same for a phosphate buffer.
9. Explain how the instability of carbonic acid helps it function in preventing a major upset in the acid-base balance of blood and write the equations for all of the relevant equilibria.
10. Explain how hyperventilation or hypoventilation helps the body regulate the pH of blood.
11. Calculate the equivalent weight of an ion.
12. Calculate how many equivalents or milliequivalents of an ion are present in a sample.
13. Describe what the "anion gap" refers to and how it is used in certain diagnoses.
14. Explain the difference between "neutralizing capacity" and "pH."
15. Describe a titration.
16. Describe the difference between "end point" and "equivalence point."
17. Calculate the equivalent weight of an acid or of a base.
18. Convert equivalents into milliequivalents.
19. Calculate how much acid (or base) should be weighed out to prepare a specified volume of a solution that is to have a certain normality.
20. Use the equation $N_a \times mL_a = N_b \times mL_b$ in situations involving data obtained from acid-base titrations.

21. Define all of the terms listed in the Glossary.

GLOSSARY

Acidic Solution. A solution in which the molar concentration of hydronium ions is greater than that of hydroxide ions; a solution in which the pH is less than 7.00 (at 25 $^{\circ}$C).

Acidosis. A condition in which the pH of the blood is below normal.

Alkalosis. A condition in which the pH of the blood is above normal.

Anion Gap. $\text{anion gap} = \dfrac{\text{meq of Na}^+}{\text{L}} - \left(\dfrac{\text{meq of Cl}^-}{\text{L}} + \dfrac{\text{meq of HCO}_3^-}{\text{L}} \right)$

Basic Solution. A solution in which the molar concentration of hydroxide ions is greater than the molar concentration of hydronium ions; a solution in which the pH is greater than 7.00 (at 25 $^{\circ}$C).

Buffer. A combination of solutes that holds the pH of a solution relatively constant even if small amounts of acids or bases are added.

Carbonate Buffer. A mixture or a solution that includes bicarbonate ions and carbonic acid (dissolved carbon dioxide) in which the bicarbonate ion can neutralize added acid and carbonic acid can neutralize added base.

End Point. The stage in a titration when the operation is stopped.

Equivalence Point. The stage in a titration when the reactants have been mixed in the exact molar proportions represented by the balanced equation; in an acid-base titration, the stage when the moles of hydrogen ions furnished by the acid match the moles of hydroxide ions (or other proton acceptor) supplied by the base.

Equivalent. For an acid, its mass in grams that can neutralize one mole of hydroxide ion; for a base, its mass in grams that can neutralize one mole of hydrogen ion; for an ion, usually its mass in grams divided by the amount of its electrical charge.

Equivalent Weight. A synonym for Equivalent.

Hydrolysis of Salts. Any reaction in which a cation (other than H^+) or an anion (other than OH^-) changes the ratio of the molar concentrations of hydrogen and hydroxide ions in an aqueous solution.

Hyperventilation. Breathing considerably faster and deeper than normal.

Hypoventilation. Breathing more slowly and less deeply than normal; shallow breathing.

Indicator. A dye that, in solution, has one color below a measured pH range and a different color above this range.

Ion Product Constant of Water (K_w). The product of the molar concentrations of hydrogen ions and hydroxide ions in water at a given temperature.

$$K_w = [H^+][OH^-]$$
$$= 1.0 \times 10^{-14} \qquad \text{(at 25 °C)}$$

K_w. (See Ion Product Constant of Water.)

Milliequivalent (meq). A quantity of substance equal to one-thousandth of an equivalent.

Neutralizing Capacity. The capacity of a solution or a substance to neutralize an acid or a base – expressed as a molar concentration.

Neutral Solution. A solution in which the molar concentration of hydronium ions exactly equals the molar concentration of hydroxide ions.

Normality. The concentration of a solution in units of equivalents per liter.

pH. The negative power to which the base 10 must be raised to express the molar concentration of hydrogen ions in an aqueous solution.

$$[H^+] = 1 \times 10^{-pH}$$
$$\text{or} \qquad -\log [H^+] = pH$$

Phosphate Buffer. Usually a mixture or a solution that contains dihydrogen phosphate ions ($H_2PO_4^-$) to neutralize OH^- and monohydrogen phosphate ions (HPO_4^{2-}) to neutralize H^+.

pOH.   The   negative   power to which   the base 10   must be   raised   to
express the concentration of hydroxide ions in an aqueous solution
in mol/L.

$$[OH^-] = 1 \times 10^{-pOH}$$

At 25 °C, $pH + pOH = 14.00$

Respiration. The intake and chemical use of oxygen by the body and
the release of carbon dioxide.

Standard Solution.   Any   solution for which   the   concentration   is
accurately known.

Titration. An analytical procedure that carefully combines a solution
of unknown concentration with a standard solution until an
indicator changes color or some other signal shows that equivalent
quantities have reacted. One solution is usually added from a
buret.

Ventilation. Breathing.

SELF-TESTING QUESTIONS

<u>COMPLETION</u>

1. The pH of a solution for which $[H^+] = 1.00 \times 10^{-5}$ mol/L is _____.

2. A solution with a pH of 6.50 is _____.
   (acidic, basic, neutral)

3. The concentration of hydroxide ions in pure water at room
   temperature is (in moles per liter) _____.

4. If the pH of a solution is 7.42, then the concentration of
   hydrogen ions (in moles per liter) is between 1.0×10^{-7} and 1.0 x
   _____.

5. If $[OH^-] = 4.0 \times 10^{-9}$ mol/L, the solution is _____.
   (acidic or basic)

6. If K_2CO_3 were added to water:

(a) Which ions would it liberate in the solution? (Formulas)
_____ and _____

(b) Could either ion react with water to produce some OH^-? _____
If yes, which ion? (Formula) _____

(c) Could either ion react with water to produce some H^+? _____
If yes, which ion? (Formula) _____

(d) In what way will K_2CO_3 change the pH – leave it the same, lower it, or raise   it? _____

7. Will the addition of Na_3PO_4 to water make the solution more basic, more acidic, or will it   not affect the pH? _____

8. Write the ionic equation for the buffering of an acid by the carbonate buffer. _____

9. Write the ionic equation for the buffering of a base by the carbonate buffer. _____

10. If too much acid invades a solution buffered by the carbonate buffer, the buffer will be destroyed as _____
breaks down to _____ (which leaves the solution) and water.

11. At a considerable sacrifice in accuracy, parts (a) through (d) ignore the fact that the following equilibrium and many others exist in the solution described in part (a):

$$PO_4^{3-} + H_2PO_4^- \rightleftharpoons 2HPO_4^{2-}$$

(a) If you had a solution that contained both NaH_2PO_4 and Na_3PO_4 as solutes, would the pH of that solution be less than or greater than 7? _____

(b) Whatever the pH is, would the solution of part (a) be buffered? _____

(c) If so, what ionic equation represents the reaction that will occur if some acid is added to the solution of part (a)?
_____

(d) If so, what ionic equation represents the reaction that will occur if some alkali is added to the solution of part (a)? _____

12. The concentration of potassium ion in a blood sample was 0.0041 mol K^+/L. This is the same as _____ meq/L.

13. How many milligrams of K^+ ion are in 3.80 meq of K^+ ion?

14. The results of a blood analyses were: 180 meq Na^+/L, 136 meq Cl^-/L, and 19 meq HCO_3^-/L. The anion gap here is _____, and this suggests that there _____ a disturbance in metabolism.   (is or is not)

15. Give the names of two acid-base indicators. _____ _____

16. If you titrate acetic acid with sodium hydroxide, what salt is present at the equivalence point? (Name) _____

17. Would the pH of this solution (question 16) be less than 7, greater than 7, or exactly equal to 7? _____

18. What is the normality of a solution of hydrochloric acid known to have a concentration of 0.50 M? _____

19. What is the normality of a solution of sulfuric acid known to have a concentration of 0.10 M? _____

20. What is the normality of an acid if 42 mL of it exactly neutralize 46 mL of a 0.10 N base? _____

MULTIPLE-CHOICE

1. The ion-product constant for water is given by the equation

(a) $K_w = \dfrac{[H^+][OH^-]}{[H_2O]}$
(b) $K_w = \dfrac{[H_3O^+][OH^-]}{[H_2O]^2}$

(c) $K_w = [H^+][OH^-]$
(d) $K_w = [\log pH][\log pOH]$

2. The equivalent weight of HCl_{aq} (if the atomic weight of H is 1.01); and that of Cl is 35.45) is

(a) 36.46 g   (b) 36.46 mg   (c) 0.3646 mg   (d) 18.23 g

3. Which of the following make an acidic solution when added to pure water:
   (a) H_2SO_4 (c) NaCl   (e) $AlCl_3$
   (b) NH_3 (d) Na_2CO_3 (f) Na_2HPO_4

4. Which of the choices in question 3 make a basic solution when added to pure water?

5. Dissolving NaOH in water produces a solution that
   (a) turns blue litmus red
   (b) turns phenolphthalein red
   (c) has a tart taste
   (d) turns red litmus blue

6. The pH of a solution is 6.00. Hence,
   (a) it is more acidic than a solution whose pH is 4.00
   (b) its concentration of hydrogen ions is 1.00×10^{-8} moles per liter
   (c) its concentration of hydrogen ions is 1.00×10^{-6} moles per liter
   (d) its concentration of hydrogen ions is 1.00×10^{6} moles per liter

7. If the pH of an aqueous solution cannot be changed by adding small amounts of a strong acid or a strong base, the solution contains
   (a) an indicator
   (b) a buffer
   (c) a protective colloid
   (d) a strong acid and a strong base already

8. A solution of $KC_2H_3O_2$ in water has a pH value
   (a) of 7.00   (b) less than 7.00
   (c) greater than 7.00   (d) less than 1.00.

9. A blood specimen was found to have an anion gap of 11 meq/L. It also had $Cl^- = 138$ meq/L and $HCO_3^- = 18$ meq/L. Therefore the concentration of Na^+ was probably close to
   (a) 167 meq/L   (b) 156 meq/L   (c) 145 meq/L   (d) 11 meq/L

10. The pH of a solution with a concentration of 0.001 M nitric acid is closest to

(a) 10^{-3} (b) 10^{-11} (c) 3   (d) 11

11. If a 0.10 M solution of the monoprotic acid, HX, is known to have a pH of 4.50, we know that the acid HX is
    (a) strong   (b) weak   (c) neutral   (d) slightly soluble

12. The end point in a titration of any acid by any base is reached when (pick the one best choice)
    (a) the pH of the solution is 7.00 (at room temperature)
    (b) the equivalence point is reached
    (c) the indicator is halfway through its characteristic change in color
    (d) an equal volume of acid has been mixed with an equal volume of base

13. If the formula weight of a monopratic acid is 165, then one milliequivalent of this acid has a mass of
    (a) 0.165 g   (b) 1.65 g   (c) 16.5 g   (d) 165 mg

14. A solution labeled as 0.1 N HCl contains
    (a) 0.1 equivalent of HCl per liter
    (b) 0.1 mole of HCl per liter
    (c) 100 milliequivalents of HCl per liter
    (d) 10 milliequivalents of HCl per liter

15. To prepare 250 mL of 0.250 $\underline{N}$ H_2SO_4, you would need how many grams of H_2SO_4?
    (a) 6.14 g   (b) 3.07 g   (c) 61.4 g   (d) 0.614 g

ANSWERS

ANSWERS TO SELF-TESTING QUESTIONS

Completion

1. 5.00
2. acidic
3. 1.00×10^{-7} moles/liter
4. 10^{-8}
5. acidic
6. (a) K^+, CO_3^{2-}
   (b) Yes; CO_3^{2-}

(c) No

(d) Raise it. (If CO_3^{2-} ties up some of the H^+ ions from the water, then there will be a slight excess of OH^- ions; the solution will have become slightly basic and, therefore, the pH will have gone up.)

7. more basic

8. $$HCO_3^- + H^+ \rightarrow H_2CO_3$$

9. $H_2CO_3 + OH^- \rightarrow HCO_3^- + H_2O$

10. carbonic acid (H_2CO_3); carbon dioxide (CO_2)

11.

(a) greater than 7

(b) Yes

(c) The most likely event for neutralizing H^+ is $H^+ + PO_4^{3-} \rightarrow HPO_4^{2-}$

(d) The most likely event for neutralizing OH^- is

$$OH^- + H_2PO_4^- \rightarrow HPO_4^{2-} + H_2O$$

12. 4.1 meq/L

13. 149 mg K^+

14. 25 meq/L; is

15. litmus and phenolphthalein (or any others in Figure 9.2. The two given as answers here are probably the most common and their characteristic colors in acids and bases should be learned.)

16. sodium acetate

17. greater than 7 (Sodium acetate hydrolyzes in water.)

18. 0.50 N

19. 0.20 N

20. 0.11 N

Multiple-Choice

1. c

2. a

3. a and e

4. b, d, and f

5. b and d

6. c

7. b

8. c

9. a

10. c   (Since HNO_3 is a strong acid we may assume that it is 100% ionized and that $0.001 M = [H^+]$. Hence, pH = 3 because $0.001 = 10^{-3}$.)

11. b   (If it were strong, its pH would be much closer to 1 since $0.10 = 1.0 \times 10^{-1}$.)

12. c

13. a and d

14. a, b, and c

15. b

10 Radioactivity and Nuclear Chemistry

OBJECTIVES

After you have studied this chapter and have worked the Practice Exercises and Review Exercises, you should be able to do the following. The first nine objectives are basic to the study of radioactivity and nuclear chemistry. Objectives 10 through 12 are most important for those who may one day work with radioisotopes. Finally, objectives 13 through 15 will help you in reading and studying current issues involving nuclear power and radioctive pollution.

1. Name three kinds of atomic radiations from natural sources.
2. Contrast a chemical change with a nuclear change.
3. Balance a nuclear equation when given all but one of the particles involved.
4. Contrast alpha, beta, and gamma rays according to their composition and their relative penetrating power.
5. Explain the term "half-life."
6. Do inverse square law calculations.
7. Describe the steps one can take to protect oneself from radiations.
8. Name three ways of detecting radiations.
9. Describe in general terms why radiations are damaging to cells.
10. Discuss five desirable properties of a radionuclide that is to be used in medical diagnosis.
11. Explain how technetium-99 is used in the diagnosis of disease.

12. Explain how I-131 or I-123 can be used in the diagnosis of disease.
13. Not all of the many units of radiation measurement (Section 10.3) may be assigned in your course. For those that are, name the unit used to describe each of the following (as assigned) and give its symbol (if it has one).
    (a) how radioactive is a sample;
    (b) how intense is an exposure to X rays (or gamma rays);
    (c) how much energy does a particular kind of radiation release in a given mass of tissue (irrespective of the kind of tissue);
    (d) how much energy is associated with a specific radiation;
    (e) how relatively stable is a particular radioactive isotope.
14. Describe the kinds of atomic events that occur in the burner reactor of a nuclear power plant.
15. Name three radioactive isotopes that can be released into the environment from a nuclear power plant.
16. Define each of the terms given in the Glossary.

GLOSSARY

Alpha (α) Particle. The nucleus of a helium atom; $_2^4\text{He}$.

Alpha (α) Radiation. A stream of high-energy alpha particles.

Background Radiation. Cosmic rays plus the natural atomic radiation emitted by the traces of radioactive isotopes in soils and rocks.

Becquerel (Bq). The SI unit for the activity of a radioactive source; one nuclear disintegration (or other transformation) per second.

$$1 \text{ curie } = 3.7 \times 10^{10} \text{ Bq}$$

Beta (β) Particle. The electron; $_{-1}^0 e$

Beta (β) Radiation. A stream of high-energy electrons.

Cosmic Radiation. A stream of ionizing radiations, from the sun and outer space, that consists mostly of protons but also includes alpha particles, electrons, and the nuclei of atoms up to atomic number 28.

Curie (Ci). A unit of activity of a radioactive source. $1 \text{ Ci } = 3.70 \times 10^{10}$ disintegrations/s

Electron Volt (eV). A very small unit of energy used to describe the
  energy of radiation.

$$1 \text{ eV} = 1.6 \times 10^{-19} \text{ joule}$$
$$1 \text{ eV} = 3.8 \times 10^{-20} \text{ calorie}$$
$$1000 \text{ eV} = 1 \text{ KeV (1 kiloelectron volt)}$$
$$1000 \text{ KeV} = 1 \text{ MeV (1 megaelectron volt)}$$

Fission. The splitting of the nucleus of a heavy atom approximately
  in half and that is accompanied by the release of one or a few
  neutrons and energy.

Gamma Radiation. A natural radiation similar to but more powerful
  than X rays.

Gray (Gy). The SI unit of absorbed dose of radiation equal to one
  joule of energy absorbed per kilogram of tissue.

Half-Life. The time needed for half of the atoms in a sample of a particular radioactive isotope to undergo radioactive decay.

Inverse Square Law. The intensity of radiation varies inversely with
  the square of the distance from its source.

Ionizing Radiation. Any radiation that can create ions from molecules
  within the medium that it enters, such as alpha, beta, gamma, X,
  and cosmic radiation.

Nuclear Chain Reaction. The mechanism of nuclear fission by which one
  fission event makes enough fission initiators (neutrons) to cause
  more than one additional fission event.

Nuclear Equation. A representation of a nuclear transformation in
  which the chemical symbols of the reactants and products include
  mass numbers and atomic numbers.

Rad. One rad equals 100 ergs (1×10^{-5} J) of energy absorbed per gram
  of tissue as a result of ionizing radiations.

Radiation. A process whereby light or heat is emitted; also the
  emitted light or heat. In atomic physics, the emission of some
  ray such as an alpha, beta, or gamma ray.

Radiation Sickness. The set of symptoms that develops following exposure to heavy doses of ionizing radiations.

Radical. A particle with one or more unpaired electrons.

Radioactive. The property of unstable atomic nuclei whereby they emit alpha, beta, or gamma rays.

Radioactive Decay. The change of a radioactive isotope into another isotope by the emission of alpha rays or beta rays.

Radioactive Disintegration Series. A series of isotopes selected and arranged such that each isotope but the first is produced by the radioactive decay of the preceding isotope and the last isotope is nonradioactive.

Radionuclide. A radioactive isotope.

Rem. One rem is the quantity of a radiation that produces the same effect in humans as one roentgen of X rays or gamma rays.

Roentgen. One roentgen is the quantity of X rays or gamma radiation that generates ions with an aggregate of 2.1×10^9 units of charge in 1 mL of dry air at normal pressure and temperature.

Threshold Exposure. The level of exposure to some toxic agent below which no harm is done.

Transmutation. The change of an isotope of one element into an isotope of a different element.

SELF-TESTING QUESTIONS

COMPLETION

Questions About General Aspects of Radioactivity

1. A substance that will ruin photographic film even when it is stored at room temperature in an air tight black wrapping is probably _____.

2. The nuclei of helium atoms are often called _____ particles.

3. If a substance emits beta radiation, its atoms have lost _____ from their _____, and their atomic number has changed by (how many) _____ unit(s) to a _____ value.
   (higher or lower)

4. If a substance emits gamma radiation, its nuclei have lost _____ and they have experienced a change in _____ unit(s) of atomic number and _____ unit(s) of atomic mass.

5. Traces of radionuclides in soils and rocks contribute to our _____

6. If a radioactive source initially weighing 8 grams had only 4 grams of that isotope left after 20 days, then the _____ of that source is 20 days.

7. If a radioactive source initially weighing 12 grams had only 3 grams of that isotope left after 8 days, its half-life is _____ days.

8. If a radionuclide emits an alpha particle, its atomic number changes by _____ unit(s) _____ ands its atomic mass
   (up or down)
   changes by _____ unit(s) _____.
   (up or down)

9. Radiations that closely resemble gamma radiation are called _____.

10. Because alpha radiation emitted from radioactive sources has enough energy to knock electrons from neutral molecules of air, we say that this radiation is an _____.

11. Complete and balance these nuclear equations.
    (a) $^{210}_{84}Po \rightarrow {}^{4}_{2}He +$ _____
    (b) $^{212}_{86}Rn \rightarrow {}^{208}_{84}Po +$ _____
    (c) $^{212}_{82}Pb \rightarrow {}^{0}_{-1}e +$ _____

Questions About Medical and Health Aspects of Radiations

12. Potentially, the most serious damage that a radiation can do to a cell is to hit its _____.

13. When a high-energy particle hits a water molecule, the molecule can lose first an electron and then a proton leaving an unstable particle with the formula _____.

14. Exposure to high intensity radiations can cause a set of symptoms that collectively are called _____.

15. Four kinds of serious, long-term consequences of exposure even to low levels of ionizing radiations are _____, _____, _____, and _____.

16. The most penetrating radiations are _____ and _____.

17. Two strategies for protecting oneself from exposure to the radiations of some radioactive source are _____ and _____.

18. The inverse square law states that _____ _____.

19. If the intensity of radiation, in arbitrary units, is 25.0 units at a point 5.00 m from a source, then at a point 10.0 m from the source the intesity will be _____ units.

20. If the intensity of radiation, in arbitrary units, is 100 units at a point 1.0 m from a source, then how many meters from the source would you have to move to reduce the exposure to 1.00 unit? _____

21. Because there are some naturally occurring radionuclides in such common substances as rocks and soil, we can never escape exposure to _____ radiation.

22. Three radionuclides produced as wastes in the operation of atomic reactors and the tissues they "seek" in the body are

<u>Radionuclide</u>	<u>Tissue It Seeks</u>
_____	_____
_____	_____
_____	_____

23. Which is more likely to be used in diagnosis and which in cancer therapy?

    X rays of 100 keV energy   _____

    X rays of 1.3 MeV energy   _____

24. When a film badge dosimeter is used, the amount of exposure to radiations is correlated with the degree of _____ on the film.

25. If a nucleus of a molybdenum-98 atom successfully captures a neutron, the product will be which radionuclide? _____

26. When selecting a radionuclide for use in diagnosis, the radiologist must find one with chemical properties that are compatible with the body, and besides this are five other properties:

    (1) _____

    (2) _____

    (3) _____

    (4) _____

    (5) _____

27. Complete the following table in the manner in which it is now partly filled. The first column contains the names of radioactive isotopes used in medicine and therapy. The second column gives the chemical form in which the isotope is administered. The third column is for a very brief statement describing the purpose(s) of the use of the isotope.

	Isotope	Form	Purpose(s)
(a)	technetium-99^m	TcO_4^-	_____
(b)	iodine-131	_____	_____
(c)	iodine-123	_____	_____

(d) cobalt-60 _____ _____

(e) phosphorus-32 _____ _____

Questions About Units of Radiation Measurements

28. Complete the following table relating the unit of measurement used to the kind of measurement taken.

Kind of Measurement	Common Unit Used	SI Unit
(a) The energy of the particles in a stream of alpha or beta rays or the energy of gamma rays	_____	_ _ _ _
(b) Exposure to X rays or gamma rays	_____	_ _ _ _
(c) Absorbed dose	_____	_____
(d) Activity of a radioactive source	_____	

29. The energy released in tissue when it is exposed to ionizing radiations is called the _____, which is commonly described in units of the _____.

30. A radioactive source with as many disintegrations per second as a 0.5 gram sample of radium has an activity, in common units, of _____.

31. To allow variations in the responses of different types of radiations to different types of tissues, the absorbed dose may be multiplied by various fractions (modifying factors or quality factors) to give a number called the _____, which is expressed in units of _____.

32. If a 0.10-gram sample of some radioactive source is rated as 5.0 microcuries in activity, a 0.20-gram sample would have an activity of _____.

33. If some radioactive source has a half-life of 15 years, then a 0.2-gram sample of this substance would have a half-life that is _____ as a 0.1-gram sample.
    (half, the same, or double)

MULTIPLE-CHOICE

Questions About General Aspects of Radioactivity

1. If a radiation from a radioactive element consists of high-energy helium nuclei, the radiation is called
   (a) an alpha ray
   (b) a beta ray
   (c) a gamma ray
   (d) cosmic rays

2. In the nuclear reaction $^{210}_{83}Bi \rightarrow ^{210}_{84}Po +$ _____ the other product is
   (a) an alpha particle
   (b) a beta particle
   (c) a neutron
   (d) a proton

3. In the nuclear reaction $^{230}_{90}Th \rightarrow ^{}_{88}Ra +$ alpha particle the blank line (the atomic mass of Ra-88) would contain
   (a) 230   (b) 226   (c) 234   (d) 229

Questions About Medical and Health Aspects of Radiations

4. The best material for shielding yourself from ionizing radiation is
   (a) concrete   (b) plexiglass   (c) glass   (d) lead

5. If you can move away from a radioactive source so that you have doubled your distance from it, your exposure to its radiations will be reduced by a factor of
   (a) two   (b) three   (c) four   (d) sixteen

6. If you were forced to live where you were exposed to a radioactive source, you would be wisest in selecting one with
   (a) a very short half-life
   (b) a very long half-life
   (c) gamma-ray emissions
   (d) beta-ray emissions

7. The principal reason for the danger associated with radioactivity is that the radiations
   (a) liberate a great amount of energy in tissue
   (b) cannot be shielded
   (c) generate ions and radicals in tissue
   (d) provoke a rapid rise in the white-cell count

Questions About Units of Radiation Measurements

8. The unit of the dose equivalent is the
   (a) curie   (b) roentgen   (c) rad   (d) rem

9. A radiologist selecting an isotope for use in diagnosis would try to find one that
   (a) is quickly eliminated by the body
   (b) has a short half-life
   (c) can do the job in the least concentration
   (d) is a gamma emitter only

10. A unit of the activity of a radioactive source is the
    (a) curie   (b) roentgen   (c) rad   (d) half-life

11. If exposure to radiation results in the absorption of 10 joule of energy per kilogram of tissue, then the tissue has absorbed
    (a) 1 rad   (b) 10 Gy   (c) 1 rem   (d) 100 rems

12. A radioactive source rated at 1 millicurie would be equivalent to
    (a) 100 curie   (c) 1,000 g of radium
    (b) 1 milligram of radium   (d) a half-life of 1 millisecond

ANSWERS

ANSWERS TO SELF-TESTING QUESTIONS

Completion

1. radioactive
2. alpha
3. electrons, nuclei, one, higher
4. energy (no particle is lost), no, no
5. background radiation
6. half-life
7. four
8. two, down, four, down
9. X rays
10. ionizing radiation
11. The missing formulas are   (a) $^{206}_{82}Pb$ (b) $^{4}_{2}He$ (c) $^{212}_{83}Bi$
12. nucleus

13. $\cdot \ddot{O}H$

14. radiation sickness
15. cancer, tumor, mutation, birth defect
16. X rays and gamma radiation
17. Use a dense shielding material like lead and get as far from the source as practical.
18. The intensity of radiation is inversely proportional to the square of the distance from the source.
19. 6.25
20. 10.0 m
21. background
22. strontium-90, bone seeker
    iodine-131, thyroid seeker
    cesium-137, general circulation
23. 100 keV - diagnosis
    1.3 MeV - therapy
24. fogging
25. molybdenum-99
26. (1) short $t_{1/2}$
    (2) decays to a nonradioactive product (or one with a long $t_{1/2}$)
    (3) a value of $t_{1/2}$ long enough for preparation and use
    (4) gamma-emitter only
    (5) give a hot spot or a cold spot

27.

	Isotope	Form	Purpose(s)
(a)	technetium-99^m	TcO_4^-	brain scanning
(b)	iodine-131	I^-	testing of thyroid function; treatment of thyroid cancer
(c)	iodine-123	I^-	same as for I-131
(d)	cobalt-60	Co	gamma ray source for cancer treatment
(e)	phosphorus-32	PO_4^{3-}	chronic leukemia treatment

28.

Common Unit Used	SI Unit
(a) electron volt	none
(b) rem	none
(c) rad	gray
(d) curie	becquerel

29. absorbed dose (or radiation absorbed dose), rad
30. 0.5 curie
31. dose equivalent, rems
32. 10 microcuries
33. the same

Multiple-Choice

1. a	5. c	9. a, b, c, d
2. b	6. b	10. a
3. b	7. c	11. b
4. d	8. d	12. b

11 Introduction to Organic Chemistry

OBJECTIVES

After you have studied this chapter and worked the Practice Exercises and Review Exercises in it, you should be able to do the following. Objectives 6 through 9 are particularly important because they concern the symbols that we'll use in later work with organic compounds.

1. State what the vital force theory was and how it affected research in its day.
2. Describe Wöhler's experiment and explain how it helped to overthrow the vital force theory.
3. List the main differences between organic and inorganic compounds.
4. Give the ways in which carbon is a unique element.
5. State how a molecular formula and a structural formula are alike and how they are different.
6. Explain why each possible conformation of a carbon chain does not represent a different compound.
7. Give an example (both names and structures) of two compounds that are related as isomers.
8. Write condensed structures from full structures.
9. Examine a pair of structures and tell if they are identical, are related as isomers, or are different.
10. Describe the molecular geometry of a carbon atom bonded to four other atoms.
11. Describe how an sp^3 hybrid orbital forms.

12. Explain how four sp^3 hybrid orbitals must point to the corners of a regular tetrahedron.
13. State the principle of maximum overlap and explain how it applies to the geometry of methane.
14. Give definitions for each of the terms listed in the Glossary.

GLOSSARY

Branched Chain. A sequence of atoms to which additional atoms are attached at points other than the ends.

Condensed Structure. (See Structure.)

Conformation. One of the infinite number of contortions of a molecule that are permitted by free rotations around single bonds.

Free Rotation. The absence of a barrier to the rotation of two groups with respect to each other when they are joined by a single, covalent bond.

Functional Group. An atom or a group of atoms in a molecule that is responsible for the particular set of reactions that all compounds with this group have.

Hybrid Orbital. An atomic orbital obtained by mixing two or more pure orbitals (those of the s, p, d, or f types).

Inorganic Compound. Any compound that is not an organic compound.

Isomerism. The phenomenon of the existence of two or more compounds with identical molecular formulas but different structures.

Isomers. Compounds with identical molecular formulas but different structures.

Nonfunctional Group. A section of an organic molecule that remains un-changed during a chemical reaction at a functional group.

Orbital Hybridization. The mixing of two or more ordinary atomic orbitals to give an equal number of modified atomic orbitals, called hybrid orbitals, each of which possesses some of the characteristics of the originals.

Sigma Bond (σ-Bond).   A covalent bond associated with a molecular orbital whose shape is symmetrical about the bonding axis.

sp^3 Hybrid Orbital.   One of four equivalent hybrid orbitals formed by the mixing of one s orbital and three p orbitals and whose axes point to the corners of a regular tetrahedron.

Straight Chain.   A continuous, open sequence of covalently bound carbon atoms from which no additional carbon atoms are attached at interior locations of the sequence.

Structure.   Synonym for structural formula.   (See Structural Formula.)

Vital Force Theory.   A discarded theory that held that organic compounds could be made in the laboratory only if the chemicals possessed a vital force contributed by some living thing.

DRILL EXERCISES

I. EXERCISES ON CONDENSED STRUCTURES

The rules for converting from a full structure to a condensed structure are:

1.   H
      |
  H — C —   becomes CH_3 (sometimes H_3C)
      |
      H

2.   H
        | |
      H — C — H or — C — becomes CH_2
        | |
                            H

3.   H
     | |
  H — C — or — C — becomes CH
     | |

4. All bonds of hydrogen to carbon, oxygen, nitrogen, or sulfur may be "understood."

5. Any single bond between heavy atoms (C — C, C — O, C — N, C — S, for example) may be shown or it may be understood, provided that it would normally be drawn horizontally and, therefore, is part of the main chain.

6. Any single bond between heavy atoms that hold substituents onto the main chain must be shown. (In other words, we always show a substituent group attached to a main chain by writing the group either above the chain or below it. and by connecting the group to the main chain by a line representing the bond.)

7. All double and triple bonds must be shown. (A modification of this rule will appear in a later chapter when we study carbonyl compounds.)

Practice converting between full and condensed structures with these additional exercises.

SET A.   Condense each of the following structures.

```
        H   H   H
        |   |   |
1.  H — C — C — C — H          _____
        |   |   |
        H   H   H
```

```
        H   H   H   H
        |   |   |   |
2.  H — C — C — C — C — H      _____
        |   |   |   |
        H   H   H   H
```

```
        H   H   H   H   H
        |   |   |   |   |
3.  H — C — C — C — C — C — H  _____
        |   |   |   |   |
        H   H   H   H   H
```

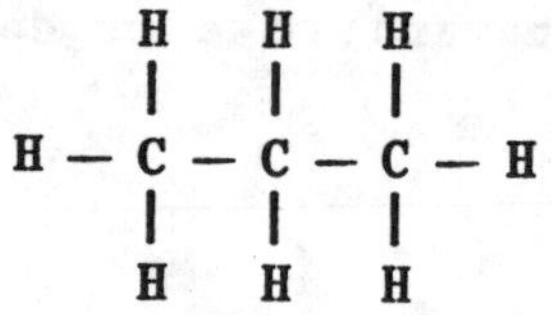

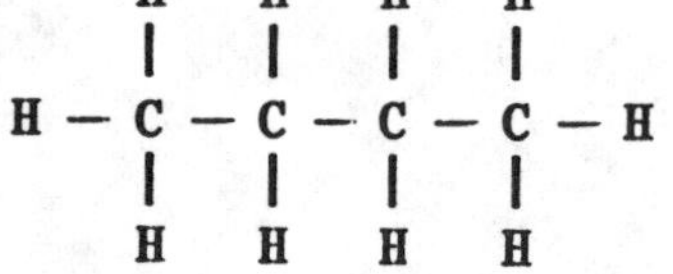

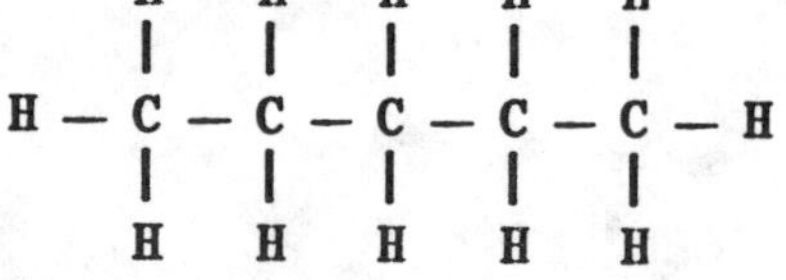

```
                        H
                        |
                    H — C — H
                        |
     H   H   H   H      H
     |   |   |   |      |
4. H—C — C — C — C — C — C — H   _____
     |   |   |   |      |
     H   |   H   H      H
         |              |
     H — C — H      H — C — H
         |              |
         H              H
```

```
          H
          |
      H — C — H
          |
      H   |      H   H   H
      |   |      |   |   |
5. H—C — C — O — C — C — C — H   _____
      |   |      |   |   |
      H   H      H   H   H
                 |
             H — C — H
                 |
                 H
```

```
          H
          |
      H — C — H                  H
          |                      |
      H   |      H   H   H   O   H
      |   |      |   |   |   |   |
6. H—C — C — C — C — C — C — C — H   _____
      |   |      |   |   |   |   |
      H   |      H   H   H   H   H
          |
      H — C — H
          |
          H
```

7.
$$
\begin{array}{ccccc}
 & H & H & H & \\
 & | & | & | & \\
H - & C & - C & - C & - O - H \\
 & | & | & | & \\
 & H & H & H &
\end{array}
$$
_____

8.
$$
\begin{array}{ccccccc}
 & H & H & H & H & H & \\
 & | & | & | & | & | & \\
H - O - & C & - C & - C & = C & - C & - H \\
 & | & | & & & | & \\
 & H & H & & & H &
\end{array}
$$
_____

9.
$$
\begin{array}{ccccccccc}
 & H & & H & H & H & H & O & \\
 & | & & | & | & | & | & \| & \\
H - & C & - O - & C & = C & - C & - C & - C & - O - H \\
 & | & & & & | & | & & \\
 & H & & & & H & | & &
\end{array}
$$

$$
\begin{array}{c}
H - C - H \\
| \\
H
\end{array}
$$
_____

10.
$$
\begin{array}{cccccc}
 & H & H & H & H & H & \\
 & | & | & | & | & | & \\
H - & C & - C & - C & - C & - C & - N - H \\
 & | & | & | & | & | & \\
 & H & O & H & H & H &
\end{array}
$$

$$
\begin{array}{c}
H - C - H \\
| \\
H
\end{array}
$$
_____

SET B. Write the full structures that correspond to these condensed structures. Show all single bonds.

1. CH_3CH_3

2.
$$
\begin{array}{c}
CH_3 \\
| \\
CH_3CCH_2CH_2CH_3 \\
| \\
CH_3
\end{array}
$$

3.

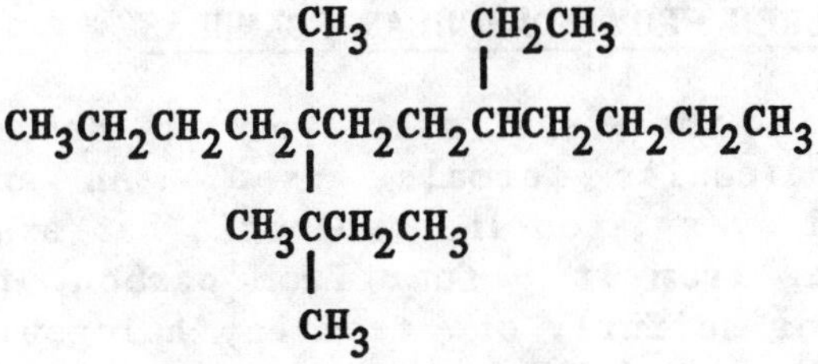

$$\overset{\overset{\displaystyle CH_3}{|}}{\underset{\underset{\displaystyle CH_3}{|}}{\underset{\displaystyle CH_3CCH_2CH_3}{|}}} \quad \overset{\overset{\displaystyle CH_2CH_3}{|}}{}$$

CH₃CH₂CH₂CH₂CCH₂CH₂CHCH₂CH₂CH₂CH₃

CH₃CCH₂CH₃

CH₃

4. $CH_3CH = CH - CH_3$

5. $HOCH_2CH_3$

6. 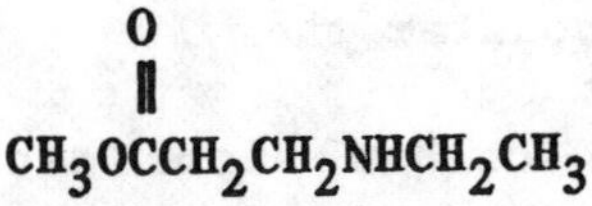 $CH_3O\overset{\overset{\displaystyle O}{\|}}{C}CH_2CH_2NHCH_2CH_3$

7. 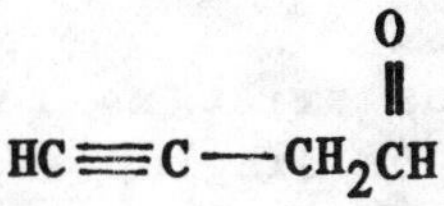 $HC \equiv C - CH_2\overset{\overset{\displaystyle O}{\|}}{C}H$

8. $NH_2CH_2CH_2NH_2$

II. EXERCISES ON DEVISING STRUCTURES FROM MOLECULAR FORMULAS

Review Exercise 11.12 in the text asks you to devise a structural formula from each molecular formula given. An answer cannot be correct unless each and every atom in a molecule has exactly the correct number of bonds going from it – four from carbon, three from nitrogen, two from oxygen (or sulfur), one from any halogen, and one from hydrogen. If the question in the text gives you trouble, see how the following semi-systematic approach works. The number of rules may seem a bit awesome, but you will quickly advance beyond needing them.

Rule 1. Identify those atoms in the given molecular formula that must have two, three, or four bonds – all of the atoms other than a halogen or hydrogen. (We'll work with these to establish a "skeleton" for the structure upon which the hydrogen and halogen atoms will be hung. Think of each covalence of every atom (4 for carbon, etc.) as an unused arm. A correct structure will be one in which all the arms are in some way linked. (There can be "no unjoined arms" could be another way of stating it.)

Rule 2. The number of unused valences ("free arms") left over on the skeleton – **all being single bonds** – must equal the number of hydrogens or halogens to be appended.

Rule 3. Don't put a halogen on anything but carbon – not on oxygen, not on sulfur, and not on nitrogen. (Although such possibilities do exist, we'll never encounter them.)

Rule 4. In writing a structure, assemble the carbon atoms along a horizontal line as much as possible and pin the substituents to the resulting chain of carbon atoms. (This rule has nothing to do with rules of valence, only with looks.)

When atoms have more than one bond, options exist as to how they may occur in structures. All of the bonds may be single, or some may be single and some double or even triple. Here are all the possible options for carbon, nitrogen, oxygen, the halogens, and hydrogen.

$$\text{For carbon} \qquad -\overset{\displaystyle |}{\underset{\displaystyle |}{C}}- \qquad \overset{\diagdown}{\underset{\diagup}{C}}= \qquad -C\equiv$$

$$\overset{\displaystyle |}{\text{For nitrogen}} \quad -N- \qquad -N= \qquad\qquad N\equiv$$

For nitrogen $-\overset{\displaystyle |}{N}-$ $-N=$ $N\equiv$

For oxygen $-O-$ $O=$

For halogen $-X$ (X = F, Cl, Br, or I)

For hydrogen $-H$

Now let's work a few sample exercises.

<u>Sample</u> <u>Exercise</u> <u>1</u>:   Write a structural formula for CH_4O.

Step 1. The multivalent atoms are C and O.   (Rule 1)

Step 2. Assemble the options:

For carbon   For oxygen

$$-\overset{\displaystyle |}{\underset{\displaystyle |}{C}}- \qquad\qquad\qquad -O-$$

$$\overset{\diagdown}{\underset{\diagup}{C}}= \qquad\qquad\qquad O=$$

$$-C\equiv$$

Step 3. Try the possible combinations of those in the first column
with those in the second.

(a) $-\overset{\displaystyle |}{\underset{\displaystyle |}{C}}--O-$ (Leaves only single bonds and leaves four, what we need)

(b) $-O\diagdown$ ... $C=$ (Leaves a double bond. Discard; rule 2.)

(c) $C==O$ (Leaves only two spots for hydrogens; not enough. Discard.)

(d) $-O-C\equiv$ (Leaves a triple bond. Discard, rule 2.)

Step 4. Identify the skeleton, or skeletons, (rules 1 and 2) and append the hydrogens.

In this sample exercise only one skeleton meets the rules, the one generated by combination (a).

$$
-\overset{|}{\underset{|}{C}}--O- \ +\ 4H- \ \longrightarrow \ H--\overset{H}{\underset{H}{\underset{|}{C}}}--O--H \quad \text{or} \quad H-\overset{H}{\underset{H}{\underset{|}{C}}}-O-H
$$

(the answer)

We may condense the answer to CH_3OH.

Step 5. Run a mental check; does the carbon have 4 bonds; the oxygen, 2; and the hydrogens 1 each? They do, therefore, the structure obeys the rules of covalence.

<u>Sample Exercise 2</u>:   Convert C_2H_4 to a structural formula.

Step 1. The multivalent atoms are just the two Cs.
Step 2. Assemble the options:

First Carbon   Second Carbon

$$
-\overset{|}{\underset{|}{C}}- \qquad\qquad -\overset{|}{\underset{|}{C}}-
$$

$$
\overset{\backslash}{\underset{/}{C}}= \qquad\qquad \overset{\backslash}{\underset{/}{C}}=
$$

$$
-C\equiv \qquad\qquad -C\equiv
$$

Step 3. Try combinations from each column. Remember that the remaining unused bonds must equal four in number, for the four hydrogens, and that these remaining bonds must all be single bonds. Mentally discarding all combinations that leave double or triple bonds "open," we have these possibilities:

$$
-\overset{|}{\underset{|}{C}}--\overset{|}{\underset{|}{C}}- \qquad\qquad \overset{\backslash}{\underset{/}{C}}\!==\!\overset{/}{\underset{\backslash}{C}} \qquad\qquad -C\equiv C-
$$

The first leaves room for six hydrogens, too many; the last for two hydrogens, too few. The middle will work; room is left for four hydrogens.

Step 4. Put the pieces together:

$$\begin{array}{ccc} \diagdown \quad \diagup \\ C{=\!=}C \\ \diagup \quad \diagdown \end{array} \quad + \quad 4H - \quad \longrightarrow \quad \begin{array}{ccc} H \qquad H \\ \diagdown \quad \diagup \\ C{=}C \\ \diagup \quad \diagdown \\ H \qquad H \end{array}$$

The condensed structure is: $CH_2{=\!=}CH_2$

Step 5. Mentally check for obedience to the rules of covalence. Each carbon has four bonds; each hydrogen has one.

<u>Sample Exercise</u> <u>3</u>: Convert C_2H_4O into a structural formula that obeys the rules of covalence.

Step 1. The multivalent atoms are two carbons and one oxygen.
Step 2. Assemble the options:

Carbon	Carbon	Oxygen
$\begin{array}{c} \mid \\ -\,C\,- \\ \mid \end{array}$	$\begin{array}{c} \mid \\ -\,C\,- \\ \mid \end{array}$	$-\,O\,-$
$\begin{array}{c} \diagdown \\ C{=\!=} \\ \diagup \end{array}$	$\begin{array}{c} \diagdown \\ C{=} \\ \diagup \end{array}$	$O{=}$
$-\,C\!\equiv$	$-\,C\!\equiv\!\!\!\!\equiv$	

Step 3. Try the possible combinations, remembering that the skeleton must have no double or triple bond left "hanging" and must have, via single bonds only, room for four hydrogens. Here are the possibilities that have only single bonds remaining. (Beneath each is a number equalling the spaces for hydrogens.)

$$\begin{array}{cccc} \mid\;\; \mid & \mid\;\; \mid & \diagdown\;\; \mid & \mid\;\; \mid \\ -\,C{-}C{-}O\,- & -\,C{-}O{-}C\,- & C{=\!=}C{-}O\,- & -\,C{-}C{=\!=}O \\ \mid\;\; \mid & \mid\;\; \mid & \diagup & \mid \\ (6) & (6) & (4) & (4) \end{array}$$

Any others you may have drawn, that avoid leaving open double or triple bonds, will be duplicates of one or more of these.

(For instance,

$$-\overset{|}{O}--\overset{|}{C}--\overset{|}{C}-\text{ is the same as }-\overset{|}{C}--\overset{|}{C}--\overset{|}{O}-;\quad O{=}{=}\overset{|}{C}--\overset{|}{C}-\text{ is the same}$$

as $\qquad -\overset{|}{C}--\overset{|}{C}{=}{=}O.)$ The last two combinations have four

spaces for hydrogens. We must consider them both.

Step 4. Put the pieces together:

$$\underset{/}{\overset{\backslash}{C}}{=}{=}\overset{|}{C}--O-\ +\ 4H-\ \rightarrow\ \underset{\underset{H}{|}}{\overset{\overset{H}{|}}{\overset{\backslash}{C}}}{=\!\!=}\overset{\overset{H}{|}}{C}-O-H\ \text{ or }\ CH_2{=\!\!=}CHOH$$

$$-\overset{|}{C}--\overset{|}{C}{=}{=}O\ +\ 4H-\ \rightarrow\ H-\underset{\underset{H}{|}}{\overset{\overset{H}{|}}{C}}-\overset{\overset{H}{|}}{C}{=\!\!=}O\ \text{ or }\ CH_3-CH{=\!\!=}O$$

Usually, you will see carbon-oxygen double bonds aligned
vertically, and the last structure will usually be seen as

$$CH_3-\overset{\overset{O}{\|}}{C}-H.$$ (Remember that not all of the single bonds in the
main chain need be "understood.")

Step 5. Check the answer: four bonds from the carbons; two from the
oxygens. Either structure is correct; the two are isomers.

Write structural formulas for the following molecular formulas.

1. H_2O_2 _____

2. N_2H_4 _____

3. CH_4S _____

4. C_2H_5Br _____

5. C_2H_3Cl _____

SELF-TESTING QUESTIONS

<u>COMPLETION</u>

1. If compound X boils at 176 ^{O}C, it is almost certainly in the
   family of _____ compounds. We would assign it to
                    (ionic or molecular)
   the organic class if, when analyzed, it was found to contain
   _____.

2. The structure

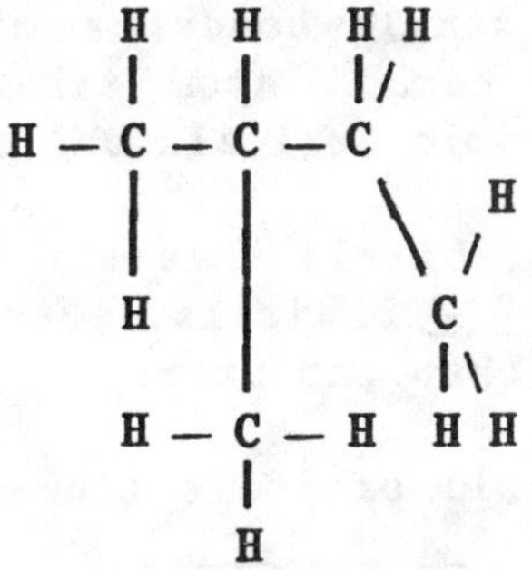

```
        H   H   H H
        |   |   |/
   H — C — C — C
        |   |    \  H
        |   |     / 
        H   |    C
            |    |\
   H — C — H   H H
        |
        H
```

   could be most neatly condensed to _____.

3. The structure

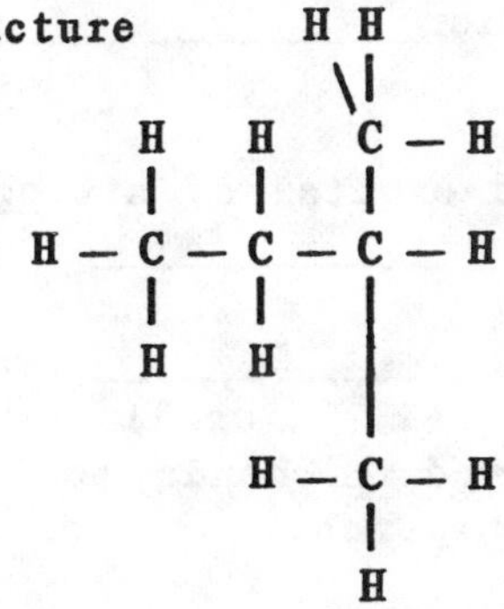

```
             H H
              \|
     H   H    C — H
     |   |    |
H — C — C — C — H
     |   |    |
     H   H    |
              H — C — H
                  |
                  H
```

   could be most neatly condensed to _____.

4. Are the compounds whose formulas are given questions 2 and 3
   isomers or are they identical? _____

5. Write full structural formulas for each of the following:

_____   _____   _____   _____
(a) CH_5N (b) CH_2Br_2 (c) C_3H_8 (d) C_3H_4

6. Write the condensed structure of at least one isomer of $CH_3CH_2CH_2OH$ (C_3H_8O). _____

7. The electron configuration of an isolated atom of carbon is _____.

8. When a carbon atom forms four single bonds, a mixing of some of the orbitals of the isolated carbon atom takes place to give _____ new, hybrid atomic orbitals called _____
(1, 2, 3, or 4)
hybrid orbitals. The original orbitals that mix are the _____ orbital and the three _____ orbitals. Each hybrid orbital has _____ lobes, one larger than the other.

9. The individual axes of the hybrid orbitals (question 8) point to the corners of a _____ and these axes make angles of _____ with each other.

10. The electron configuration of an atom of nitrogen when it is forming _____ single bonds is _____.
            (1, 2, or 3)

11. In ammonia, one valence level hybrid orbital of nitrogen does not form a bond to hydrogen, but it holds _____ electrons.

12. In the water molecule, there are _____ valence level
                                    (1, 2, or 3)
hybrid orbitals that are not involved in bonds to hydrogen but that contain _____ electrons each.

MULTIPLE-CHOICE

1. The compounds in which of the following pairs are related as isomers?   HO
                                                |
(a) $CH_2 = CH - CH_2 - OH$ and $CH_2 - CH = CH_2$

(b) H_2 and D_2

(c) $CH_3 - CH_2 - NH - CH_3$ and $CH_3 - CH_2 - CH_2 - NH_2$

(d) $CH_3 - C \equiv C - CH_2NH_2$ and $CH_3 - CH_2 - CH_2 - C \equiv N$

2. According to all the rules, which compound(s) should not be possible?

(a) $AlPO_4$ (b) $CH_3CH = C = CH_2$ (c) $H - O - \overset{\overset{\displaystyle O}{\|}}{C} - O - H$ (d) $CH = NH$

3. A structural formula for C_5H_{12} would be

(a) $CH_3(CH_2)_3CH_3$

(b) $CH_3CH_2CH_2CH_2CH_3$

(c) $CH_3\underset{\underset{\displaystyle CH_3}{|}}{CH}CH_2CH_3$

(d) $CH_3\underset{\underset{\displaystyle CH_3}{|}}{\overset{\overset{\displaystyle CH_3}{|}}{C}}CH_3$

4. Which (if any) of the choices in question 3 are identical compounds?

5. Which (if any) of the choices in question 3 are isomers?

6. The orbitals used by carbon in forming the bonds in CH_4 are

(a) $2s^2$ and $2p_x^2$

(b) $2s^2$, $2p_x^1$ and $2p_y^1$

(c) $1s^1$, $2s^1$, sp_x^1 and $2p_y^1$

(d) $2(sp^3)^1$, $2(sp^3)^1$, $2(sp^3)^1$, and $2(sp^3)^1$

7. The bond angle in H_2O is closest to
(a) 180° (b) 105° (c) 90° (d) 65.5^o

8. Using hybrid orbitals, instead of the half-filled 2p orbitals, of nitrogen to form the bonds to the three hydrogen atoms in NH_3 results in stronger bonds because
(a) the large lobe of the hybrid orbital extends farther and can

overlap better with hydrogen's 1s orbital.
(b) the hybrid orbitals have two pair of electrons instead of only one, and thus they can form stronger bonds with half-filled 1s orbitals of hydrogen atoms.
(c) the hybrid orbitals permit the N — H bonds to have axes that are closer to each other than they can be in any alternative bonding network.
(d) none of the above

ANSWERS

ANSWERS TO DRILL EXERCISES

I. Exercises on Condensed Structures

Set A
1. $CH_3CH_2CH_3$ or $CH_3 - CH_2 - CH_3$ (Single bonds along the main chain

may be shown or they may be understood. You will see both practices often.
2. $CH_3CH_2CH_2CH_3$ or $CH_3 - CH_2 - CH_2 - CH_3$

3. $CH_3CH_2CH_2CH_2CH_3$ or $CH_3 - CH_2 - CH_2 - CH_2 - CH_3$

$$\begin{matrix} & CH_3 & & \\ & | & & \end{matrix}$$
4. $CH_3CHCH_2CH_2CCH_3$
   with CH_3 and CH_3 branches

$$CH_3$$
5. $CH_3CH - O - CH_2CHCH_3$ with CH_3 branch

6. $CH_3CCH_2CH_2CH_2CHCH_3$ with CH_3, OH, CH_3 substituents

7. $CH_3CH_2CH_2OH$ (often seen as $CH_3CH_2CH_2 - O - H$)

8. $HOCH_2CH_2CH = CHCH_3$ (Do not write this as $OHCH_2CH_2CH = CHCH_3$.
It implies $O - H - CH_2 -$ etc., which is wrong.)

9. $CH_3 - O - CH = CHCH_2\underset{\underset{CH_3}{|}}{CH}\overset{\overset{O}{\|}}{C}OH$

10. $CH_3\underset{\underset{\underset{CH_3}{|}}{\underset{O}{|}}}{CH}CH_2CH_2CH_2NH_2$ or $CH_3 - O - \underset{\underset{CH_3}{|}}{CH}CH_2CH_2CH_2NH_2$

Set B

1.
$$
\begin{array}{ccc}
 & H & H \\
 & | & | \\
H - & C - & C - H \\
 & | & | \\
 & H & H
\end{array}
$$

2.
$$
\begin{array}{c}
H \\
| \\
H - C - H \\
\end{array}
$$

$$
\begin{array}{ccccc}
H & H & H & H \\
| & | & | & | \\
H - C - C - C - C - C - H \\
| & | & | & | \\
H & H & H & H
\end{array}
$$

$$
\begin{array}{c}
H - C - H \\
| \\
H
\end{array}
$$

3.

```
                      H           H   H
                      |           |   |
                  H — C — H   H — C — C — H
                      |           |   |
                      |           |   H
                      |           |
  H   H   H   H       |   H   H   |   H   H   H   H
  |   |   |   |       |   |   |   |   |   |   |   |
H—C — C — C — C — C — C — C — C — C — C — C — C—H
  |   |   |   |       |   |   |   |   |   |   |   |
  H   H   H   H       |   H   H   H   H   H   H   H
                      |
                      |   H   H
                  H   |   |   |
                  |   |   |   |
              H — C — C — C — C — H
                  |   |   |   |
                  H   |   H   H
                      |
                      |
                  H — C — H
                      |
                      H
```

4.

```
  H   H   H   H                      H           H
  |   |   |   |                      |           |
H—C — C = C — C—H   or   H — C — C = C — C — H      (The two are
  |       |                      |   |   |   |        equivalent.)
  H       H                      H   H   H   H
```

5.

```
          H   H
          |   |
H — O — C — C — H
          |   |
          H   H
```

6.

```
      H       O   H   H   H   H
      |       ‖   |   |   |   |
H — C — O — C — C — C — N — C — C — H
      |           |   |       |   |
      H           H   H       H   H
```

7.

```
              H   O
              |   ‖
H — C ≡ C — C — C — H
              |
              H
```

8.

```
      H   H   H   H
      |   |   |   |
H — N — C — C — N — H
          |   |
          H   H
```

II. Exercises on Devising Structures from Molecular Formulas

1. H — O — O — H

2. H — N — N — H
       | |
       H   H

3.
       H
       |
  H — C — S — H
       |
       H

4.
      H H
      | |
  H — C — C — Br   (Satisfy yourself that
      | |
      H H

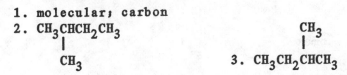

are identical.)

5.
      H H
      | |
  H — C = C — Cl

ANSWERS TO SELF-TESTING QUESTIONS

Completion

1. molecular; carbon

2. $CH_3CHCH_2CH_3$
      |
     CH_3

3.
             CH_3
             |
  $CH_3CH_2CHCH_3$

(Your answers to 2 and 3 may be correct but still not look exactly like those given here. All that matters is that they show the correct nucleus-to-nucleus sequence while obeying the accepted conventions for condensing. One of these conventions is that as much of the structure as possible is written out on one line. Should you have questions about your answers, consult your instructor or one of the assistants.)

4. identical

5. (a) 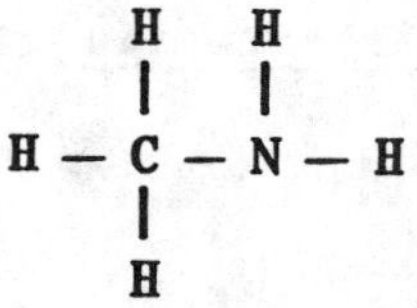
$$\begin{array}{c} \text{H} \quad \text{H} \\ | \quad | \\ \text{H} - \text{C} - \text{N} - \text{H} \\ | \\ \text{H} \end{array}$$

(c)
$$\begin{array}{c} \text{H} \quad \text{H} \quad \text{H} \\ | \quad | \quad | \\ \text{H} - \text{C} - \text{C} - \text{C} - \text{H} \\ | \quad | \quad | \\ \text{H} \quad \text{H} \quad \text{H} \end{array}$$

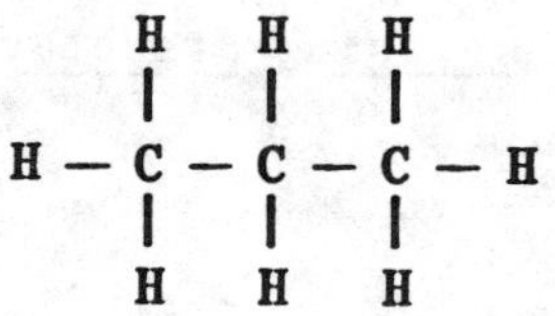

(b) 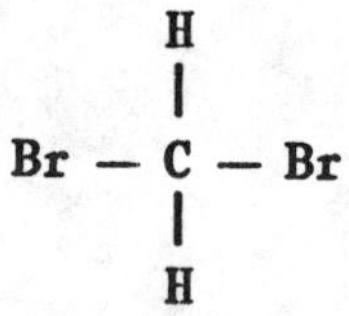
$$\begin{array}{c} \text{H} \\ | \\ \text{Br} - \text{C} - \text{Br} \\ | \\ \text{H} \end{array}$$

(d) 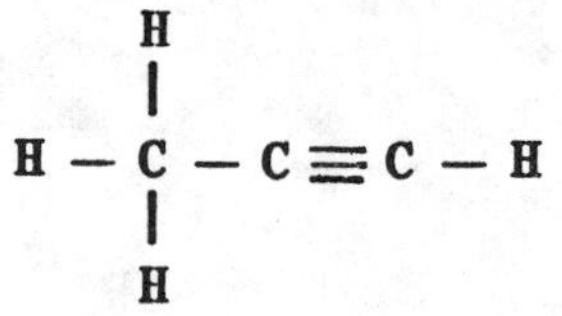
$$\begin{array}{c} \text{H} \\ | \\ \text{H} - \text{C} - \text{C} \equiv \text{C} - \text{H} \\ | \\ \text{H} \end{array}$$

6. There are two isomers: $CH_3 - O - CH_2CH_3$ and CH_3CHCH_3.

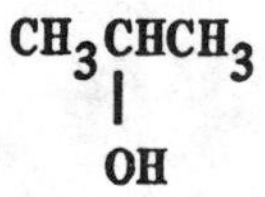

7. $1s^2 2s^2 2p_x^1 2p_y^1$

$\qquad\qquad\qquad\qquad\qquad\qquad\qquad\qquad$ OH

8. 4; sp^3; 2s; 2p; two
9. regular tetrahedron; 109.5°

10. 3; $1s^2 2(sp^3)^2 2(sp^3)^1 2(sp^3)^1 2(sp^3)^1$
11. two
12. 2; two

Multiple-Choice

1. c and d
2. d
3. a, b, c, and d
4. a and b
5. a (or b), c, and d are isomers.
6. d
7. b
8. a

12 Saturated Hydrocarbons Alkanes and Cycloalkanes

OBJECTIVES

In this chapter, we continue the very important task of learning how to write organic structures and how to name organic compounds. We also learn two valuable "map signs" that will help us to relate the properties of substances to their structures:

1. All substances whose molecules are wholly or mostly hydrocarbons are insoluble in water; and
2. Alkanes or alkane-like regions of molecules of other compounds do not react with strong alkalis, strong acids, water, reducing agents, or most oxidizing agents.

After you have studied this chapter and worked the Practice Exercises and Review Exercises in it, you should be able to do the following:

1. Recognize if a substance is a saturated hydrocarbon from its structure.
2. Write the IUPAC names of all of the straight-chain alkanes (through decane) and give their carbon content (the number of carbon atoms per molecule).
3. Write the name and structure of the next higher homolog of any straight-chain alkane (through nonane).
4. Write the names and structures of all alkyl groups through the C_4 set.

5. Write the common names of all alkanes through the C_4 set.
6. Write the IUPAC names of alkanes and cycloalkanes from their structures.
7. Write the structures of alkanes and cycloalkanes from their IUPAC names.
8. Recognize the primary, secondary, and tertiary carbons in a structure.
9. Write an equation for the complete combustion of any hydrocarbon.
10. Write an equation for the chlorination of any given alkane showing the structures of all the possible isomeric monochloro products.
11. Define and, where applicable, give an example of each of the terms in the Glossary.

GLOSSARY

Alkane. Any saturated hydrocarbon, one that has only single bonds. A normal alkane is any whose molecules have straight chains.

Alkyl Group. A substituent group that is an alkane minus one H atom.

Homolog. Any member of a homologous series of organic compounds.

Homologous Series. A series of organic compounds in the same family whose successive members differ by individual CH_2 units.

Hydrocarbon. Any organic compound that consists entirely of carbon and hydrogen.

International Union of Pure and Applied Chemistry System (IUPAC System). A set of systematic rules for naming compounds and designed to give each compound one unique name and for which only one structure can be drawn; the Geneva system of nomenclature.

IUPAC System. (See International Union of Pure and Applied Chemistry System.)

Like–Dissolves–Like Rule. Polar solvents dissolve polar or ionic solutes and nonpolar solvents dissolve nonpolar or weakly polar solutes.

Nomenclature. The system of names and the rules for devising such names, given structures, or for writing structures, given names.

Nonfunctional Group.   A section of an organic molecule that remains
   unchanged during a chemical reaction at a functional group.

Ring Compound.   A compound whose molecules contain three or more atoms
   joined in a ring.

Primary Carbon.   In a molecule, a carbon atom that is joined directly
   to just one other carbon, such as the end carbons in $CH_3CH_2CH_3$.

Saturated Compound. A compound whose molecules have only single bonds.

Secondary Carbon.   Any carbon atom in an organic molecule that has two
   and only two bonds to other carbon atoms, such as the middle
   carbon atom in $CH_3CH_2CH_3$.

Tertiary Carbon.   Any carbon in an organic molecule that has three and
   only three bonds to adjacent carbon atoms.

Unsaturated Compound.   Any compound whose molecules have a double or a
   triple bond.

DRILL EXERCISES

EXERCISES ON STRUCTURAL FEATURES
   The following set of structures illustrates several features
discussed in the text. Supply the information requested by writing
the identifying number(s) of the structure(s).

A. $CH_3CH_2CH_2CH_2CH_2CH_3$

$$
\text{B. } CH_3CH \overset{\overset{\displaystyle CH_3}{|}}{-} \underset{\underset{\displaystyle CH_3}{|}}{CHCH_3}
$$

C. $CH_3CH = CHCH_2CH_2CH_3$

D. $CH_3CH_2CH_2OH$

E.

F.

1. Give the letters of
   (a) the saturated compounds _____
   (b) the straight–chain compounds _____
   (c) the branch–chain compounds _____
   (d) any pairs of compounds related as isomers _____
   (e) any compound with a 3° carbon _____

2. Write the IUPAC name of A _____
                        of B _____
                        of E _____

3. Write the structures of E and F using the geometric figure method
   illustrated in the discussion of "Ring Compounds" in Section 12.1
   of the text.

   _____   _____
              E   F

4. Write the structure and give
   the IUPAC name for the next
   lower homolog of A.   _____

5. Write the full structure of CH₃

   _____

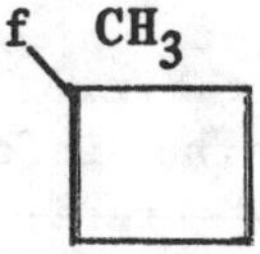

6. Write the full structure of Cl

   _____

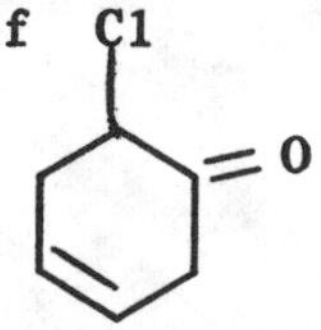

Examine the following compounds and supply the information requested.

A. CH₃ CH₃   B.   CH₃   C.   CH₃
                                     CH₃   CH₃

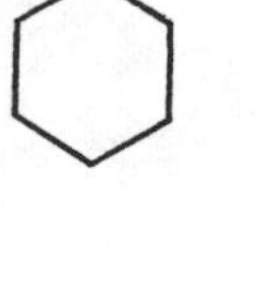

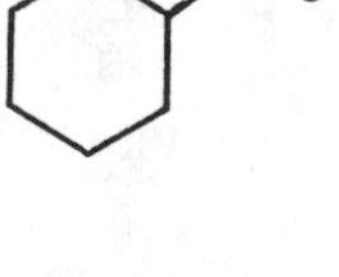

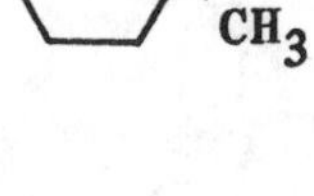

D.

E.   CH$_3$

F.   CH$_3$

CH$_3$

CH$_3$

CH$_3$

CH$_3$

CH$_3$

7. Group in sets the letters of structures representing identical
   compounds.   _____

8. Write IUPAC names for A _____
                       for E _____

9. To develop skill in the rapid recognition of alkyl groups regardless of how they are oriented in space, place the name of each
   group on the line beneath its structure.

(a)   CH$_3$
          /
     — CH$_2$

(b)   CH$_3$
             |
      — CH$_2$ — CH — CH$_3$

(c)   CH$_3$
       \
        CH —
       /
      CH$_3$

_____

_____

_____

(d)   — CH$_2$ CH$_3$
           | |
         CH$_2$ — CH$_2$

(e)   CH$_3$
          |
   CH$_3$ — CH — CH$_2$ —

(f)   CH$_3$
          |
   CH$_3$ — C — CH$_3$
          |

_____

_____

_____

(g)   CH$_3$
             |
     CH$_3$ — C —
             |
            CH$_3$

(h)   CH$_3$
          |
    — CH — CH$_3$

(i)   CH$_2$—
          |
   CH$_3$ — CH — CH$_3$

_____

_____

_____

(j) |
     CH$_3$ — CH — CH$_3$

(k) CH$_3$ — CH$_2$ — CH — CH$_3$
                      |

(l) |
     CH$_3$ — CH — CH$_2$ — CH$_3$

_____

_____

_____

(m) 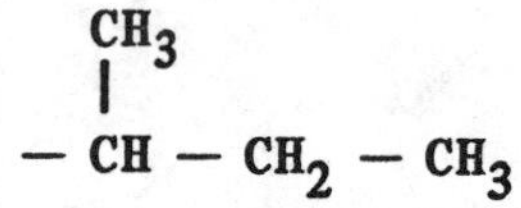
$$CH_3$$
$$|$$
$$- CH - CH_2 - CH_3$$

(n) 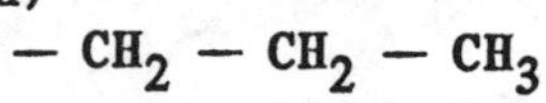
$$- CH_2 - CH_2 - CH_3$$

(o) 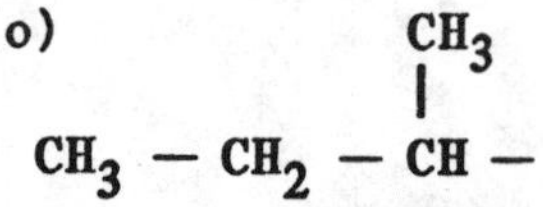
$$CH_3$$
$$|$$
$$CH_3 - CH_2 - CH -$$

_____

(p) 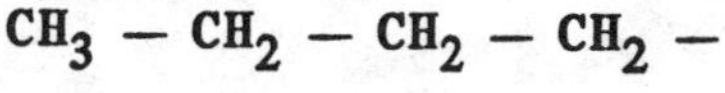
$$CH_3 - CH_2 - CH_2 - CH_2 -$$

(q) 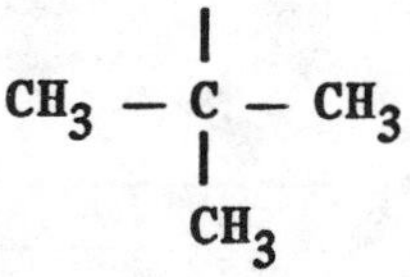
$$|$$
$$CH_3 - C - CH_3$$
$$|$$
$$CH_3$$

(r) 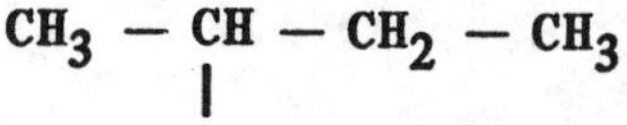 $CH_3 - CH - CH_2 - CH_3$
        $|$

_____

10. To develop the skill of looking at a structure and visualizing it with its longest continuous chain on one horizontal line, write the IUPAC names of each of the following.

(a) 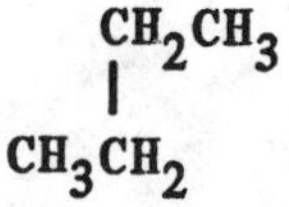
$$CH_2CH_3$$
$$|$$
$$CH_3CH_2$$

(b) 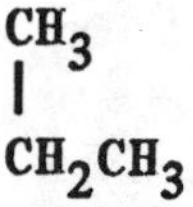
$$CH_3$$
$$|$$
$$CH_2CH_3$$

(c) 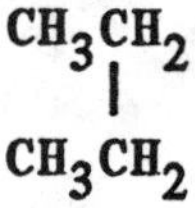
$$CH_3CH_2$$
$$|$$
$$CH_3CH_2$$

_____

(d) 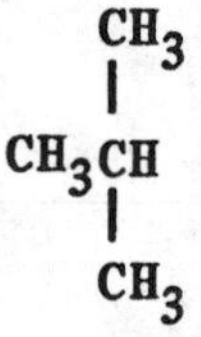
$$CH_3$$
$$|$$
$$CH_3CH$$
$$|$$
$$CH_3$$

(e) 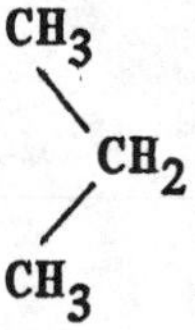
$$CH_3$$
$$\backslash$$
$$CH_2$$
$$/$$
$$CH_3$$

_____

        Before going on to the Self–Testing Question, be sure to work all of the Practice Exercises and Review Exercises in the text first. As always, use the Self–Testing Questions as a "final exam" for the chapter.

SELF-TESTING QUESTIONS

COMPLETION

1. The IUPAC name of $CH_3CH_2CH_2CH_3$ is _____.

2. The structure of 1-chloro-2,3-dimethylbutane is

_____

3. The structure of isobutane is

_____

4. Write the structure of each compound on the line above the name.

(a) _____   (b) _____
    isopropyl chloride   t-butyl bromide

(c) _____   (d) _____
    isobutyl chloride   sec-butyl bromide

(e) _____   (f) _____
    butyl iodide   propyl chloride

5. Give the common names of the following compounds.
   (a) the isomer of butane with a 3° carbon   _____
   (b) the isomer of butane with two 2^O carbons   _____
   (c) the isomer of butane with three 1^O carbons _____

6. When hexane burns completely in air the balanced equation is:

_____

7. The IUPAC names and structures of all the monobromo derivatives
   of pentane are:

8. Pentane has an isomer that can form only one monochloro derivative. Write the structure of this isomer of pentane.

_____

9. The structure of 1,3-dimethylcyclohexane is

_____

MULTIPLE-CHOICE

1. A family of organic compounds containing only carbon and hydrogen and having only single bonds are the
   (a) alkenes   (b) alkanes   (c) alkynes   (d) cycloalkanes

2. The combustion of butane produces
   (a) butylene and water   (c) an alcohol
   (b) carbon dioxide and water   (d) isobutane

3. The compound CH_3 ⬡ is

   (a) methylcyclopentane   (c) cycloheptane
   (b) a saturated compound   (d) cyclohexane

4. The name of

$$CH_3CH_2CH_2\underset{\underset{CH_3-CH-CH_3}{|}}{\overset{\overset{CH_2CH_2CH_3}{|}}{CH}}CHCH_2CH_2CH_2CH_3$$

   (a) 4-isobutyl-5-propylnonane
   (b) 2-methyl-3,4-dipropyloctane
   (c) 5-propyl-6-isopropylnonane
   (d) 4-isopropyl-5-propylnonane

5. The chlorination of 2-methylpentane would produce how many isomeric monochloro compounds?
   (a) 6   (b) 5   (c) 4   (d) 3

6. Because carbon and hydrogen have very similar electronegativities, hydrocarbons are generally
   (a) electropositive   (c) polar
   (b) nonpolar   (d) hydrogen-bonded

7. The compound shown has how many hydrogens per molecule?
   (a) 8   (b) 7   (c) 12   (d) 10

8. The common name of $CH_3 - CH - CH_2 - Cl$ is

    $\qquad\qquad\qquad\qquad\qquad\qquad\overset{|}{\underset{CH_3}{}}$

    (a) butyl chloride   (c) t-butyl chloride
    (b) isobutyl chloride   (d) sec-butyl chloride

9. The common name of 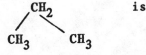 is

    (a) cyclopropane   (c) propane
    (b) isopropane   (d) dimethylmethane

10. The compound $CH_3 - C \equiv C - H$ is
    (a) soluble in water   (c) soluble in gasoline
    (b) a hydrocarbon   (d) unsaturated

ANSWERS

ANSWERS TO DRILL EXERCISES

Exercises on Structural Features
  1. (a) A, B, D, and E
     (b) A, C, and D   ("The term straight-chain" cannot apply to
                        rings. Among open-chain compounds, as long as
                        the atoms C, O, N, or S are all in a continuous
                        sequence with or without double or triple bonds,
                        the chain is "straight." If the atoms O, N, or
                        S are appended as part of substituents, the
                        chain is still straight if all the carbons occur
                        in a continuous sequence.)
     (c) B (Ring compounds are not classified as branched, either.)
     (d) A and B; C and E
     (e) B (Only saturated carbons are designated as 1^O, 2^O, or 3°.)
  2. A hexane
     B 2,3-dimethylbutane
     E cyclohexane

  3. E   F   O

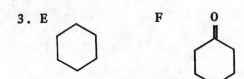

4. $CH_3CH_2CH_2CH_2CH_3$ pentane

5.
```
CH3
  \
CH —— CH2
 |      |
CH2 —— CH2
```

6.
```
      Cl
      |
      CH
     /    \
  CH2      C = O
  |        |
  CH      CH2
     \    /
      CH
```

7. A and C; B and D

8. A 1,1-dimethylcyclohexane
   E 1,3-dimethylcyclohexane   (**NOT**   1,5-dimethylcyclohexane, from counting around the ring in the wrong way. You count from position one, picked as the location of one CH_3 group, along the shortest path around the ring to the next group.)

9. (a) ethyl   (Actually, ethyl group, but we may omit "group.")
   (b) isobutyl (Any four-carbon alkyl group must be one of the four butyl groups. The two derived from isobutane are either the isobutyl or the t-butyl group.)
   (c) isopropyl   (Any three-carbon alkyl group must be one of the two propyl groups - propyl or isopropyl.)
   (d) butyl   (Straighten out the chain and you have a group based on butane, not isobutane. The two butyl groups based on butane are the butyl and the sec-butyl groups.)
   (e) isobutyl   (Compare b and e)
   (f) t-butyl   (It's the only butyl group where the unused bond is at a 3° carbon.)
   (g) t-butyl   (This is simply f rotated through part of a circle.)
   (h) isopropyl   (A C_3-alkyl group; it must be either propyl or isopropyl.)
   (i) isobutyl   (Compare b, e, and i.)
   (j) isopropyl   (Compare c, h, and j.)
   (k) sec-butyl   (It's the only butyl group where the unused bond is at a 2^O carbon.)
   (1) sec-butyl   (Compare with k.)
   (m) sec-butyl   (Carefully compare k, 1, and m. After chain straightening, the chain is straight in all three and the free bond comes from a 2^O carbon.)
   (n) propyl

(o) sec-butyl   (Compare with m and the accompanying note.)
(p) butyl
(q) t-butyl   (Just f tipped upside down. They have to be the same. Do you become someone else if you stand on your head?)
(r) sec-butyl

10. (a) butane   (d) isobutane or 2-methylpropane
    (b) propane   (e) propane
    (c) butane

ANSWERS TO SELF-TESTING QUESTIONS

Completion

1. butane

2.
$$CH_3$$
$$|$$
$$Cl - CH_2CHCHCH_3$$
$$|$$
$$CH_3$$

3. CH_3CHCH_3
   $|$
   CH_3

4. (a) CH_3CHCH_3 (b) $CH_3\overset{\displaystyle CH_3}{\underset{\displaystyle CH_3}{C}} - Br$ (c) $CH_3CHCH_2 - Cl$ with CH_3 above
        $|$
        Cl

   (d) $CH_3CHCH_2CH_3$ (e) $CH_3CH_2CH_2CH_2 - I$ (f) $CH_3CH_2CH_2 - Cl$
        $|$
        Br

5. (a) isobutane   (b) butane   (c) isobutane

6. $2C_6H_{14} + 19O_2 \rightarrow 12CO_2 + 14H_2O$

7. $CH_3CH_2CH_2CH_2CH_2 - Br$ $CH_3CH_2CH_2CHCH_3$ $CH_3CH_2CHCH_2CH_3$
                                              $|$ $|$
                                              Br Br

   1-bromopentane   2-bromopentane   3-bromopentane

8.
$$CH_3$$
$$|$$
$$CH_3 - C - CH_3$$
$$|$$
$$CH_3$$

9.

Multiple-Choice

1. b and d	6. b
2. b	7. d
3. b	8. b
4. d	9. c
5. b	10. b, c, d

13 Unsaturated Hydrocarbons

OBJECTIVES

The two topics in this chapter that are most important in our development of the molecular basis of life are:

1. The addition reactions of the carbon–carbon double bond, particularly those with hydrogen and water. (Double bonds occur in all fats and oils. They also occur in the intermediate stages of the chemical breakdown of sugars, fatty acids, and amino acids as well as in the enzymes for these reactions.)
2. The fact that benzene rings in aromatic hydrocarbons do not behave as alkenes, in spite of being highly unsaturated. These rings take part in substitution reactions rather than in addition reactions. (The benzene ring occurs in a few amino acids as well as in many drugs.)

Except for the loss of completeness, the chemistry of the triple bond may be omitted, particularly if the study of organic chemistry in your course is intended to serve only the needs of biochemistry. Likewise, the details of how benzene rings undergo substitution reactions may be left out. The two things you do need from aromatic chemistry are the ability to recognize an aromatic system, especially the benzene ring, and the knowledge that it does not add water, does not react with dilute acids or bases, and is quite stable to reducing agents and to oxidizing agents (provided the ring does not hold an — OH or an — NH_2 group).

The details concerning the special covalent bonds that occur in alkenes and benzene rings — bonds made from hybrid orbitals and the pi bond — are necessary background for understanding why the double bond is reactive toward many chemicals that leave the carbon–carbon single bond alone.

The specific objectives for this chapter are the following. Return to these to test yourself after you have completed studying all of the chapter, including the Practice Exercises and Review Exercises.

1. Recognize whether a substance is an alkene, an alkyne, or an aromatic compound from its molecular structure.
2. Write the structures of alkenes from their IUPAC names.
3. Write the IUPAC names of alkenes.
4. Predict whether a given alkene can exist as cis and trans isomers.
5. Write equations for the reactions of a given alkene with: (a) hydrogen; (b) a halogen; (c) a hydrogen halide; (d) sulfuric acid; and (e) water (in the presence of acid).
6. Use Markovnikov's Rule to predict the correct products in the addition reactions of alkenes.
7. Explain why Markovnikov's Rule works.
8. Correctly use terms associated with polymerizations, like polymer, monomer, polymerization, and promoter.
9. Describe the structure of benzene in molecular orbital terms.
10. Explain why benzene takes part in substitution reactions rather than in addition reactions.
11. Write equations for the reactions of benzene that result in halogenation, nitration, and sulfonation.
12. Write the common names of all of the monosubstituted benzenes given in this chapter.
13. Devise names for derivatives of benzene having two or more substituents.
14. Describe how sp^2 orbitals arise.
15. Describe how the sigma and pi bonds of a double bond are constructed.
16. Define each of the terms in the Glossary and give examples or illustrations where applicable.

GLOSSARY

Addition Reaction. Any reaction in which two parts of a reactant molecule add to a double or a triple bond.

Aliphatic Compound. Any organic compound whose molecules lack a benzene ring or a similar structural feature.

Aromatic Compound. Any organic compound whose molecules have a benzene ring (or a feature very similar to this).

Carbocation. Any cation in which a carbon atom has just six outer level electrons; a carbonium ion.

Geometric Isomerism. Isomerism caused by restricted rotations that give different geometries to the same structural organization; cis-trans isomerism.

Geometric Isomers. Isomers whose molecules have identical atomic organizations but different geometries; cis-trans isomers.

Markovnikov's Rule. In the addition of an unsymmetrical reactant to an unsymmetrical double bond of a simple alkene, the positive part of the reactant molecule (usually H^+) goes to the carbon that has the greater number of hydrogen atoms and the negative part goes to the other carbon of the double bond.

Monomer. Any compound that can be used to make a polymer.

Pi Bond (π Bond). A covalent bond formed when two electrons fill a molecular orbital created by the side-to-side overlap of two p orbitals.

Pi (π) Electrons. The pair of electrons in a pi bond.

Polymer. Any substance with a very high formula weight whose molecules have a repeating structural unit.

Polymerization. A chemical reaction that makes a polymer from a monomer.

sp^2 Hybrid Orbital. A hybrid orbital made by mixing one s orbital with two p orbitals to form three new, identical orbitals whose axes are in one plane and point to the corners of an equilateral triangle.

Substitution Reaction. A reaction in which one atom or group replaces another atom or group in a molecule.

KEY MOLECULAR "MAP SIGNS"

Key Molecular "Map Signs" In Organic Molecules	The Associated Chemical and Physical Properties
If the molecule is largely (or, of course, entirely) hydrocarbonlike	Expect the substance to be relatively insoluble in water and relatively soluble in organic solvents. Expect the substance to be less dense than water.
Alkanelike portions of all molecules	In these portions, expect no chemical changes involving water, acids, alkalies, or chemical oxidizing agents or reducing agents.
An alkene double bond Markovnikov's Rule applies – – – –	Adds H — H; becomes saturated Adds X — X; forms di-halo compounds (X = Cl, Br) { Adds H — X; forms alkyl halides (X = F, Cl, Br, I) Adds H — OH; forms alcohols Adds H — OSO_3H; forms alkyl hydrogen sulfates Adds its own kind; polymers form
Benzene ring	Not like an alkene - gives substitution, not addition reactions. Ring can nitrate, sulfonate, and halogenate (with iron(III) salts as catalysts)

DRILL EXERCISES

The chapter in the text has a number of drills on the reactions of alkenes and benzene. Do these Exercises first.

EXERCISE I

To review condensed structures, write out the following as full structures.

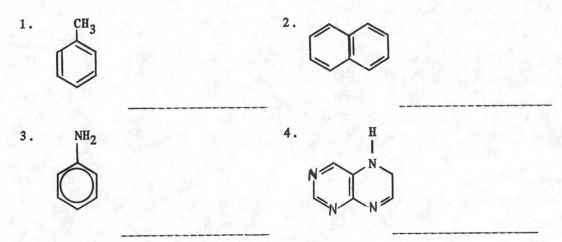

1. CH$_3$

2.

3. NH$_2$

4. H

EXERCISE II

As drill in several things — names, cis-trans isomers, and the recognition of nonequivalent positions — write the structures and the IUPAC names of all of the monochloro derivatives of all of the alkenes through the C$_4$ alkenes. Be sure to show cis and trans forms with correct geometry; for example, the structure of cis-2-chloro-2-butene is:

$$
\begin{array}{ccc}
CH_3 & & CH_3 \\
\diagdown & & \diagup \\
& C = C & \\
\diagup & & \diagdown \\
H & & Cl
\end{array}
$$

There are a total of 16 compounds.

EXERCISE III

Memorize the names and structures of the monosubstituted benzenes given in the text. Then study how to use the designations o-,
m- and p- for disubstituted benzenes. When you have done this study,
test yourself with these. Beneath each structure write the correct
name.

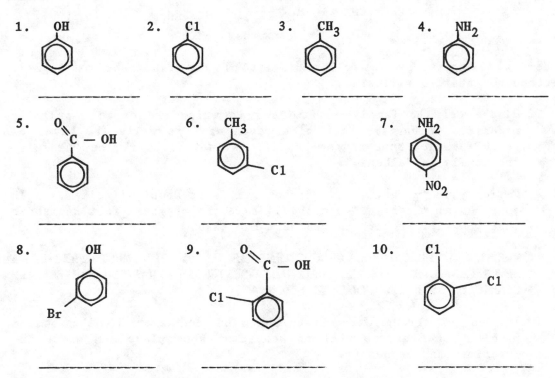

How to Work Problems in Organic Reactions

The overall goal of the chapters on organic chemistry in the
text is to learn the chemical properties of functional groups. You
know you've reached this goal when you can write the products made by
the reaction of a given set of starting materials. In other words, it
is not enough to be able to recognize the names and structural
features of functional groups – although you can't do anything else
without starting here. It's also not enough to memorize a sentence
that summarizes a chemical fact, such as "alkenes react with hydrogen
in the presence of a metal catalyst, heat, and pressure to give
alkanes" – although memorizing such sentences is also absolutely

essential to the goal. Such sentences are fundamental facts about the world in which we live. What you have to be able to do is apply such knowledge to a specific set of reactants. A typical problem, for example, is

"What forms, if anything, in the following reaction:

$$CH_3 - CH = CH_2 \ + \ H_2 \ \xrightarrow[\substack{\text{heat,}\\ \text{pressure}}]{\text{Ni}} \ ?"$$

Here's the strategy to follow in all of the reactions of organic compounds that we will study.

1. Figure out the family or families to which the organic starting materials belong(s). You might even want to write the names of these families beneath the specific structures. (In our example, the family is "alkene.")

2. Review your "memory list" of chemical facts about all alkenes. In your mental "storage," you should have the chemical-fact sentences that you will prepare in the next Exercise.

   FOR EVERY FUNCTIONAL GROUP THAT WE STUDY YOU MUST PREPARE A MEMORIZED LIST OF CHEMICAL-FACT SENTENCES THAT SUMMARIZE THE CHEMICAL PROPERTIES THAT WE STUDY.

3. If you go through the list and find no match to the stated problem, then assume that no reaction takes place and write "no reaction" as the answer.

4. When you find the match between the specific problem and one of the listed chemical properties, stop and construct the structure of the answer. The many worked examples in the text develop the patterns for doing this, and study them thoroughly first.

EXERCISE IV. SUMMARIZING THE CHEMICAL PROPERTIES OF ALKENES

When you have completed this exercise, you will have a complete sentence summarizing each kind of chemical reaction of the carbon-carbon double bond that we have studied. These statements are the chemical properties of this double bond that must be memorized, but they have to be learned in such a way that you can apply them to specific situations. Hence, following this Exercise there are several drill problems to give you practice. As you work each specific exercise among these drills, repeat to youself the sentence statement that summarizes the property being illustrated. This kind of repeated reinforcement will soon give you a surprisingly good working knowledge of these organic reactions, and you will be able to apply what you have learned to much more complicated situations with ease.

1. The alkene double bond reacts with hydrogen (in the presence of a metal catalyst and under pressure and heat) to give _____ _____.

2. The alkene double bond reacts with chlorine to give _____ _____.

3. The alkene double bond reacts with bromine to give _____ _____.

4. The alkene double bond reacts with hydrogen chloride to give _____ _____.

5. The alkene double bond reacts with hydrogen bromide to give _____ _____.

6. The alkene double bond reacts with concentrated sulfuric acid to give _____.

7. The alkene double bond reacts with water (in the presence of an acid catalyst) to give _____.

8. The complete combustion of any alkene gives _____ and _____.

Another way to organize chemical facts about a functional group such as the carbon-carbon double bond is by means of a 5 x 8" note card. An example is given on the following page, but for the remaining functional groups it is vitally important that you prepare the cards yourself. Part of the learning process is in this preparation, and having someone else do it for you robs you of that benefit.

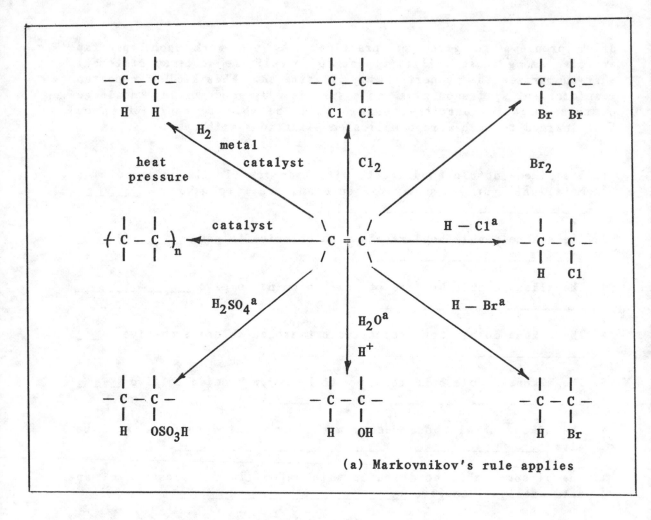

(a) Markovnikov's rule applies

Notice that the functional group is put in the center and its chemical properties are arranged about it. All of the arrows point outward. Later on, it will be useful to prepare cards summarizing key reactants, like H_2O or H_2 on which you'll list all of the reactions studied involving these substances.

EXERCISE V.   DRILL ON THE ADDITION OF HYDROGEN TO ALKENES

Write the structures of the products of the following reactions. If no reaction occurs, write "no reaction." (See Example 13.3 in the text.)

1. $\overset{\overset{\displaystyle CH_3}{|}}{CH_2 = CCH_2CH_3}$ + H_2 $\xrightarrow[\text{heat, pressure}]{\text{Ni}}$ _____

2. $CH_3CH = CHCH_3$ + H_2 $\xrightarrow[\text{heat, pressure}]{\text{Ni}}$ _____

3. 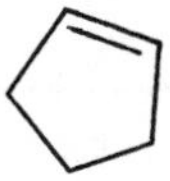   + H_2 $\xrightarrow[\text{heat, pressure}]{\text{Ni}}$ _____

4. $\overset{\overset{\displaystyle CH_3}{|}}{CH_2 = CCH = CH_2}$ + $2H_2$ $\xrightarrow[\text{heat, pressure}]{\text{Ni}}$ _____

5. 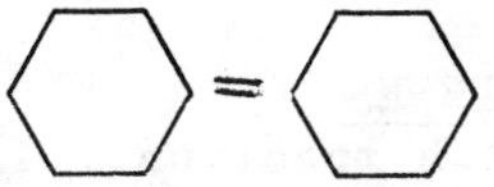   + H_2 $\xrightarrow[\text{heat, pressure}]{\text{Ni}}$ _____

EXERCISE VI.   DRILL ON THE ADDITION OF CHLORINE OR BROMINE TO ALKENES

Write the structures of the products of the following reactions. If no reaction occurs, write "no reaction." (See Example 13.4 in the text.)

1. $CH_2 = CHCH_3$ + Cl_2 $\longrightarrow$ _____

2. $\overset{\overset{\displaystyle CH_3}{|}}{CH_2 = CCH_2CH_2CH_3}$ + Br_2 $\longrightarrow$ _____

3. 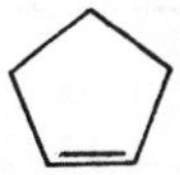   + Cl_2 $\longrightarrow$ _____

4. $CH_3CH = CHCH = CH_2$ + $2Br_2$ $\longrightarrow$ _____

5. $CH_2 =$ + Cl_2 $\longrightarrow$ _____

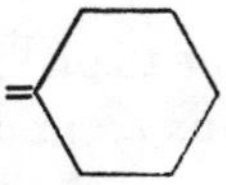

EXERCISE VII.   DRILL ON THE ADDITION OF H — Cl(g) OR H — Br(g) TO ALKENES

Write the structures of the products of each reaction. If no reaction occurs, write "no reaction." (See Example 13.5 in the text.)

1. $CH_3CH = CHCH_3$ + $H - Cl$ $\xrightarrow{\hspace{2cm}}$ _____

2. ⬠ + HBr $\xrightarrow{\hspace{2cm}}$ _____

3. CH₃ above
   $CH_2 = CCH_2CH_2CH_3$ + $H - Cl$ $\xrightarrow{\hspace{2cm}}$ _____

4. ⬡—CH_3 + HBr $\xrightarrow{\hspace{2cm}}$ _____

5. $CH_2 = CHCH_2CH_2CH_2CH = CH_2$ + $2HBr$ $\xrightarrow{\hspace{2cm}}$ _____

EXERCISE VIII.   DRILL ON THE ADDITION OF H_2O TO ALKENES

Write the structures of the products of each reaction. If no reaction occurs, write "no reaction." Note: H^+ represents an acid catalyst. (See Example 13.5 in the text.)

1. $CH_2 = CHCH_2CH_3$ + H_2O $\xrightarrow{H^+}$ _____

2. $CH_3CH = CHCH_3$ + H_2O $\xrightarrow{H^+}$ _____

3. CH_3 above
   $CH_2 = CCH_2CH_3$ + H_2O $\xrightarrow{H^+}$ _____

4. ⬠ + H_2O $\xrightarrow{H^+}$ _____

5. CH_3CH_2—⬡ + H_2O $\xrightarrow{H^+}$ _____

SELF-TESTING QUESTIONS

Use the self-testing questions as a final examination for the chapter after you have worked the exercises in the text and those given above.

COMPLETION

1. In a molecule of ethylene, the number of atoms whose nuclei are all in the same plane equals _____.

2. The structures of the two geometric isomers of 2-pentene are:

_____   _____

3. In a cis isomer, two reference groups lie on _____ side(s) of the double bond.

4. Write the overall equation for the reaction of 1,4-dimethylbenzene with each of the following reagents, assuming that only mono-substitution occurs.

(a) Br_2 ($FeBr_2$ or Fe catalyst) _____

(b) HNO_3 (in concd H_2SO_4)   _____

(c) H_2SO_4, concd   _____

5. When hydrogen chloride adds to isobutylene, the intermediate organic cation is _____ and has the structure:
                           (1°, 2°, or 3°)

_____

6. The other organic cation that, at least on paper, could also form from isobutylene is _____ and has the structure:
                           (1°, 2°, or 3°)

_____

7. Arrange these organic cations in the order of their increasing stability by arranging the letters that identify them in the correct order on this line: _____ < _____ < _____
                           least stable   most stable

A. $CH_3CH_2{}^+$ B. $CH_3CH_2\overset{\displaystyle CH_3}{\underset{\displaystyle CH_3}{\overset{|}{\underset{|}{C}}}}{}^+$ C. $CH_3CH_2{}^+CHCH_3$

8. Complete these equations by writing the structure(s) of the product(s).

(a) $CH_3C = CHCH_3 \quad + \quad HCl_{(gas)} \quad \longrightarrow \quad$ _____
  |
  CH_3

(b) $CH_2 = CCH_2CH_3 \quad + \quad H_2O \quad \xrightarrow{\ H^+\ }$ _____
         |
        CH_3

(c) $= CH_2 \quad + \quad H_2 \quad \xrightarrow[\text{pressure}]{\text{Ni, heat}}$ _____

(d) $-CH = CH_2 \quad + \quad Br_2 \quad \longrightarrow$ _____

(e) $CH_3 -$ $+ \quad H_2SO_4 \quad \longrightarrow$ _____

(f) $CH_3 -$ $+ \quad HCl_{aq} \quad \longrightarrow$ _____

(g) $CH_2 = CH—CH_2 — CH = CH_2 + 2H_2 \xrightarrow[\text{pressure}]{\text{Ni, heat}}$ _____

9. The two types of covalent bonds in a double bond are _____ _____ and _____ bonds.

10. The pi bond results from the side-to-side overlap of two _____ orbitals on adjacent carbons.

MULTIPLE-CHOICE

1. A family of organic compounds containing only carbon and hydrogen and having only single bonds are the
   (a) alkenes　　(b) alkanes　　(c) alkynes　　(d) cycloalkanes

2. A family of organic compounds whose molecules will add water (under an acid catalysis) and change into alcohols are the
   (a) alkenes　　　　　　　　(c) aromatic hydrocarbons
   (b) alkanes　　　　　　　　(d) cycloalkenes

3. A chemist, handed a sample of an organic compound, was told that

it was either $CH_2 = CHCH_2CH_3$ or $CH_3CH_2CH_2CH_3$. He could decide
which one it was by determining if the sample would react with
(a) sodium hydroxide
(b) hydrogen (with nickel present, heated under pressure)
(c) sodium chloride
(d) water (in the presence of an acid catalyst)

4. An aromatic hydrocarbon can be expected to undergo
   (a) substitution reactions   (c) addition reactions
   (b) reaction with water   (d) no reactions

5. The combustion of 1-butene will produce
   (a) butylene and water   (c) an alcohol
   (b) carbon dioxide and water   (d) cyclobutane

6. An isomer of

$$CH_3 \diagdown \diagup CH_3 \\ CH = CH \quad is$$

(a)
$$CH_3 \diagdown \diagup CH_3 \\ CH_2 - CH_2$$

(b)
$$CH_3 \diagup \\ CH_2 - CH_2 \\ CH_3$$

(c)
$$CH_3 \diagup \\ CH = CH \\ CH_3$$

(d)
$$CH_3 \diagdown \\ CH = CH_2$$

7. CH_3^+ is the
   (a) methyl radical   (c) methyl anion
   (b) methyl group   (d) methyl cation

8. The substance
$$CH_3 - \overset{\overset{\displaystyle CH_3}{|}}{\underset{\underset{\displaystyle CH_3}{|}}{C}} - OH$$
could be made by the addition of
   water to

(a) $CH_3CH = CHCH_3$

(b) $CH_2 = C \overset{\diagup CH_3}{\diagdown CH_3}$

(c) $CH_2 = CHCH_2CH_3$ (d) none of these

9. The substance $Cl - CH_2 - \overset{\overset{\displaystyle CH_3}{|}}{CH} - CH_3$ could be made by the addition

of HCl to

(a) $CH_3CH = CHCH_3$

(b) $CH_2 = C$

(c) $CH_2 = CHCH_2CH_3$

(d) none of these

10. The substance 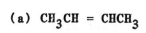 is

(a) benzene   (b) aromatic   (c) a diene   (d) none of these

11. Action of aqueous alkali on CH_3-⬡ could be expected to produce

(a) CH_3-⬡-OH

(b) CH_3-⬡ CH

(c) $HO - CH_2$-⬡

(d) no reaction

12. Action of aqueous hydrochloric acid on CH_3-⬡ could be expected to produce

(a) CH_3-⬡-Cl

(b) CH_3-⬡ Cl

(c) CH_3-⬡ Cl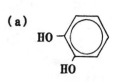

(d) no reaction

13. Action of aqueous potassium permanganate on benzene could be expected to produce

(a) HO-⬡ HO

(b) HO-⬡ OH

(c) HO-⬡

(d) no reaction

14. The trans isomer of 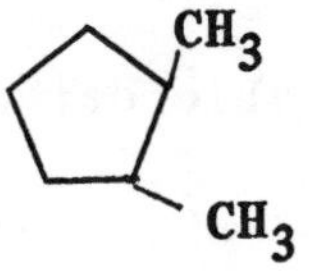 is

(a) 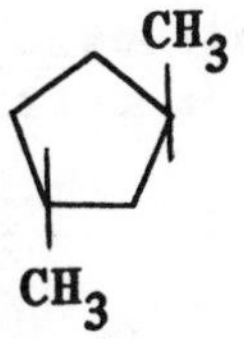   (b) 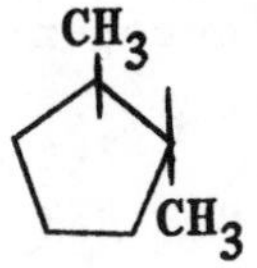   (c)    (d)

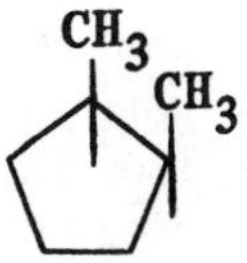

ANSWERS

<u>ANSWERS</u> <u>TO</u> <u>DRILL</u> <u>EXERCISES</u>
<u>Exercise</u> <u>I</u>

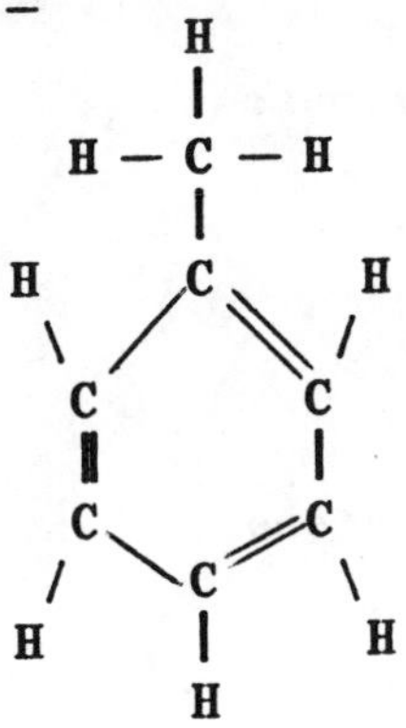

1.

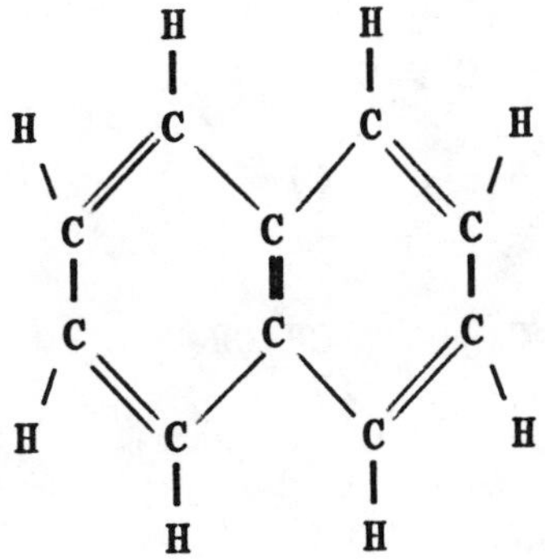

2.

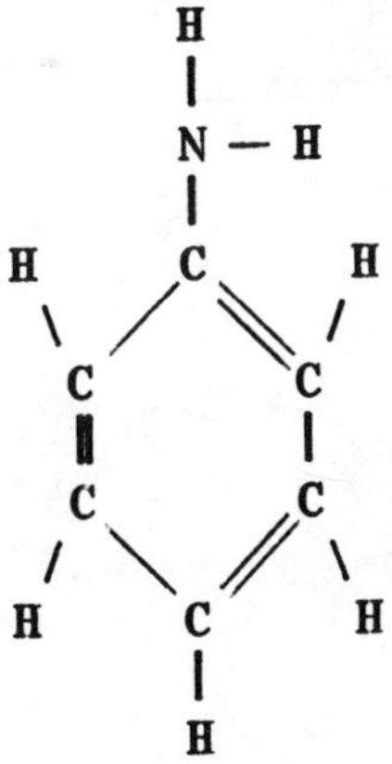

3.

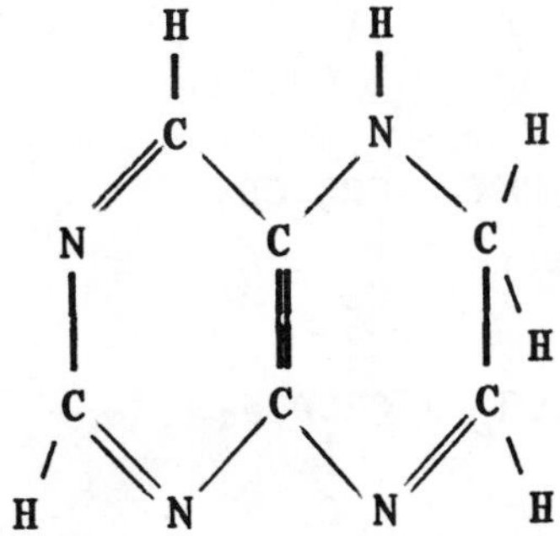

4.

Exercise II

C_2 $CH_2 = CH - Cl$ chloroethane

C_3 $Cl - CH_2 - CH = CH_2$ 3-chloro-1-propene

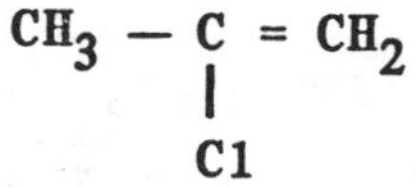

$CH_3 - \underset{\underset{Cl}{|}}{C} = CH_2$ 2-chloro-1-propene (in this case 2-chloropropene would be correct, too.)

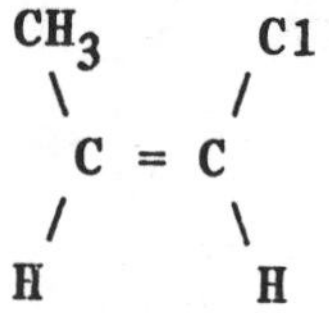

$$\underset{\underset{H}{/}}{\overset{\overset{CH_3}{\backslash}}{C}} = \underset{\underset{H}{\backslash}}{\overset{\overset{Cl}{/}}{C}}$$ cis-1-chloropropene

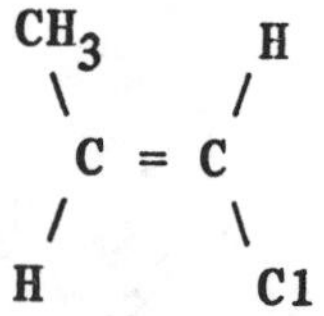

$$\underset{\underset{H}{/}}{\overset{\overset{CH_3}{\backslash}}{C}} = \underset{\underset{Cl}{\backslash}}{\overset{\overset{H}{/}}{C}}$$ trans-1-chloropropene

C_4

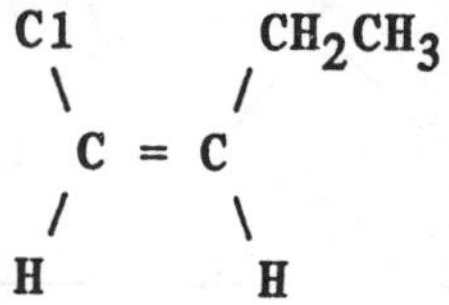

$$\underset{\underset{H}{/}}{\overset{\overset{Cl}{\backslash}}{C}} = \underset{\underset{H}{\backslash}}{\overset{\overset{CH_2CH_3}{/}}{C}}$$ cis-1-chloro-1-butene

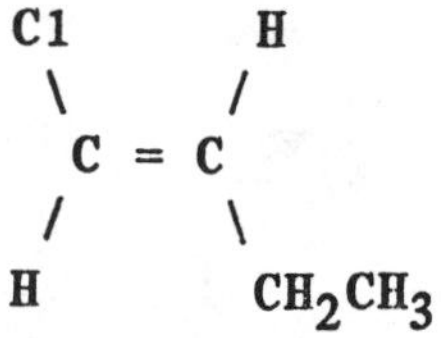

$$\underset{\underset{H}{/}}{\overset{\overset{Cl}{\backslash}}{C}} = \underset{\underset{CH_2CH_3}{\backslash}}{\overset{\overset{H}{/}}{C}}$$ trans-1-chloro-1-butene

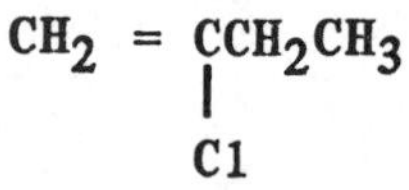

$CH_2 = \underset{\underset{Cl}{|}}{C}CH_2CH_3$ 2-chloro-1-butene

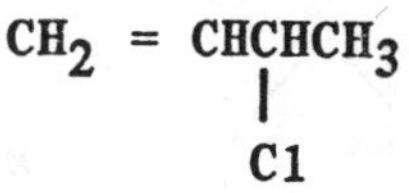

$CH_2 = CH\underset{\underset{Cl}{|}}{C}HCH_3$ 3-chloro-1-butene

$CH_2 = CHCH_2CH_2Cl$ 4-chloro-1-butene

```
ClCH2      CH3
    \      /
     C  =  C              cis-1-chloro-2-butene
    /      \
   H        H


  CH3      CH3
    \      /
     C  =  C              cis-2-chloro-2-butene (The CH3 groups
    /      \                               are cis.)
  C1        H


ClCH2      H
    \      /
     C  =  C              trans-1-chloro-2-butene
    /      \
   H        CH3


  CH3      H
    \      /
     C  =  C              trans-2-chloro-2-butene
    /      \
  C1        CH3


                 CH3
                 /
 C1 — CH  =  C           1-chloro-2-methylpropene   (-1-pro-
                 \                      pene is unnecessary)
                 CH3


                 CH3
                 /
 CH2  =  C              3-chloro-2-methylpropene   (-1-pro-
                 \                      pene is unnecessary)
                 CH2C1
```

Exercise III
 1. phenol   6. m-chlorotoluene
 2. chlorobenzene   7. p-nitroaniline
 3. toluene   8. m-bromophenol
 4. aniline   9. o-chlorobenzoic acid
 5. benzoic acid   10. o-dichlorobenzene

Exercise IV

1. an alkane (or more generally, a saturated site)
2. a 1,2-dichloro compound (where 1 and 2 refer to relative locations)
3. a 1,2-dibromo compound
4. an alkyl chloride
5. an alkyl bromide
6. an alkyl hydrogen sulfate
7. an alcohol
8. carbon dioxide and water

Exercise V

1. 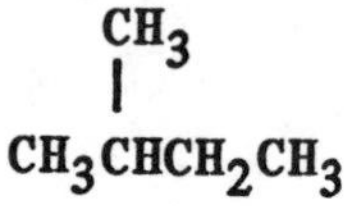 $\underset{\displaystyle \overset{|}{CH_3}}{CH_3CHCH_2CH_3}$

2. $CH_3CH_2CH_2CH_3$

3.

4. 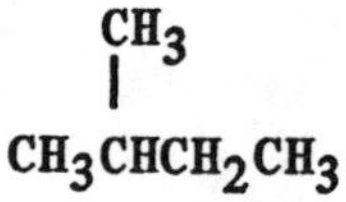 $\underset{\displaystyle \overset{|}{CH_3}}{CH_3CHCH_2CH_3}$

5.

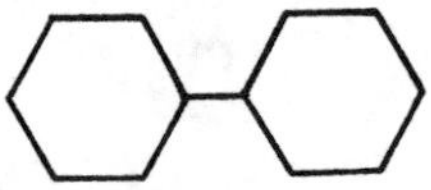

Exercise VI

1. 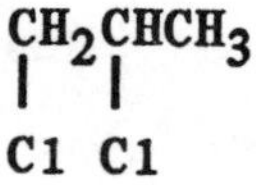 $\underset{\displaystyle \underset{Cl\ \ Cl}{|\ \ |}}{CH_2CHCH_3}$

2. 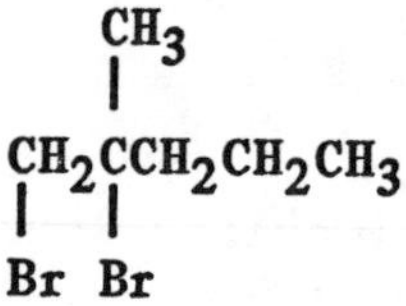 $\underset{\displaystyle \underset{Br\ Br}{|\ \ |}}{CH_2CCH_2CH_2CH_3}$ with CH_3 above

3.

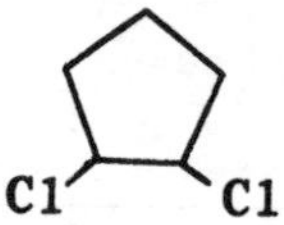

4. 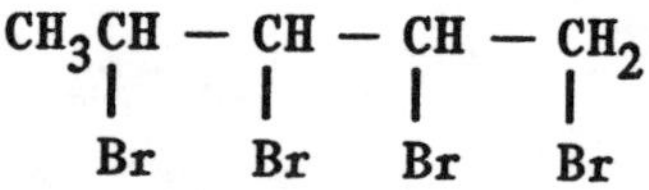 $\underset{\displaystyle \underset{Br\ \ \ Br\ \ \ Br\ \ \ Br}{|\ \ \ \ |\ \ \ \ |\ \ \ \ |}}{CH_3CH - CH - CH - CH_2}$

5. 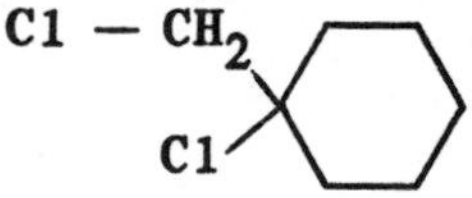 $Cl - CH_2$

Exercise VII

1. 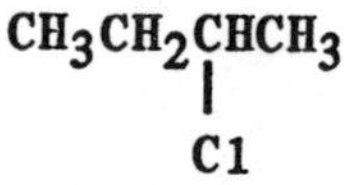 $\underset{\displaystyle \overset{|}{Cl}}{CH_3CH_2CHCH_3}$

2.

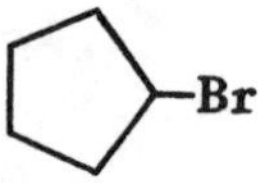

3. 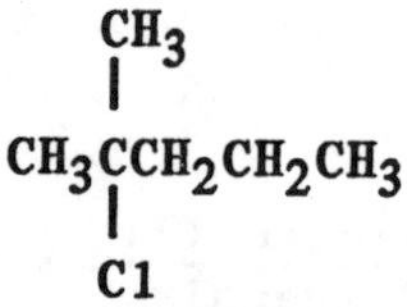 $\underset{\displaystyle \underset{Cl}{|}}{CH_3CCH_2CH_2CH_3}$ with CH_3 above

4.

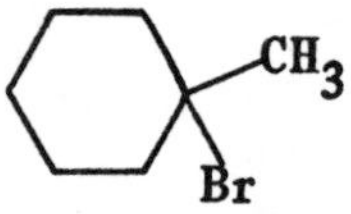

5. 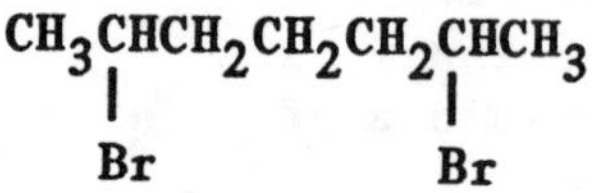 $\underset{\displaystyle \underset{Br\ \ \ \ \ \ \ \ \ \ \ \ \ Br}{|\ \ \ \ \ \ \ \ \ \ \ \ \ |}}{CH_3CHCH_2CH_2CH_2CHCH_3}$

Exercise VIII

1. $CH_3CHCH_2CH_3$
    |
    OH

2. $CH_3CH_2CHCH_3$
    |
    OH

3. CH_3
    |
  $CH_3CCH_2CH_3$
    |
    OH

4. —OH

5. CH_3CH_2 HO

ANSWERS TO SELF–TESTING QUESTIONS

Completion

1. six

2.
$$\underset{H}{\overset{CH_3}{\diagdown}} C = C \underset{H}{\overset{CH_2CH_3}{\diagup}}$$
cis

$$\underset{H}{\overset{CH_3}{\diagdown}} C = C \underset{CH_2CH_3}{\overset{H}{\diagup}}$$
trans

3. the same

4. (a) CH_3–⟨⟩–CH_3 + Br_2 $\xrightarrow[\text{FeBr}_3]{\text{Fe or}}$ CH_3–⟨⟩–CH_3 + HBr
                                                              Br

   (b) CH_3–⟨⟩–CH_3 + HNO_3 $\xrightarrow[\text{H}_2\text{SO}_4]{}$ CH_3–⟨⟩–CH_3 + H_2O
                                                              NO_2

   (c) CH_3–⟨⟩–CH_3 + H_2SO_4 $\xrightarrow[\text{heat}]{}$ CH_3–⟨⟩–CH_3 + H_2O
                                                              SO_3H

5. 3°, $CH_3 - \overset{\overset{\displaystyle CH_3}{|}}{\underset{\underset{\displaystyle CH_3}{\diagdown}}{C^+}}$

6. 1°, $^+CH_2 - \overset{\overset{\displaystyle CH_3}{\diagup}}{\underset{\underset{\displaystyle CH_3}{\diagdown}}{CH}}$

7. A $<$ C $<$ B

8. (a) 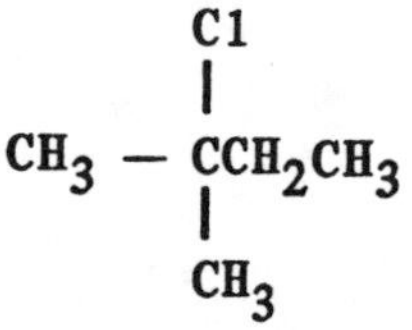
$$CH_3 - \overset{\displaystyle Cl}{\underset{\displaystyle CH_3}{C}}CH_2CH_3$$

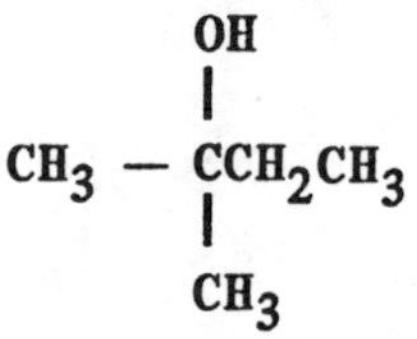

(b)
$$CH_3 - \overset{\displaystyle OH}{\underset{\displaystyle CH_3}{C}}CH_2CH_3$$

(c) ⬡—CH_3

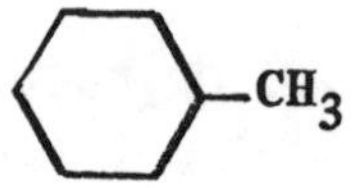

(d) 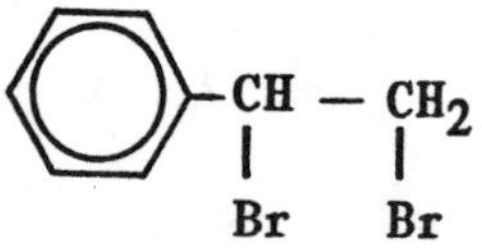 ⬡—$\overset{\displaystyle CH}{\underset{\displaystyle Br}{}} - \overset{\displaystyle CH_2}{\underset{\displaystyle Br}{}}$ (Br_2 needs Fe or $FeBr_3$ to attack the ring.)

(e) no reaction (Sulfuric acid does not attack alkanes or cyclo-alkanes.)

(f) no reaction (Aqueous acids do not attack the benzene ring or alkyl groups.)

(g) $CH_3CH_2CH_2CH_2CH_3$

9. sigma, pi
10. p

Multiple-Choice

1. b and d
2. a and d
3. b and d   (Neither would react with a or c.)
4. a
5. b
6. c
7. d
8. b
9. d   (Markovnikov's Rule prevents use of any of the other choices.)
10. c
11. d   (Benzene rings and alkyl groups are stable to aqueous alkali.)
12. d   (Benzene rings and alkyl groups are stable to aqueous acids.)
13. d   (Benzene is exceptionally stable toward oxidizing agents.)
14. b

14 Alcohols, Thioalcohols, Phenols, and Ethers

OBJECTIVES

The three most important topics in this chapter relating to the molecular basis of life are:

1. Hydrogen bonds – which occur in nearly all of the biologically important molecules, such as carbohydrates, proteins, and nucleic acids.
2. Electron-transfer events (oxidations and reductions) – which occur at one or more of the stages in the breakdown of all substances used in the body's cells; and
3. Dehydrations of alcohols – which also occur several times in the reactions of metabolism.

We shall need the nomenclature of alcohols, ethers, and phenols if only to be able to talk and write about them.

In greater detail, you should be able to do the following specific objectives after you have carefully studied this chapter and worked the Practice Exercises and Review Exercises in it.

1. Recognize the alcohol, thioalcohol, disulfide, ether, and phenol groups in structures.
2. Write the common names for alcohols (up through C_4).
3. Write the IUPAC names for alcohols.
4. Classify alcohols as 1°, 2°, or 3° and as mono-, di-, or

polyhydric.

5. Write structures that illustrate how alcohol molecules form hydrogen bonds between each other or with water molecules in aqueous solutions.

6. Explain how alcohols have much higher boiling points and solubilities in water than do hydrocarbons of comparable formula weight.

7. Contrast phenols and alcohols in their acidity.

8. Give one reaction characteristic of phenols.

9. Given the structure of an alcohol, write the structure of the alkene or the ether that could be made from it by acid-catalyzed dehydration.

10. Given the structure of an alcohol, write the structure of the aldehyde and carboxylic acid that could be made from it (if it is a primary alcohol) or the ketone that could be made from it (if it is a secondary alcohol) by oxidation.

11. Write the structure of the disulfide that could be made by the oxidation of a thioalcohol.

12. Write the structure of the thioalcohol(s) that could be made by the reduction of a disulfide.

13. Contrast phenols and alcohols in their abilities to neutralize a strong base (like OH$^-$).

14. Write sentences that summarize each of the chemical reactions of alcohols, thioalcohols, disulfides, phenols, and ethers that are studied in this chapter.

15. Prepare 5 x 8" cards for each functional group studied on which are summarized the chemical properties they give.

16. Make lists of the functional groups we have so far studied that are chemically affected by (a) oxidizing agents, (b) reducing agents, (c) hot acid catalysts, (d) neutralizing agents (at room temperature) such as aqueous acids and aqueous hydroxides.

17. Define each term in the Glossary. Where appropriate, illustrate a definition with a structure or a reaction.

GLOSSARY

Alcohol. Any organic compound whose molecules have the — OH group attached to a saturated carbon; R — OH.

Alcohol Group. The — OH group when it is joined to a saturated carbon.

Dihydric Alcohol. An alcohol with two — OH groups; a glycol.

Disulfide. A compound whose molecules have the sulfur—sulfur unit, — S — S — as in R — S — S — R.

Ether. An organic compound whose molecules have an oxygen attached by single bonds to separate carbon atoms neither of which is a carbonyl carbon atom; R — O — R'.

Mercaptan. A thioalcohol; R — S — H.

Monohydric Alcohol. An alcohol whose molecules have one — OH group.

Phenol. Any organic compound whose molecules have an — OH group attached to a benzene ring.

Primary Alcohol.   An alcohol in whose molecules an — OH group is attached to a primary carbon, as in RCH_2OH.

Secondary Alcohol.   An alcohol in whose molecules an — OH group is attached to a secondary carbon atom; $R_2CH — OH$.

Tertiary Alcohol.   An alcohol in whose molecules an — OH group is held by a carbon from which three bonds extend to other carbon atoms; $R_3C — OH$.

Thioalcohol. A compound whose molecules have the — SH group attached to a saturated carbon atom; a mercaptan.

Trihydric Alcohol. An alcohol with three — OH groups per molecule.

KEY MOLECULAR "MAP SIGNS"

We have likened the functional groups to "map signs" that enable us to "read" the structural formulas of complicated systems. The principal map signs in this chapter are the following.

Key Molecular "Map Signs" in Organic Molecules	What to Expect When This Functional Group is Present

---

$$\begin{array}{c} | \\ -\,C\,-\,O\,-\,H \\ | \end{array}$$

Alcohol group

Influence on physical properties:
· The — OH group is a good hydrogen-bond donor and hydrogen-bond acceptor.
· Molecules with the — OH group tend to be much more polar (e.g., they have higher boiling points) and much more soluble in water than are molecules without any group that can participate in hydrogen bonding (formula weights being about the same).

Influence on chemical properties:
A molecule with the alcohol group is vulnerable to
· dehydrating agents (acids + heat); either double bonds are introduced or ethers are made.
· oxidizing agents (those that can pull out the pieces of the element hydrogen):

1^0 alcohols $\rightarrow$ aldehydes $\rightarrow$ acids
2^0 alcohols $\rightarrow$ ketones
3^0 alcohols $\rightarrow$ (no reaction)

(The list for alcohols will be completed in later chapters.)

— S — H

Thioalcohol (sulfhydryl) group

Easy oxidation to — S — S — ;   this property, which we need to understand in its relation to proteins, is the only property that concerns us.

— S — S —

Disulfide

Easy reduction to 2 — S — H; again, this property which we also need to understand in its relation to proteins is the only property that concerns us.

Phenol system

A phenolic — OH group is a weak acid, but strong enough to neutralize the hydroxide ion.

$$-\overset{|}{\underset{|}{C}} - O - \overset{|}{\underset{|}{C}}$$

Ether group

(To be a simple ether, the carbons shown here must hold either Hs or Rs.)

Influence on physical properties:
The oxygen of an ether molecule can accept hydrogen bonds;   ethers there- fore are slightly more soluble in water than alkanes; they are also slightly more polar.

Influence on chemical properties:
None.   (As far as our study goes, the simple ether group is not at- tacked by water, dilute acids   or bases,   or oxidizing or reducing agents.   We have to be able   to recognize it, but   then we can   ignore it.)

DRILL EXERCISES

I. EXERCISES IN HYDROGEN BONDS

We define an H-bond donor as a molecule with a δ+ on a hydrogen attached to oxygen or nitrogen (— O — H or — N — H).

An H-bond acceptor is a molecule with a δ- on an oxygen or a tri-substituted nitrogen in any functional group.

H-bond donors can establish H-bonds not only to their own kind but also to water molecules and to any other H-bond acceptors. H-bond donors are invariably H-bond acceptors.

Some molecules, such as ethers, ketones, aldehydes, and esters, can only be H-bond acceptors. They have oxygens (or nitrogens, as in examples not yet studied), but they do not have — OH or — NH (i.e., a hydrogen with a δ+ and held by oxygen or nitrogen).

Examine these structures and answer the questions that follow.

A. $CH_3CH_2O — H$

B. $CH_3CH_2 — O — CH_2CH_3$

C. ⬡ — O — H

$$D. \quad CH_3 — \overset{\overset{\textstyle O}{\|}}{C} — O — CH_3$$

$$E. \quad CH_3 — \overset{\overset{\textstyle O}{\|}}{C} — O — H$$

F. $CH_3CH_2CH_2CH_3$

1. Which are H-bond donors? _____

2. Which are H-bond acceptors? _____

3. Which are H-bond acceptors only? _____

4. Which would be completely insoluble in water? _____

5. Which would be more soluble in water, A or B? _____

6. Which would be more soluble in water, C or E? _____

II. DRILL IN WRITING THE PRODUCTS OF THE DEHYDRATION OF ALCOHOLS

After you have studied Example 14.2 in the text and have tried Practice Exercise 5 of Chapter 14, you might feel the need for further drill. Write the products of the dehydration of the following alcohols.

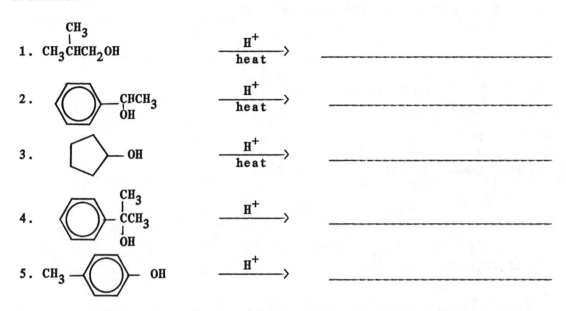

1. $CH_3\overset{\overset{\displaystyle CH_3}{|}}{CH}CH_2OH$ $\xrightarrow[\text{heat}]{H^+}$ _____

2. $C_6H_5\overset{\underset{\displaystyle OH}{|}}{CH}CH_3$ $\xrightarrow[\text{heat}]{H^+}$ _____

3. cyclopentyl—OH $\xrightarrow[\text{heat}]{H^+}$ _____

4. $C_6H_5\overset{\overset{\displaystyle CH_3}{|}}{\underset{\underset{\displaystyle OH}{|}}{C}}CH_3$ $\xrightarrow{H^+}$ _____

5. CH_3—C_6H_4—OH $\xrightarrow{H^+}$ _____

III. DRILL IN WRITING THE PRODUCTS OF THE OXIDATION OF ALCOHOLS

Examples 14.3 and 14.4 plus Practice Exercises 6, 7, and 8 of Chapter 14 are the places to begin this study. For more drill, write the products of the oxidation of the following alcohols. If oxidation in the sense being studied cannot occur, then write "no reaction." If the initial product can be oxidized further, then write the next product, too.

1. $CH_3CH_2CH_2CH_2OH$ + (O) $\longrightarrow$ _____

2. Cl—C_6H_4—CH_2OH + (O) $\longrightarrow$ _____

3. $CH_3\overset{\overset{\displaystyle CH_3}{|}}{\underset{\underset{\displaystyle OH}{|}}{C}}CH_2CH_3$ + (O) $\longrightarrow$ _____

4. 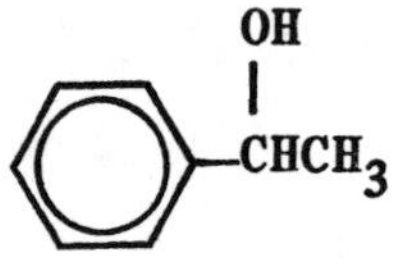
$$\underset{\text{(phenyl ring)}}{\bigcirc}-\overset{\overset{\displaystyle OH}{|}}{C}HCH_3 \quad + \quad (O) \quad \longrightarrow$$

_____

5. 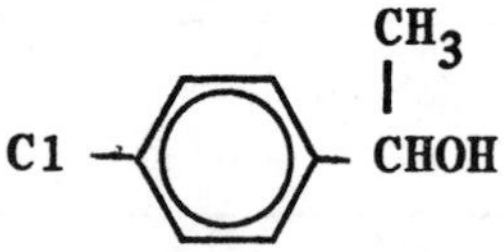 $Cl \longrightarrow \bigcirc \longrightarrow \overset{\overset{\displaystyle CH_3}{|}}{C}HOH \quad + \quad (O) \quad \longrightarrow$

_____

6. $CH_3\underset{\underset{\displaystyle CH_3}{|}}{\overset{\overset{\displaystyle OH}{|}}{C}H}CHCH_3 \quad + \quad (O) \quad \longrightarrow$

_____

7. $CH_3CH_2\overset{\overset{\displaystyle CH_3}{|}}{C}HCH_2OH \quad + \quad (O) \quad \longrightarrow$

_____

8. 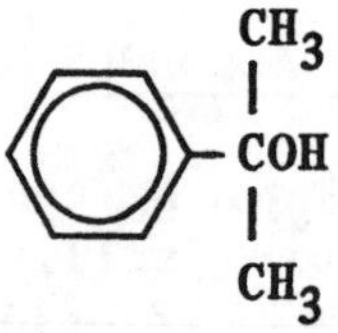 $\bigcirc - \underset{\underset{\displaystyle CH_3}{|}}{\overset{\overset{\displaystyle CH_3}{|}}{C}}OH \quad + \quad (O) \quad \longrightarrow$

_____

IV. DRILL IN WRITING THE PRODUCT OF THE OXIDATION OF A THIOALCOHOL

Example 14.5 in the text shows how to do this kind of exercise. For practice, write the products of the oxidation of the following thioalcohols.

1. $CH_3SH \quad + \quad (O) \quad \longrightarrow$ _____

2. $CH_3CH_2CH_2CH_2SH \quad + \quad (O) \quad \longrightarrow$ _____

3. 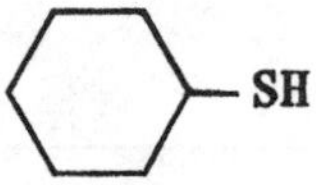 $\bigcirc - SH \quad + \quad (O) \quad \longrightarrow$ _____

4.  $\bigcirc - CH_2SH \quad + \quad (O) \quad \longrightarrow$ _____

$$CH_3$$
$$|$$
5. CH_3CHCH_2SH + (O) $\longrightarrow$ _____

V. <u>DRILL</u> <u>IN</u> <u>WRITING</u> <u>THE</u> <u>PRODUCT</u> <u>OF</u> <u>THE</u> <u>REDUCTION</u> <u>OF</u> <u>A</u> <u>DISULFIDE</u>

Example 14.6 in the text shows how to do this. Remember that disulfides aren't always symmetrical, like $R - S - S - R$. Those that we'll encounter in biochemistry usually are not; they're like $R - S - S - R'$.   The reduction of this kind gives two different thioalcohols, so both of them have to be written.

$$CH_3$$
$$|$$
1. $CH_3 - S - S - CHCH_3$ + 2(H) $\longrightarrow$ _____

2. $CH_2 - S - S - CH_2CH_3$ + 2(H) $\longrightarrow$ _____

3. $CH_3CH_2OCH_2CH_2 - S - SCH_3$ + 2(H) $\longrightarrow$ _____

$$S - S$$
$$/ \qquad \backslash$$
4. $CH_2 \qquad CH_2$ + 2(H) $\longrightarrow$
$$\backslash \qquad /$$
$$CH_2 - CH_2$$

_____

5. $- S - S -$ + 2(H) $\longrightarrow$ _____

VI. <u>DRILL</u> <u>IN</u> <u>WRITING</u> <u>THE</u> <u>STRUCTURES</u> <u>OF</u> <u>ETHERS</u> <u>THAT</u> <u>CAN</u> <u>BE</u> <u>MADE</u> <u>FROM</u> <u>ALCOHOLS</u>

Example 14.7 shows how this is done. In order to make this kind of exercise more helpful for applications in the next chapter, we will include in the drill examples in which you'll construct a structure of an unsymmetrical ether, like $R - O - R'$, from two different given alcohols. We will also make this a review of the names of alcohols. If only one alcohol is named, then the question is what symmetrical ether can be made from this alcohol? If two alcohols are named, then the question is what unsymmetrical ether can be made from the two?

1. isobutyl alcohol   _____

2. t-butyl alcohol   _____

3. 4-methylcyclohexanol   _____

4. isopropyl alcohol and methyl alcohol   _____

5. cyclopentanol and sec-butyl alcohol   _____

SELF-TESTING QUESTIONS

Treat the following questions as a final examination for the chapter. Work them only after you have done the drills with the in-chapter and end-of-chapter Review Exercises. As a review for these tests, go back to the chapter objectives, one by one, and see if you can do them.

COMPLETION

Test your knowledge of chemical properties by writing the structure(s) of the principal organic product(s) that would be expected to form in each of the following. If no reaction will occur, write "none." Reread Review Exercise 14.39 in the text, for stipulations concerning problems of this type. Notice particularly that whenever the dehydration of any given alcohol is intended to go to the corresponding ether, then the coefficient "2" will appear before the structure of that alcohol. If the "2" is absent, then the intended reaction is the internal dehydration of the alcohol to an alkene (if possible). (Normally, coefficients are not put in until after all structures of reactants and products have been written; "balancing comes last.") Where the oxidation of a 1° alcohol is indicated, show both the aldehyde and the carboxylic acid that are formed.

1. $CH_3CH_2CH_3 \xrightarrow[\text{heat}]{H_2SO_4}$ _____

2. $\overset{\overset{\displaystyle CH_3}{|}}{CH_3CHCH_2OH}$ + (O) $\longrightarrow$ _____

3. $2CH_3CH_2CH_2OH \xrightarrow[\text{heat}]{H_2SO_4}$ _____

4. $\overset{\overset{\displaystyle CH_3}{|}}{\underset{\underset{\displaystyle CH_3}{|}}{CH_3C}} - OH \xrightarrow[\text{heat}]{H_2SO_4}$ _____

5. $CH_3CH_2CH_2OH$ + (O) $\longrightarrow$ _____

6. $\overset{\overset{\displaystyle CH_3}{|}}{CH_3CHCH_2SH}$ + (O) $\longrightarrow$ _____

7. $CH_3CH_2 - \bigcirc\!\!\!\!\bigcirc - OH$ + NaOH(aq) $\longrightarrow$ _____

8. $CH_3 - \overset{\overset{\displaystyle CH_3}{|}}{\underset{\underset{\displaystyle CH_3}{|}}{C}} - CH_2 - S - S - CH_3$ + 2(H) $\longrightarrow$ _____

9.

$\overset{\displaystyle OH}{\bigcirc}$ + (O) $\longrightarrow$ _____

10. $\overset{\overset{\displaystyle OH}{|}}{CH_3CH_2CHCH_2CH_3}$ + (O) $\longrightarrow$ _____

11. $\bigcirc\!\!\!\overset{OH}{\underset{CH_3}{}}$ + (O) $\longrightarrow$ _____

12. $CH_3CH_2CH_2CH_3$ + $H_2O \xrightarrow[\text{heat}]{H^+}$ _____

13. CH_3OH + (O) $\longrightarrow$ _____

14. $HOCH_2CH_2CH_3 \xrightarrow[\text{heat}]{H_2SO_4}$ _____

15. $CH_2 = CHCH_2OH$ + $H_2 \xrightarrow[\text{pressure}]{\text{Ni, heat}}$ _____

16. $CH_3CH_2 - O - CH_2 - CH_2 - \overset{\overset{\displaystyle OH}{|}}{C}HCH_3$ + (O) $\longrightarrow$ _____

MULTIPLE-CHOICE

1. The oxidation of $CH_3 - \overset{\overset{\displaystyle OH}{|}}{C}H - CH_2 - CH_3$ could be made to produce:

   (a) $CH_3 - \overset{\overset{\displaystyle OH}{|}}{C}H - O - CH_2 - CH_3$

   (c) $CH_3 - CH_2 - \overset{\overset{\displaystyle O}{\|}}{C} - CH_3$

   (b) $H - \overset{\overset{\displaystyle O}{\|}}{C} - CH_2CH_2CH_3$

   (d) $HO - \overset{\overset{\displaystyle O}{\|}}{C}CH_2CH_2CH_3$

2. The **best** explanation for the solubility of glycerol in water is
   (a) glycerol is a small molecule
   (b) glycerol molecules are polar
   (c) glycerol molecules can donate
       and accept hydrogen bonds to
       and from water molecules in the
       solution
   (d) glycerol's ions are well-solvated by
       water

   $HO - CH_2 - \overset{\overset{\displaystyle }{|}}{C}H - CH_2 - OH$
   OH

   Glycerol

3. The substance whose structure is $CH_3 - OH$ is known as
   (a) wood alcohol   (c) methanol
   (b) grain alcohol   (d) potable alcohol

4. The substance whose structure is $CH_3CH_2OCH_2CH_3$ is
   (a) a common fuel   (c) a common anesthetic
   (b) a common antifreeze   (d) an eye irritant in smog

5. The dehydration of 3-hexanol would produce
   (a) 2-hexene only
   (b) 3-hexene only
   (c) a mixture of 2-hexene and 3-hexene
   (d) cyclohexene

6. The oxidation of 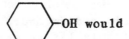 —OH would

   (a) destroy the ring   (c) produce
   (b) produce hydroquinone   (d) cyclohexene

7. The reaction of ⬡ — OH with NaOH(aq) gives

   (a) ⬡ = O

   (b) ⬡ — O⁻Na⁺ + H₂O

   (c) ⬡ — O — ⬡

   (d) none of these

8. The common name of CH_3CHCH_3 is
                                |
   (a) isopropanol   OH
   (b) 2-propyl alcohol
   (c) 2-propanol
   (d) isopropyl alcohol

Multiple-choice questions 9 through 11 refer to this structure:

$$CH_3 - O - CH_2 - CH_2 - \underset{\underset{OH}{|}}{CH} - CH_2CH = CH_2$$

with groups labeled A, B, and C

9. The group labeled Ⓐ is

   (a) an easily hydrolyzed group
   (b) an easily oxidized group
   (c) an easily reduced group
   (d) a generally unreactive group

10. The group labeled Ⓑ could be

(a) oxidized to a ketone
(b) oxidized to an aldehyde
(c) involved in an acid-catalyzed dehydration
(d) reduced to a ketone

11. The group labeled $\boxed{C}$ could be

(a) made to react with dilute sodium hydroxide
(b) made to add a water molecule (if an acid catalyst were available)
(c) reduced to a carbon-carbon single bond by hydrogen (with a catalyst, heat, and pressure)
(d) involved in hydrogen bonding

12. A substance that can neutralize aqueous sodium hydroxide is

(a)

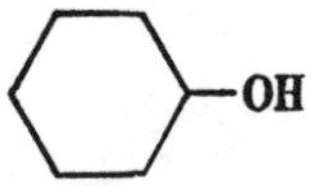

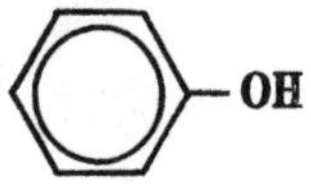

(c)

(b)

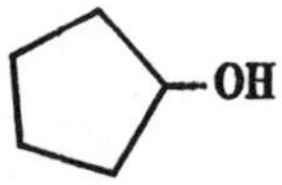

(d)

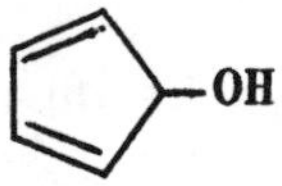

13. A mild reducing agent will react with
  (a) CH_3CH_2SH
  (b) CH_3CH_2OH
  (c) $CH_3 - S - S - CH_3$
  (d) $CH_3CH_2 - O - CH_2CH_3$

ANSWERS

ANSWERS TO DRILL EXERCISES

  I. Exercises in Hydrogen Bonds
1. A, C, and E
2. A, B, C, D, and E
3. B and D
4. F   (an alkane)
5. A   (It's of lower formula weight, and it can both donate and accept H — bonds to water, the solvent. B can only accept H — bonds from water.)
6. E   (It's less hydrocarbon like, of lower-formula weight, and has two H — bond accepting sites – the two oxygens – whereas

structure C has only one H — bond accepting site.)

II. Drill in Writing the Products of the Dehydration of Alcohols

1. 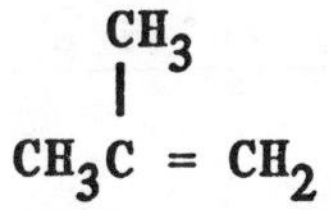
$$\underset{\underset{\displaystyle CH_3}{|}}{CH_3C} = CH_2$$

2.
— CH = CH$_2$

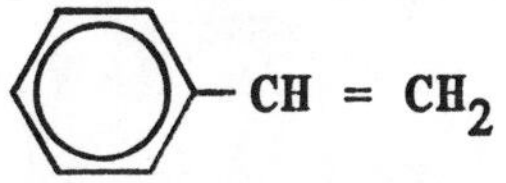

3.

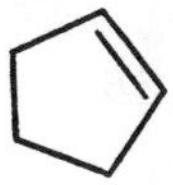

4. 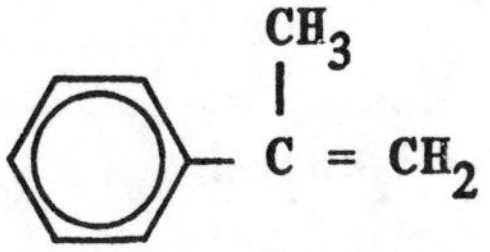
— $\underset{\underset{\displaystyle CH_3}{|}}{C} = CH_2$

5. no reaction

III. Drill in Writing the Products of the Oxidation of Alcohols

1. $CH_3CH_2CH_2CHO \xrightarrow{\text{more (O)}} CH_3CH_2CH_2CO_2H$

2. 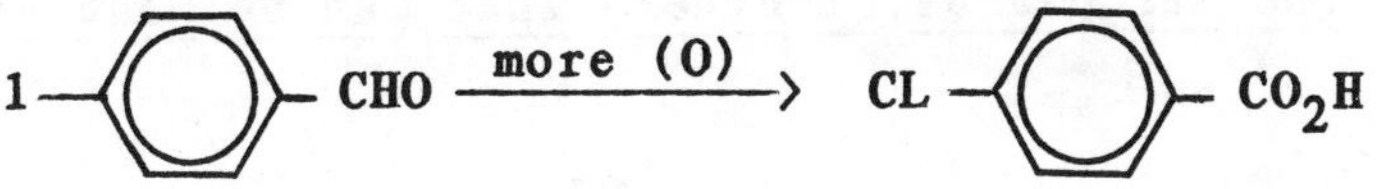 Cl—— $CHO \xrightarrow{\text{more (O)}} CL$—— CO_2H

3. no reaction

4. 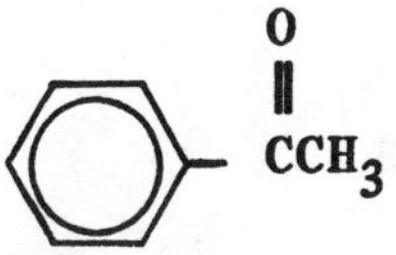 — $\overset{\overset{\displaystyle O}{\|}}{C}CH_3$

5. 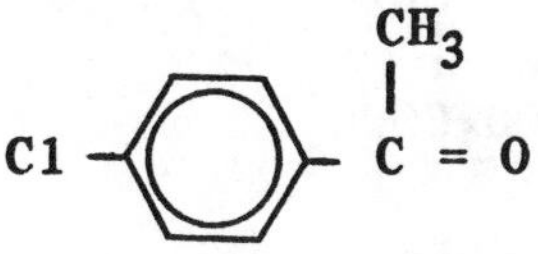 Cl —— $\underset{\underset{\displaystyle CH_3}{|}}{C} = O$

6. 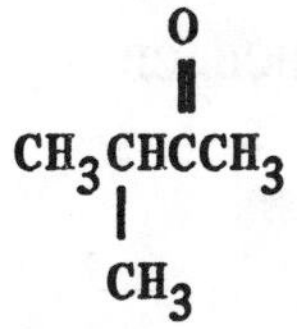 $CH_3\underset{\underset{\displaystyle CH_3}{|}}{\overset{\overset{\displaystyle O}{\|}}{CH}C}CH_3$

7. 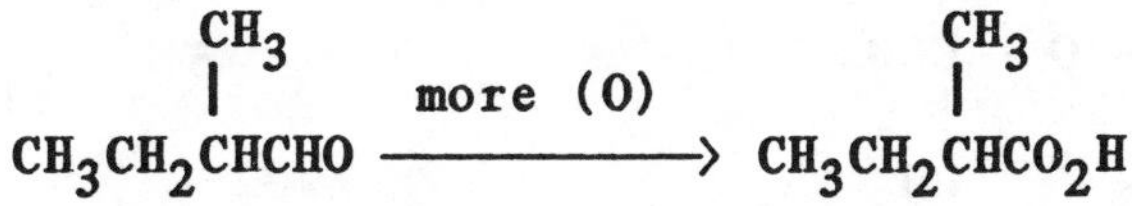 $CH_3CH_2\underset{\underset{\displaystyle CH_3}{|}}{CH}CHO \xrightarrow{\text{more (O)}} CH_3CH_2\underset{\underset{\displaystyle CH_3}{|}}{CH}CO_2H$

8. no reaction

IV. Drill in Writing the Product of the Oxidation of a Thioalcohol

1. $CH_3 - S - S - CH_3$

2. $CH_3CH_2CH_2CH_2 - S - S - CH_2CH_2CH_2CH_3$

3.  — $S - S$ —

4. 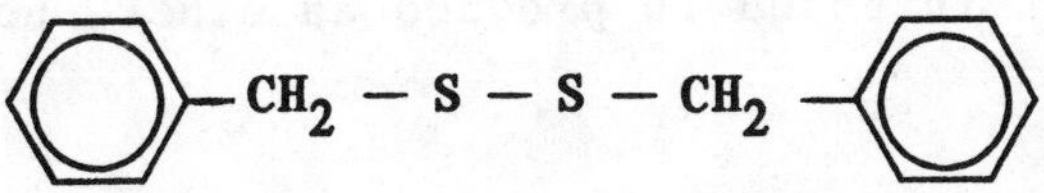 — $CH_2 - S - S - CH_2$ —

$$\overset{\displaystyle CH_3}{\underset{|}{}} \qquad \overset{\displaystyle CH_3}{\underset{|}{}}$$

5. $CH_3CHCH_2 - S - S - CH_2CHCH_3$

V. Drill in Writing the Product of the Reduction of a Disulfide

1. $CH_3SH \; + \; HSCHCH_3$

   $\qquad\qquad\qquad\;\; \overset{|}{CH_3}$

2. $- CH_2SH \; + \; HSCH_2CH_3$

3. $CH_3CH_2OCH_2CH_2SH \; + \; HSCH_3$

4. $HSCH_2CH_2CH_2CH_2SH$

5. 2 $- SH$

VI. Drill in Writing the Structures of Ethers That Can be Made From Alcohols

1. $CH_3\overset{CH_3}{\underset{|}{CH}}CH_2OCH_2\overset{CH_3}{\underset{|}{CH}}CH_3$

4. $CH_3\overset{CH_3}{\underset{|}{CH}}OCH_3$

2. $CH_3\overset{\overset{\displaystyle CH_3}{|}}{\underset{\underset{\displaystyle CH_3}{|}}{C}} - O - \overset{\overset{\displaystyle CH_3}{|}}{\underset{\underset{\displaystyle CH_3}{|}}{C}}CH_3$

5. $- O - \overset{CH_3}{\underset{|}{CH}}CH_2CH_3$

3. $CH_3 -$ $- O -$ $- CH_3$

ANSWERS TO SELF-TESTING QUESTIONS

Completion

1. none

2. $CH_3\overset{\overset{\displaystyle CH_3}{|}}{CH} - \overset{\overset{\displaystyle O}{\|}}{C} - H \xrightarrow{\;(O)\;} CH_3\overset{\overset{\displaystyle CH_3}{|}}{CH} - \overset{\overset{\displaystyle O}{\|}}{C} - OH$

3. $CH_3CH_2CH_2 - O - CH_2CH_2CH_3 \; + \; H_2O$ (Note the coefficient "2," our signal that the dehydration is intended to produce an ether, not an alkene.)

4. $CH_2 = C \overset{CH_3}{\underset{CH_3}{\Big<}}$ $+$ H_2O

5. $CH_3CH_2\overset{O}{\overset{\|}{C}} - H$ $\xrightarrow{(O)}$ $CH_3CH_2\overset{O}{\overset{\|}{C}} - OH$

6. $CH_3\overset{CH_3}{\overset{|}{CH}}CH_2 - S - S - CH_2\overset{CH_3}{\overset{|}{CH}}CH_3$ $+$ H_2O

7. $CH_3CH_2 - \bigcirc - O^-Na^+$ $+$ H_2O (The electrical charges can be
omitted.)

8. $CH_3\overset{CH_3}{\underset{CH_3}{\overset{|}{\underset{|}{C}}}}CH_2SH$ $+$ $HSCH_3$

9. $\overset{O}{\overset{\|}{\bigcirc}}$

10. $CH_3CH_2\overset{O}{\overset{\|}{C}}CH_2CH_3$

11. none ($3°$ alcohols do not oxidize)
12. none (Reactant is an alkane.)

13. $H - \overset{O}{\overset{\|}{C}} - H$ $\xrightarrow{(O)}$ $H - \overset{O}{\overset{\|}{C}} - OH$ (This will also continue to oxidize

to $HO - \overset{O}{\overset{\|}{C}} - OH$, which   is   carbonic
acid.)

14. $CH_2 = CHCH_3$ $+$ H_2O (Note the absence of the coefficient
"2," a signal that internal dehydra-
tion to an alkene is intended.)

15. $CH_3CH_2CH_2OH$ (Only   the   double   bond responds to
hydrogenation.   We studied   no re-
action   of   the   alcohol   group with
hydrogen – and   none,   in   fact,   ex–

ists; therefore, we can assume that this group is unaffected by the hydrogen and all the conditions present for hydrogenation.)

$$
16. \ CH_3CH_2 - O - CH_2CH_2\overset{\overset{\displaystyle O}{\|}}{C}CH_3 \ + \ H_2O
$$

(Only the 2° alcohol is affected by an oxidizing agent. We studied no reaction of an ether, the other functional group, with oxygen. Therefore, we can assume that it is untouched by the oxidizing agent, which is, in fact, true.)

Multiple-Choice

1. c
2. c
3. a and c
4. c
5. c (Water can split out from the carbon holding the — OH group in two directions.)
6. c (The reactant is cyclohexanol, not phenol.)
7. b (The reactant is phenol.)
8. d (Choices a and b mix the common system and the IUPAC system – a definite no-no in organic chemistry. Choice c is the IUPAC name.)
9. d
10. a and c
11. b and c
12. b
13. c

15 Aldehydes and Ketones

OBJECTIVES

After studying this chapter and working the exercises in it, you should be able to do the following. Pay particular attention to objectives 1, 5, and 8 through 13. They are very important to our future needs.

1. Recognize the following families in the structures of compounds: aldehydes, ketones, carboxylic acids, hemiacetals, hemiketals, acetals, and ketals.
2. Give the common and the IUPAC names and structures for aldehydes, ketones, and acids through C_4.
3. Compare the physical properties of aldehydes and ketones with those of alcohols or hydrocabons of comparable formula weights.
4. Write specific examples of reactions for the synthesis of simple aldehydes and ketones from alcohols.
5. Write specific examples of reactions for the oxidation of aldehydes.
6. Describe what one does and sees in a positive Tollens' Test.
7. Describe what one does and sees in a positive Benedict's Test.
8. Illustrate specific reactions for the reduction of aldehydes and ketones.
9. Give specific equations for the addition of an alcohol to an aldehyde (or ketone) to form a hemiacetal (or a hemiketal).
10. Write the structures of the original alcohol and aldehyde (or ketone) from the structure of a hemiacetal (or hemiketal).

11. Give specific examples of reactions forming acetals or ketals.
12. Give specific examples of reactions showing the hydrolysis of acetals or ketals.
13. Give definitions of the terms in the Glossary and provide illustrations where applicable.

Be sure to prepare and learn the one-sentence statements of chemical properties of aldehydes, ketones, hemiacetals, hemiketals, acetals, and ketals. Also, be sure to prepare the 5 x 8" reaction summary cards, and the cards that list the functional groups affected by various kinds of inorganic reactants.

GLOSSARY

Acetal.   Any organic compound in which two ether linkages extend from one CH unit, as in:

$$R - O - \overset{\textstyle |}{C}H - O - R'$$

Aldehyde.   An organic compound that has a carbonyl group joined to H on one side and C on the other, R — CH = O.

Aldehyde Group.   — CH = O

Benedict's Reagent.   A solution of copper(II) sulfate, sodium citrate, and sodium carbonate that is used in the Benedict's test.

Benedict's Test.   The use of Benedict's reagent to detect the presence of any compound whose molecules have easily oxidized functional groups – α–hydroxyaldehydes and α–hydroxyketones – such as those present in monosaccharides. In a positive test the intensely blue color of the reagent disappears and a reddish precipitate of copper(I) oxide separates.

Hemiacetal.   Any compound whose molecules have both an — OH group and an ether linkage coming to a — CH — unit:

$$\overset{\textstyle OH}{\underset{\textstyle |}{-CH}} - O - \overset{\textstyle |}{\underset{\textstyle |}{C}} - \qquad as\ in \qquad R - \overset{\textstyle OH}{\underset{\textstyle |}{CH}} - O - R'$$

Hemiketal. Any compound whose molecules have both an — OH group and
    an ether linkage coming to a carbon that otherwise bears no H
    atoms:

$$-\overset{\displaystyle |}{\underset{\displaystyle |}{C}} - O - \overset{\displaystyle |}{\underset{\displaystyle |}{C}} - \qquad \text{as in} \qquad R - \overset{\displaystyle OH}{\underset{\displaystyle R}{\overset{|}{\underset{|}{C}}}} - O - R'$$

Ketal. A substance whose molecules have two ether linkages joined to
    a carbon that also holds two hydrocarbon groups as in $R_2C(OR')_2$.

Keto Group. The carbonyl group when it is joined on each side to car-
    bon atoms.

Ketone. Any compound with a carbonyl group attached to two carbon
    atoms, as in $R_2C = O$.

Tollens' Reagent. A slightly alkaline solution of the diammine com-
    plex of the silver ion, $Ag(NH_3)^+$, in water.

Tollens' Test. The use of Tollens' reagent to detect an easily oxi-
    dized group such as the aldehyde group.

KEY MOLECULAR "MAP SIGNS"

Key Molecular "Map Signs" in Organic Molecules	What to Expect When This Functional Group is Present
$$-\overset{\displaystyle O}{\overset{\|}{C}} - H$$ Aldehyde group	Influence on physical properties: The aldehyde group is moderately polar; it can accept H-bonds. Influence on chemical properties: · One of the most easily oxidized groups: Changes into a carboxyl group Gives Tollens' test

- Can be reduced to a 1° alcohol group
- Adds an alcohol molecule to form a hemiacetal
- Can be converted into an acetal system

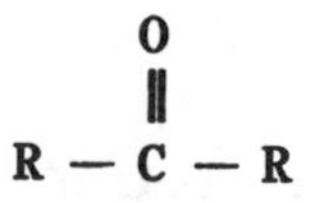

$$O$$
$$\|$$
$$R - C - R$$

Ketone

Influence on physical properties:
Same as the aldehyde group

Influence on chemical properties:
- Strongly resists oxidation (unlike the aldehydes)
- Can be reduced to a 2° alcohol group
- Adds an alcohol to form a hemiketal (although not as readily as an aldehyde gives the hemiacetal)
- Can be converted into a ketal system

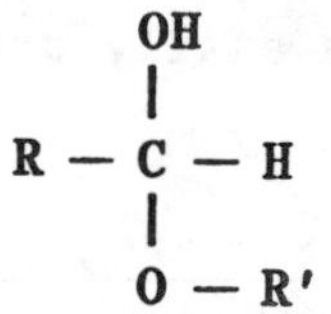

$$OH$$
$$|$$
$$R - C - H$$
$$|$$
$$O - R'$$

Hemiacetal

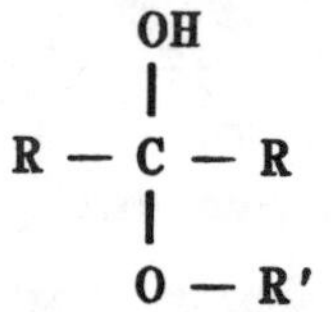

Two properties of importance:
- Both the hemiacetal and the hemiketal systems are unstable; they exist in equilibrium with the aldehyde or ketone and the alcohol (R'OH) that formed them.

$$OH$$
$$|$$
$$R - C - R$$
$$|$$
$$O - R'$$

Hemiketal

- Both systems can be changed to the acetal or ketal system by a reaction with another alcohol molecule.

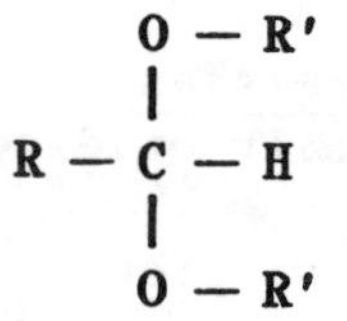

$$O - R'$$
$$|$$
$$R - C - H$$
$$|$$
$$O - R'$$ Acetal

One important property:
- Both the acetal and the ketal systems react with water when an acid catalyst is present, but they do not react in the presence of a basic

```
      O — R'                      catalyst;  in so  doing, they revert
      |                           back   to   the   original  aldehyde
  R — C — R                              ⎛   O   ⎞              ⎛   O   ⎞
      |                                  ⎜   ‖   ⎟   or ketone  ⎜   ‖   ⎟
      O — R'    Ketal                    ⎝R — C — H⎠            ⎝R — C — R⎠
                                  and alcohol  (2R'OH).
```

SELF-TESTING QUESTIONS

<u>COMPLETION</u>

1. Write the full structure of each compound on the line above its
   condensed structure.

   (a) _____   (b) _____
              CH_3CHO CH_3CHOCH_3
                                                         |
                                                         OH

   (c) _____   (d) _____
            CH_3CH_2OH CH_3CHOCH_3
                                                         |
                                                        OCH_3

2. What functional groups have we studied that will be attacked by
   oxidizing agents? (Give the name of the product, too.)

   (a) In this chapter   _____

                            _____

   (b) In previous chapters _____

                            _____

                            _____

                            _____

3. What functional groups have we studied that will be attacked by

reducing agents? (Give the name of the product, too.)

(a) In this chapter   _____

_____

(b) In earlier chapters   _____

_____

_____

_____

4. By means of short statements, summarize the chemical reactions of alcohols (from both this and earlier chapters).

Example: <u>Alcohols can be dehydrated to form alkenes.</u> _____

_____

_____

_____

_____

_____

_____

_____

_____

5. Examine the structures shown in question 1, parts a through d. Answer the following questions by writing the condensed structures of both the reactants and the products.

(a) Which of the structures in question 1
    will give a positive test with
    Tollens' Reagent?   _____

(b) Which can be hydrolyzed?   _____

(c) Which can be oxidized under basic conditions? _____
_____

(d) Which can be reduced? _____

6. Write the structure(s) of the principal **organic** product(s) that could be expected to form in each case. If no reaction occurs, write "none."

(a) $CH_3CH = O \xrightarrow{(O)}$ _____

(b) $CH_3CH = O + HOCH_3 \rightleftharpoons$ _____

(c) $CH_3\overset{\displaystyle OH}{\underset{\displaystyle OCH_3}{\overset{|}{\underset{\backslash}{CH}}}} + HOCH_3 \xrightarrow{H^+}$ _____

(d) $CH_3\overset{\displaystyle O}{\overset{||}{C}}CH_3 + H_2 \xrightarrow[\text{heat, pressure}]{Ni}$ _____

(e) $CH_3\overset{\displaystyle OCH_3}{\overset{|}{CH}}OCH_3 + H_2O \xrightarrow{H^+}$ _____ + _____

(f) $O = CHCH_2CH_3 \xrightarrow{(O)}$ _____

(g) $CH_3 - S - S - CH_2CO_2H \xrightarrow[\text{agent}]{\text{mild reducing}}$ _____ + _____

(h) $HOCH_2CH_2\overset{\displaystyle O}{\overset{||}{C}}CH_3 \xrightarrow[\text{heat}]{H_2SO_4}$ _____

7. Write the common and the IUPAC name of each of the following.

	Common	IUPAC

(a) $CH_3CH_2\overset{\displaystyle O}{\overset{||}{CH}}$ _____   _____

(b) $CH_3\overset{\displaystyle O}{\overset{||}{C}}CH_3$ _____   _____

$$\text{(c)} \quad CH_3\overset{\displaystyle O}{\overset{\|}{C}}H$$ _____ _____

$$\text{(d)} \quad CH_3CH_2\overset{\displaystyle O}{\overset{\|}{C}}CH_2CH_3$$ _____ _____

MULTIPLE-CHOICE

1. Which of these compounds could be hydrolyzed the most easily (assuming acid catalysis)?

   (a) $CH_3CH_2OCH_2CH_3$ (c) $CH_3OCH_2CH_2OCH_3$

   (b) $CH_3OCH_2OCH_3$ (d) $CH_3CH_2CH_2CH_2OH$

2. Which of these compounds could be the most easily oxidized under mild conditions:

   (a) $CH_3CH_2OCH_2CH_3$

   (b) $CH_3CH_2CH_2CH_2\overset{\displaystyle O}{\overset{\|}{C}}H$

   (c) $CH_3CH_2\overset{\displaystyle O}{\overset{\|}{C}}CH_2CH_3$

   (d) $CH_3CH_2O\overset{\displaystyle CH_3}{\overset{\|}{C}H}OCH_2CH_3$

3. Which of these compounds would be most soluble in water:

   (a) $CH_3CH_2CH_2CH_2OH$ (c) $CH_3CH_2CH_2CH_2CH_3$

   (b) $CH_3CH_2CH_2CH_2\overset{\displaystyle O}{\overset{\|}{C}}H$ (d) $CH_3CH_2CH_2OCH_2CH_3$

4. The substance that precipitates in a positive Benedict's Test is
   (a) CuO   (b) Ag   (c) Cu_2O (d) Ag^+

5. The acetal that could be hydrolyzed to CH_3CH_2OH and CH_3CHO is

   (a) $CH_3CH_2\overset{\displaystyle OH}{\overset{/}{\underset{\underset{\displaystyle O-CH_3}{\backslash}}{C}H}}$

   (c) $CH_3\overset{\displaystyle OH}{\overset{/}{\underset{\underset{\displaystyle O-CH_2CH_3}{\backslash}}{C}H}}$

(b) $CH_3CH_2OCHCH_3$
$\qquad\qquad\;\; |$
$\qquad\qquad OCH_2CH_3$

(d) none of these

6. The alcohol that could be oxidized to is

(a)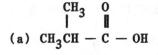

(c)
$\qquad\quad CH_3$
$\qquad\qquad |$
$\quad CH_3CHCH_2CH_2OH$

(b)
$\qquad\quad CH_3$
$\qquad\qquad |$
$\quad CH_3CH - OH$

(d) $CH_3CH - CH_2OH$
$\qquad\qquad\; |$
$\qquad\qquad CH_3$

7. Isopropyl alcohol could be made by the catalytic reduction of
   (a) acetone
   (b) propionaldehyde
   (c) methyl ethyl ketone
   (d) acetaldehyde

8. The oxidation of $CH_3CH_2CH_2\overset{\overset{\textstyle O}{\textstyle \|}}{C}H$ would give
   (a) isobutyraldehyde
   (b) butyric acid
   (c) butyraldehyde
   (d) 1-butanol

9. Which choice contains the best description of the functional group(s) present in this structure?

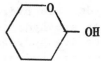

(a) a hemiacetal   (b) an acetal   (c) a hemiketal   (d) a ketal

10. If a hydride donor were used, which of the following systems would be able to accept it?

   (a) $CH_3CH_2CH_2CH_3$

   (c) $CH_3OCH_2CH_3$

   (b) $CH_3CH_2CH_2OH$

   (d) $CH_3CH_2CH_2\overset{\overset{\textstyle O}{\textstyle \|}}{C}H$

ANSWERS

ANSWERS TO SELF-TESTING QUESTIONS

Completion

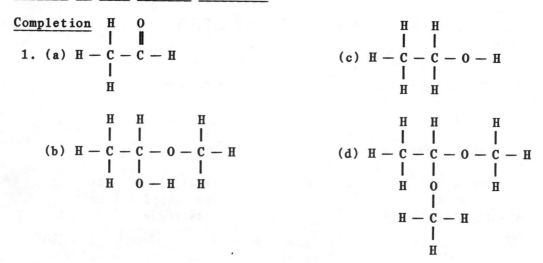

1. (a)
$$H - \overset{\overset{\displaystyle H}{|}}{\underset{\underset{\displaystyle H}{|}}{C}} - \overset{\overset{\displaystyle O}{\|}}{C} - H$$

(b)
$$H - \overset{\overset{\displaystyle H}{|}}{\underset{\underset{\displaystyle H}{|}}{C}} - \overset{\overset{\displaystyle H}{|}}{\underset{\underset{\displaystyle O-H}{|}}{C}} - O - \overset{\overset{\displaystyle H}{|}}{\underset{\underset{\displaystyle H}{|}}{C}} - H$$

(c)
$$H - \overset{\overset{\displaystyle H}{|}}{\underset{\underset{\displaystyle H}{|}}{C}} - \overset{\overset{\displaystyle H}{|}}{\underset{\underset{\displaystyle H}{|}}{C}} - O - H$$

(d)
$$H - \overset{\overset{\displaystyle H}{|}}{\underset{\underset{\displaystyle H}{|}}{C}} - \overset{\overset{\displaystyle H}{|}}{\underset{\underset{\displaystyle O}{|}}{C}} - O - \overset{\overset{\displaystyle H}{|}}{\underset{\underset{\displaystyle H}{|}}{C}} - H$$

2. (a) Aldehydes are oxidized to carboxylic acids.
   (b) Alkenes are oxidized to various products (which we did not study in any detail).
      1^O alcohols are oxidized to aldehydes or to carboxylic acids.
      2^O alcohols are oxidized to ketones.
      Mercaptans are oxidized to disulfides.
3. (a) Aldehydes are hydrogenated to 1^O alcohols.
      Ketones are hydrogenated to 2^O alcohols.
   (b) Alkenes are hydrogenated to alkanes.
      Alkynes are hydrogenated to alkenes and thence to alkanes.
      Disulfides are reduced to mercaptans.
4. Alcohols can be dehydrated to form ethers.
   1^O alcohols can be oxidized to aldehydes, and then to carboxylic acids.
   2^O alcohols can be oxidized to ketones.
   Alcohols will add to aldehydes or ketones to form hemiacetals or hemiketals (both of which are generally too unstable to be isolated).
   Alcohols will react with hemiacetals or hemiketals to form acetals or ketals.
5. (a) $CH_3CHO \longrightarrow CH_3\overset{\overset{\displaystyle O}{\|}}{C} - OH$

The hemiacetal, b, will also give a positive test because, in solution, it exists in equilibrium with acetaldehyde and methyl alcohol; acetaldehyde will react with Tollens' Reagent.

(b) "To be hydrolyzed" means to undergo a chemical reaction with water. Only the acetal, d, reacts with water:

$$CH_3\underset{\underset{OCH_3}{|}}{C}HOCH_3 \ + \ H_2O \ \xrightarrow{\ H^+\ } \ CH_3CHO \ + \ 2CH_3OH$$

(c) The aldehyde, a, the hemiacetal, b, and the alcohol, c, can be oxidized under basic conditions. The alcohol will give acetaldehyde or acetic acid, depending on other conditions. The oxidation of a and b was discussed in the answer to part a of this question. The acetal is stable in a base and will not hydrolyze; therefore, it cannot break down into oxidizable compounds.

(d) The aldehyde, a, can be reduced to CH_3CH_2OH. The hemiacetal, b, will also react because it is present in equilibrium with its parent aldehyde (CH_3CHO) and alcohol (CH_3OH). As with a, the aldehyde will be taken out by the reducing agent and changed to ethyl alcohol.

6. (a) CH_3CO_2H

                                           (e) $CH_3CH = O \ + \ 2CH_3OH$

(b) $CH_3\underset{\overset{|}{OH}}{C}HOCH_3$ (f) $CH_3CH_2CO_2H$

(c) $CH_3\underset{\underset{OCH_3}{\diagdown}}{\overset{\overset{OCH_3}{\diagup}}{C}}H$ $(+ \ H_2O)$ (g) $CH_3SH \ + \ HSCH_2CO_2H$

(d) $CH_3\underset{\overset{|}{OH}}{C}HCH_3$ (h) $CH_2 = CH\overset{\overset{O}{\|}}{C}CH_3$

7.

	Common	IUPAC
(a)	propionaldehyde	propanal
(b)	acetone	propanone (2-propanone is all right, but the number is unnecessary here)
(c)	acetaldehyde	ethanal
(d)	diethyl ketone	3-pentanone

Multiple-Choice

1. b (an acetal)   (a is an ether; c is a di-ether and not an acetal; and d is an alcohol.)

2. b (an aldehyde) (a is an ether; c is a ketone; and d is an acetal. The oxidizing reagent is neutral or basic, as in Tollens' Test or Benedict's Test; therefore, the acetal will hold together. However, if the reagent is acidic, the acetal will hydrolyze to produce some acetaldehyde (and ethyl alcohol) and the aldehyde will be oxidized.)

3. a (an alcohol)   (It is the only one that can both accept and donate hydrogen bonds.)

4. c

5. b

6. d

7. a

8. b

9. a

10. d

16 Carboxylic Acids and Esters

OBJECTIVES

After you have studied this chapter and worked the Practice Exercises and Review Exercises in it, you should be able to do all of the following. Objectives 1 through 5 are very important for understanding the applications of this chapter in biochemistry and in our study of the molecular basis of life.

1. Recognize the structural features of: the carboxylic acid group, the carboxylate ion group, the ester group and ester bond, the acid chloride group, and the anhydride group.
2. By examining a structure of a substance, determine the probable ability of the substance to neutralize either a base or an acid or to be hydrolyzed, saponified, or esterified.
3. Write equations that are specific examples of the formation of:
   (a) an acid from an alcohol or an aldehyde
   (b) a carboxylic acid salt from the acid
   (c) a carboxylic acid from its salt
   (d) an ester from an alcohol and
      (1) a carboxylic acid
      (2) a carboxylic acid chloride
      (3) a carboxylic acid anhydride
   (e) an alcohol and an acid from an ester
   (f) an alcohol and the salt of a carboxylic acid from an ester
4. Give a general explanation for the relative acidity of an acid over that of an alcohol.

5. Explain through equations and discussion how the carboxyl group may be used as a "solubility switch."
6. Write the structural features common to phosphate, diphosphate, and triphosphate esters.
7. Define the terms in the Glossary, and give illustrations where applicable.

Reaction summary cards should be made for acids, their salts, acid chlorides, acid anhydrides, and esters. The partly completed card for alcohols can now be completed. A list of sentences that summarize chemical facts should be prepared and learned. The lists of reactions organized by key reagents (e.g., acids, bases, water and so forth) should be brought up to date.

GLOSSARY

Acid Anhydride. In organic chemistry, a compound formed by splitting water out between two — OH groups of the acid function of an organic acid. The structural features are:

$$-\overset{\overset{\displaystyle O}{\|}}{C} - O - \overset{\overset{\displaystyle O}{\|}}{C} - \qquad\qquad -\overset{\overset{\displaystyle O}{\|}}{\underset{\underset{\displaystyle OH}{|}}{P}} - O - \overset{\overset{\displaystyle O}{\|}}{\underset{\underset{\displaystyle OH}{|}}{P}} -$$

Carboxylic acid   Phosphoric acid
anhydride system   anhydride system

Acid Chloride. A derivative of an acid in which the — OH group of the acid has been replaced by — Cl.

$$R - \overset{\overset{\displaystyle O}{\|}}{C} - Cl$$

Acid Derivative. Any organic compound that can be made from an organic acid or that can be changed back to the acid by hydrolysis. (Examples are acid chlorides, acid anhydrides, esters, and amides.)

Acyl Group. $$R - \overset{\overset{\displaystyle O}{\|}}{C} -$$

Acyl Group Transfer Reaction. Any reaction in which an acyl group transfers from a donor to an acceptor.

Amide. Any organic compound whose molecules have a carbonyl-nitrogen

unit,
$$-\overset{\overset{\text{O}}{\|}}{\text{C}}-\overset{|}{\text{N}}-\ .$$

Carboxylic Acid. A compound whose molecules have the carboxyl group, $-CO_2H$.

Ester. A derivative of an acid and an alcohol that can be hydrolyzed to these parent compounds. Esters of carboxylic acids and phosphoric acid occur in living systems.

$$-\overset{|}{\underset{|}{\text{C}}}-\text{O}-\overset{\overset{\text{O}}{\|}}{\text{C}}-\qquad\qquad -\overset{|}{\underset{|}{\text{C}}}-\text{O}-\overset{\overset{\text{O}}{\|}}{\underset{\underset{\text{OH}}{|}}{\text{P}}}-\text{OH}$$

System in an ester of   System in an ester of
a carboxylic acid   phosphoric acid

Esterification. The formation of an ester.

Fatty Acid. Any carboxylic acid that can be obtained by the hydrolysis of animal fats or vegetable oils.

Saponification. The reaction of an ester with sodium or potassium hydroxide to give an alcohol and the salt of an acid.

KEY MOLECULAR "MAP SIGNS"

Key Molecular "Map Signs" in Organic Molecules	What to Expect When This Functional Group is Present
$-\overset{\overset{\text{O}}{\|}}{\text{C}}-\text{O}-\text{H}$ Carboxyl group	Influence on physical properties: The carboxyl group is a very polar group; it can both donate and accept H-bonds

[Often written as CO_2H or $COOH$]

Influence on chemical properties:

- Can neutralize OH^- (or HCO_3^- or CO_3^{2-}); in so doing, the group becomes

$$-\overset{\overset{\textstyle O}{\|}}{C}-O^-$$

(the carboxylate ion discussed next).
- Can be changed into an ester by reacting with an alcohol when a mineral acid catalyst is present.

$$-\overset{\overset{\textstyle O}{\|}}{C}-O^-$$

carboxylate
ion

Influence on physical properties:
- One of the most effective groups at bringing long hydrocarbon chains into solution.
- All salts of the carboxylic acids are solids at room temperature.

Influence on chemical properties:
- Aqueous solutions will test slightly basic.
- Can neutralize mineral acids (and revert to carboxyl group).

$$-\overset{\overset{\textstyle O}{\|}}{C}-O-\overset{\overset{\textstyle |}{}}{\underset{|}{C}}-$$

Ester

Influence on physical properties: The ester group is a moderately polar group; it can accept H-bonds but cannot donate them.

Influence on chemical properties:
- Can be hydrolyzed (to the carboxylic acid and alcohol).
- Can be saponified (to the carboxylate ion and alcohol).

$$-\overset{\overset{\textstyle O}{\|}}{C}-Cl$$

Acid chloride

$$-\overset{\overset{\textstyle O}{\|}}{C}-O-\overset{\overset{\textstyle O}{\|}}{C}-$$

Acid anhydride

Properties of importance:
Acid chlorides and acid anhydrides are
- easily hydrolyzed (to the carboxylic acid).
- easily react with alcohols to give esters.

O   O
‖   ‖
R — O — P — O — P — O — H
| |
OH   ↑ OH

Influence on physical properties: Both esters exist at the pH of body fluids as negatively charged ions; therefore they are very soluble in water.

O   O   O
‖   ‖   ‖
R — O — P — O — P — O — P — O — H
| ↑ | ↑ |
OH   OH   OH

Di- and triphosphate esters

Influence on chemical properties:

· Can neutralize OH^- (to the extent they begin with $-OH$ groups on phosphorus and have not been neutralized).

· In absence of a catalyst, they only react very slowly with water at the pH of body fluids.
· In the presence of the appropriate enzymes, they react rapidly with the alcohol (and amino) groups of biochemicals and suffer breakage of the bonds indicated by the arrows (↑).

Additional Reactions of Functional Groups Introduced in Earlier Chapters

|
— C — O — H
|

Alcohol (or phenol)

· Forms esters with carboxylic acids, either by reacting with the carboxylic acid (when a mineral acid catalyst is present) or by reacting with an acid chloride or acid anhydride.

DRILL EXERCISES

I. EXERCISES IN STRUCTURES AND NAMES

1. To make sure that you understand the condensed structures that are often used with carbonyl compounds, write out the following condensed structures as full structures.

   (a) CH_3CO_2H

   (b) $HO_2CCH_2CH_3$

   (e) $CH_3CO_2CH_3$

   (f) CH_3CONH_2

(c) CH_3CHO (g) $HOOCCH_3$
(d) CH_3CH_2OH

2. It is important to be able to recognize quickly the presence of functional groups in structures; otherwise, the chemical and physical properties of such structures cannot be "read" (in the sense of "reading" a map with knowledge of "map signs"). Study each of the following structures and assign them to their correct families. Some will have more than one functional group; name them all.

Examples:

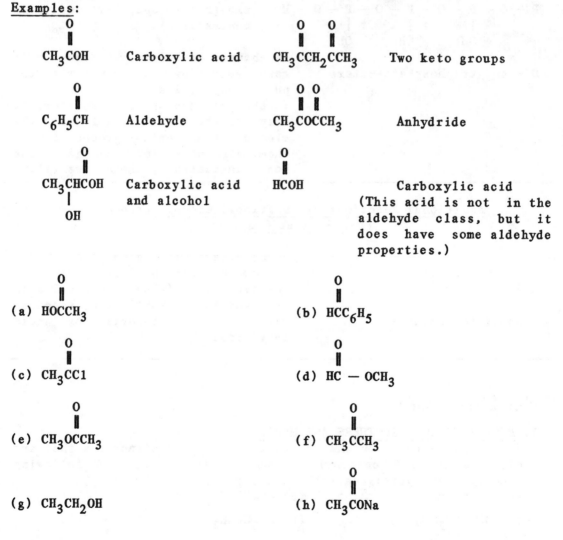

$$\overset{\text{O}}{\overset{\|}{CH_3COH}}$$ Carboxylic acid

$$\overset{\text{O}}{\overset{\|}{C_6H_5CH}}$$ Aldehyde

$$\underset{\underset{OH}{|}}{\overset{\text{O}}{\overset{\|}{CH_3CHCOH}}}$$ Carboxylic acid and alcohol

$$\overset{\text{O}\ \ \text{O}}{\overset{\|\ \ \ \|}{CH_3CCH_2CCH_3}}$$ Two keto groups

$$\overset{\text{O O}}{\overset{\|\ \|}{CH_3COCCH_3}}$$ Anhydride

$$\overset{\text{O}}{\overset{\|}{HCOH}}$$ Carboxylic acid (This acid is not in the aldehyde class, but it does have some aldehyde properties.)

(a) $\overset{\text{O}}{\overset{\|}{HOCCH_3}}$

(c) $\overset{\text{O}}{\overset{\|}{CH_3CCl}}$

(e) $\overset{\text{O}}{\overset{\|}{CH_3OCCH_3}}$

(g) CH_3CH_2OH

(b) $\overset{\text{O}}{\overset{\|}{HCC_6H_5}}$

(d) $\overset{\text{O}}{\overset{\|}{HC - OCH_3}}$

(f) $\overset{\text{O}}{\overset{\|}{CH_3CCH_3}}$

(h) $\overset{\text{O}}{\overset{\|}{CH_3CONa}}$

(i) $\overset{\overset{\textstyle O}{\|}}{HCH}$

(j)

(k) $\overset{\overset{\textstyle O}{\|}}{C_6H_5COH}$

(l) $\overset{\overset{\textstyle O}{\|}}{C_6H_5OCH}$

(m) $\overset{\overset{\textstyle O}{\|}}{HOCH_2CH_2CCH_3}$

(n) $\overset{\overset{\textstyle O}{\|}}{CH_3\underset{\underset{\textstyle OH}{|}}{C}HCCH_3}$

(o) $(CH_3)_2CHCOOH$

(p) $\overset{\overset{\textstyle O}{\|}}{HCCH_2}CH_2\overset{\overset{\textstyle O}{\|}}{C}CH_3$

(q) $CH_3O\overset{\overset{\textstyle O}{\|}}{C}CH_2CH_3$

(r) $CH_3OCH_2\overset{\overset{\textstyle O}{\|}}{C}CH_3$

(s) $CH_3O\overset{\overset{\textstyle O}{\|}}{C}CH_2\overset{\overset{\textstyle O}{\|}}{C}H$

(t) $CH_3\overset{\overset{\textstyle O}{\|}}{C}O\overset{\overset{\textstyle O}{\|}}{C}CH_2CH_3$

(u) $CH_3CH_2CO_2H$

(v) $CH_3CH_2OCH\underset{\underset{\textstyle CH_3}{|}}{}\overset{\overset{\textstyle O}{\|}}{C}CH_3$

(w) $CH_3O\overset{\overset{\textstyle O}{\|}}{C}CH_2\overset{\overset{\textstyle O}{\|}}{C}OCH_3$

(x) $C_6H_5\overset{\overset{\textstyle O}{\|}}{C}O\overset{\overset{\textstyle O}{\|}}{C}C_6H_5$

(y) $C_6H_5\overset{\overset{\textstyle O}{\|}}{C}Cl$

(z) $HO-\underset{}{\bigcirc}-\overset{\overset{\textstyle O}{\|}}{C}H$

(aa) $CH_2 = CHO\overset{\overset{\textstyle O}{\|}}{C}CH_3$

(bb) $CH_3CH_2O\overset{\overset{\textstyle O}{\|}}{C}CH = CH_2$

$$(cc) \quad CH_3\overset{O}{\overset{\|}{C}}OCH_2CH_2O\overset{O}{\overset{\|}{C}}CH_3 \qquad\qquad (dd) \quad CH_3\overset{O}{\overset{\|}{C}}CH_2CH_2\overset{O}{\overset{\|}{C}}OH$$

3. If after studying the ways to name esters and salts discussed in the text, you continue to have trouble, try the following. Prefixes in the common names for esters and acid salts (and aldehydes) relate to the acid portion of the structure. You may need some drill simply in recognizing what the acid portion is. It is that part of the structure that contains the carbonyl group plus whatever else consists solely of carbon and hydrogen. In other words, the acid portion is the part of the structure of the acid derivative that came from the parent acid.

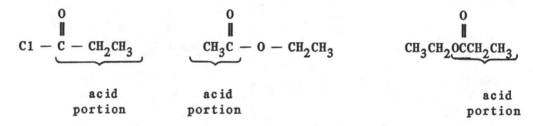

| acid portion | acid portion | acid portion |

For the following structures, circle the acid portion in each structure and then write the prefix associated with its common name.

Examples:

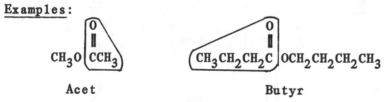

Acet   Butyr

Note that the oxygen that has only single bonds and the R-group attached to it are not included.

(a) $HC\overset{O}{\overset{\|}{}}OCH_3$

(b) $CH_3CH_2\overset{O}{\overset{\|}{C}}O^-Na^+$

(c) $CH_3CH_2O\overset{O}{\overset{\|}{C}}H$

(d) $CH_3CH_2O\overset{O}{\overset{\|}{C}}CH_3$

$$\text{O}$$
$$\|$$
(e) $CH_3CH_2COCH_3$

$$\text{O}$$
$$\|$$
(f) $CH_3OCCH_2CH_2CH_3$

4. Complete the following table according to the example given.

STRUCTURE	FAMILY	COMMON NAME
$CH_3\overset{\overset{\displaystyle O}{\|}}{C}OCH_3$	ester	methyl acetate
(a) $(CH_3)_2CH O\overset{\overset{\displaystyle O}{\|}}{C}CH_3$	_____	_____
(b) $CH_3CH_2\overset{\overset{\displaystyle O}{\|}}{C}O^-Na^+$	_____	_____
(c) $CH_3O\overset{\overset{\displaystyle O}{\|}}{C}CH_2CH_2CH_3$	_____	_____
(d) $(CH_3)_2CHCH_2O\overset{\overset{\displaystyle O}{\|}}{C}CH_2CH_2CH_3$	_____	_____
(e) $H-\overset{\overset{\displaystyle O}{\|}}{C}CH_2CH_3$	_____	_____

5. Write the IUPAC names for compounds (a) – (e) of the previous exercise.

(a) _____

(b) _____

(c) _____

(d) _____

(e) _____

6. Write the condensed structural formulas for each of the following.

(a) methyl ethanoate

(b) 2,3-dimethylbutanal

(c) propanoic anhydride

(d) sec-butyl ethanoate

(e) 2-methylbutanoyl chloride

(f) sodium 3-chloropropanoate

II. DRILL ON THE REACTIONS OF CARBOXYLIC ACIDS WITH STRONG BASES

Write the structures of the salts that form in the following situations. Assume in every case that the aqueous base is being used at room temperature. You may write the structures showing the electrical charges (e.g., $CH_3CO_2^-Na^+$) or not (e.g., CH_3CO_2Na), but you should never draw a line between the two parts of the salts (e.g., $CH_3CO_2 - Na$) because a line means a covalent bond, which is not present here.

1. $CH_3CH_2CH_2CO_2H$ + KOH(aq) ———> _____

2. $CH_3(CH_2)_6CO_2H$ + NaOH(aq) ———> _____

3. $CH_3 - O - CH_2CO_2H$ + NaOH(aq) ———> _____

4. $HOCH_2CH_2CO_2H$ + KOH(aq) ———> _____

5. $HO_2CCH_2CH_2CO_2H$ + 2NaOH(aq) ———> _____

III. DRILL ON THE REACTIONS OF CARBOXYLIC ACID SALTS WITH STRONG ACIDS

Write the structures of the acids that form in the following situations. Assume in every case that the aqueous acid is being used at room temperature.

1. $HCO_2^-Na^+$ + $HCl(aq)$ $\longrightarrow$ _____

2. $K^{+-}O_2CCH_2CH_2CH_3$ + $HCl(aq)$ $\longrightarrow$ _____

3. $CH_3(CH_2)_8CO_2^-Na^+$ + $HCl(aq)$ $\longrightarrow$ _____

4. CH_3 —⬡— $CO_2^-K^+$ + $HCl(aq)$ $\longrightarrow$ _____

5. $CH_2CH_2OCH_2CH_2CO_2^-Na^+$ + $HCl(aq)$ $\longrightarrow$ _____

IV. <u>DRILL</u> <u>ON</u> <u>WRITING</u> <u>THE</u> <u>STRUCTURES</u> <u>OF</u> <u>ESTERS</u> <u>THAT</u> <u>CAN</u> <u>FORM</u> <u>FROM</u> <u>GIVEN</u> <u>ACIDS</u> <u>AND</u> <u>ALCOHOLS</u>

      Example 16.2 in the text provides the pattern. If after doing Review Exercises 6 and 7 you feel the need for more practice, try the following. Write the structures of the esters that can form between the following pairs of compounds. This will also serve as a review of the names of acids and alcohols. On the lines above the names write the structures of the reactants, and then form the structures of the products.

1. _____ + _____ $\longrightarrow$ _____
    butyric acid   methyl alcohol

2. _____ + _____ $\longrightarrow$ _____
    formic acid   propyl alcohol

3. _____ + _____ $\longrightarrow$ _____
    isobutyl alcohol   propionic acid

4. _____ + _____ $\longrightarrow$ _____
    t-butyl alcohol   acetic acid

5. $HO_2CCH_2CH_2CH_2CO_2H$ + $2CH_3OH$ $\longrightarrow$ _____
                                   (show the di-ester)

V.   DRILL ON WRITING THE PRODUCTS OF THE HYDROLYSIS OF ESTERS

Example 16.4 shows how to do this. For more drill beyond Review Exercise 10, try the following. Write the structures of the products of the hydrolysis of the following esters.

1. $CH_3CH_2\overset{\overset{\displaystyle O}{\|}}{C}OCH_3$ + H_2O $\xrightarrow[\text{heat}]{H^+}$ _____

2. $CH_3O\overset{\overset{\displaystyle O}{\|}}{C}CH_3$ + H_2O $\xrightarrow[\text{heat}]{H^+}$ _____

3. CH_3-⬡-$\overset{\overset{\displaystyle O}{\|}}{C}O\overset{\overset{\displaystyle CH_3}{|}}{C}HCH_3$ + H_2O $\xrightarrow[\text{heat}]{H^+}$ _____

4. $CH_3CH_2\overset{\overset{\displaystyle CH_3}{|}}{C}H - O - \overset{\overset{\displaystyle O}{\|}}{C}\underset{\underset{\displaystyle CH_3}{|}}{C}HCH_3$ + H_2O $\xrightarrow[\text{heat}]{H^+}$ _____

5. $CH_3CH_2O\overset{\overset{\displaystyle O}{\|}}{C}CH_2CH_2CH_2\overset{\overset{\displaystyle O}{\|}}{C}OCH_3$ + $2H_2O$ $\xrightarrow[\text{heat}]{H^+}$

_____

VI. DRILL IN WRITING THE PRODUCTS OF THE SAPONIFICATION OF ESTERS

Example 16.5 in the text describes how to do this kind of exercise. For additional drill, write the structures of the products of the complete saponification of the esters of the preceding drill exercise. Use NaOH(aq) as the saponifying agent.

1. _____   4. _____

2. _____   5. _____

3. _____

SELF-TESTING QUESTIONS

Use the following questions as your own final examination for the chapter. As a review, go back to the chapter objectives and find out if you can do them.

COMPLETION

1. Write the structure(s) of the organic product(s) that would form in each reaction. If no reaction occurs, write "none." Some of the reactions will involve a review of earlier chapters.

(a) CH_3CO_2H + NaOH $\xrightarrow[\text{water}]{}$ _____

(b) CH_3CO_2H + CH_3OH $\xrightarrow[\text{heat}]{H^+}$ _____

(c) $CH_3CH_2CH_2CH_2CO_2^-Na^+$ + HCl(aq) $\longrightarrow$ _____

(d) $HOC \overset{O}{\underset{\|}{}} \!\!-\!\! \bigcirc \!\!-\!\! \overset{O}{\underset{\|}{}} COH$ + $2CH_3CH_2OH$ $\xrightarrow[\text{heat}]{H^+}$ _____

(e) $CH_3O\overset{O}{\underset{\|}{C}}CH_3$ + H_2O $\xrightarrow[\text{heat}]{H^+}$ _____ + _____

(f) $CH_3\underset{\underset{OH}{|}}{C}HCO_2H$ $\xrightarrow{(O)}$ _____

(g) $CH_3O\overset{O}{\underset{\|}{C}}CH_2CH_2\overset{O}{\underset{\|}{C}}OCH_3$ $\xrightarrow[\text{heat}]{H_2O,\ H^+}$ _____ + _____

(h) $CH_3O\overset{O}{\underset{\|}{C}}CH_2CH_2\underset{\underset{O}{\|}}{C}OCH_3$ $\xrightarrow[\text{heat}]{H_2O,\ H^+}$ _____ + _____

(i) $CH_3CH_2CH_2\overset{O}{\underset{\|}{C}} - Cl$ + CH_3CH_2OH $\longrightarrow$ _____

(j) $CH_3S - S - CH_2 - \overset{\overset{\displaystyle O}{\|}}{C} - O^-Na^+$ + HCl $\xrightarrow[\text{water}]{}$ _____

(k) CH_3OCCH_3 + NaOH $\xrightarrow[\text{heat}]{}$ _____ + _____

(where the C bears $\overset{\displaystyle O}{\|}$)

2. Which of the reactions in question 1 illustrate

(a) esterification   _____

(b) saponification   _____

MULTIPLE-CHOICE

1. The compound whose structure is: $CH_3 - O - CH_2 - \overset{\overset{\displaystyle O}{\|}}{C} - O - CH_3$
   has which functional group(s)?
   (a) two ether groups
   (b) an ether and an ester group
   (c) two ester groups
   (d) two ether and one ketone group

2. The structure of sodium butyrate is

(a) $Na^{+-}O\overset{\overset{\displaystyle O}{\|}}{C}OCH_2CH_2CH_3$

(b) $Na^{+-}\overset{\overset{\displaystyle O}{\|}}{C} - O - CH_2CH_2CH_2CH_3$

(c) $CH_3CH_2CH_2 - \overset{\overset{\displaystyle O}{\|}}{C} - O - Na$

(d) $CH_3CH_2CH_2\overset{\overset{\displaystyle O}{\|}}{C}O_2^-Na^+$

3. The name of $CH_3\overset{\overset{\displaystyle CH_3}{|}}{C}HCO_2CH_2\overset{\overset{\displaystyle CH_3}{|}}{C}HCH_3$ is

(a) isopropyl isobutyrate
(b) isobutyl 2-methylpropanoate
(c) isobutyl butyrate
(d) isopropyl 3-methylbutanoate

4. The acid derivative that is most reactive toward water is
   (a) the ester   (c) the acid chloride
   (b) the acid salt   (d) both (b) and (c)

5. The action of aqueous hydrochloric acid on $CH_3CH_2CH_2CH_2CO_2^-K^+$ gives
   (a) $HO_2CCH_2CH_2CH_2CH_3$ + KCl

   (b) $CH_3CH_2CH_2CH_3$ + CO_2 + KCl

   (c) $CH_3CH_2CH_2CH_2COCl$ + KOH

   (d) none of these

6. The action of aqueous potassium hydroxide at room temperature on $CH_3CH_2 - O - CH_2CH_2CO_2H$ gives
   (a) CH_3CH_2OH + $HOCH_2CH_2CO_2^-K^+$

   (b) $CH_3CH_2OCH_2CH_2CO_2^-K^+$

   (c) $K^{+-}OCH_2CH_2OCH_2CH_2CO_2H$ + H_2O

   (d) no reaction

7. In the following structure, the numbered arrows point toward functional groups. What are the numbers of the arrows pointing toward groups readily attacked by relatively mild oxidizing agents?
   (a) 1
   (b) 2
   (c) 3
   (d) 4
   (e) 5

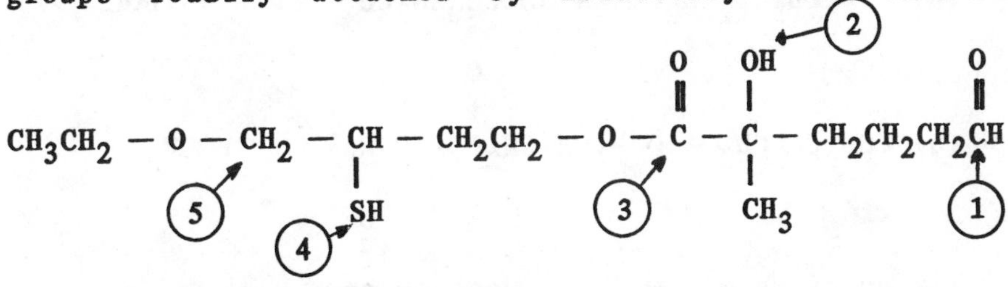

   $CH_3CH_2 - O - CH_2 - CH - CH_2CH_2 - O - \overset{\overset{O}{\|}}{C} - \overset{\overset{OH}{|}}{\underset{CH_3}{C}} - CH_2CH_2CH_2\overset{\overset{O}{\|}}{CH}$

   ⑤   ④ SH   ③   ①   ②

8. In the structure of question 7, which groups, if any, are subject to acid-catalyzed hydrolysis?
   (a) 1   (b) 2   (c) 3   (d) 4   (e) 5   (f) none

9. In the structure of question 7, which groups, if any, will neutralize aqueous sodium hydroxide at room temperature?
   (a) 1   (b) 2   (c) 3   (d) 5   (e) none of these

10. The action of one mole of water (containing a trace of acid catalyst) on one mole of compound Y produces one mole of CH_3CO_2H and one mole of CH_3CH_2OH. The structure of Y is

(a) $CH_3CH_2OCH_2CH_3$

(c) $CH_3CH - O - CH_2CH_3$
               $|$
               OH

(b) $CH_3CH_2O\overset{\displaystyle O}{\overset{\|}{C}}CH_3$

(d) $CH_3\overset{\displaystyle O}{\overset{\|}{C}} - O - \overset{\displaystyle O}{\overset{\|}{C}}CH_2CH_3$

11. The esterification of propionic acid by ethyl alcohol would produce

(a) $CH_3CH_2CH_2O\overset{\displaystyle O}{\overset{\|}{C}}CH_3$

(c) $CH_3CH_2\overset{\displaystyle O}{\overset{\|}{C}}CH_2CH_3$

(b) $CH_3CH_2\overset{\displaystyle O}{\overset{\|}{C}}OCH_2CH_3$

(d) $CH_3CH_2\overset{\displaystyle OH}{\overset{|}{C}}OCH_2CH_3$
                              $|$
                              H

12. The saponification of $CH_3CH_2CH_2O - CH_2CH_2\overset{\displaystyle O}{\overset{\|}{C}} - O - CH_3$ by sodium hydroxide would produce

(a) $CH_3CH_2CH_2OH$ + $HOCH_2CH_2\overset{\displaystyle O}{\overset{\|}{C}}O - Na^+$ + $HOCH_3$

(b) $CH_3CH_2CH_2 - O - CH_2CH_2\overset{\displaystyle O}{\overset{\|}{C}} - O^-Na^+$ + $HOCH_3$

(c) $CH_3CH_2CH_2O - Na^+$ + $HOCH_2CH_2\overset{\displaystyle O}{\overset{\|}{C}}OCH_3$

(d) $CH_3CH_2CH_2OCH_2CH_2\overset{\displaystyle O}{\overset{\|}{C}}OH$ + Na^+OCH_3

ANSWERS

ANSWERS TO DRILL EXERCISES

I. Exercises in Structures and Names

1. (a)
$$H - \underset{\underset{H}{|}}{\overset{\overset{H}{|}}{C}} - \overset{\overset{O}{\|}}{C} - O - H$$

(b)
$$H - O - \overset{\overset{O}{\|}}{C} - \underset{\underset{H}{|}}{\overset{\overset{H}{|}}{C}} - \underset{\underset{H}{|}}{\overset{\overset{H}{|}}{C}} - H$$

(c)
$$H - \underset{\underset{H}{|}}{\overset{\overset{H}{|}}{C}} - \overset{\overset{O}{\|}}{C} - H$$

(d)
$$H - \underset{\underset{H}{|}}{\overset{\overset{H}{|}}{C}} - \underset{\underset{H}{|}}{\overset{\overset{H}{|}}{C}} - O - H$$

(e)
$$H - \underset{\underset{H}{|}}{\overset{\overset{H}{|}}{C}} - \overset{\overset{O}{\|}}{C} - O - \underset{\underset{H}{|}}{\overset{\overset{H}{|}}{C}} - H$$

(f)
$$H - \underset{\underset{H}{|}}{\overset{\overset{H}{|}}{C}} - \overset{\overset{O}{\|}}{C} - \overset{\overset{H}{|}}{N} - H$$

(g)
$$H - O - \overset{\overset{O}{\|}}{C} - \underset{\underset{H}{|}}{\overset{\overset{H}{|}}{C}} - H$$

2. (a) acid
   (c) acid chloride
   (e) ester
   (g) alcohol
   (i) aldehyde
   (k) acid
   (m) alcohol, ketone
   (o) acid
   (q) ester
   (s) ester, aldehyde
   (u) acid
   (w) diester
   (y) acid chloride
   (aa) alkene, ester
   (cc) diester

   (b) aldehyde
   (d) ester
   (f) ketone
   (h) acid salt
   (j) ketone
   (l) ester
   (n) alcohol, ketone
   (p) aldehyde, ketone
   (r) ether, ketone
   (t) anhydride
   (v) ether, ketone
   (x) anhydride
   (z) phenol, aldehyde
   (bb) ester, alkene
   (dd) ketone, acid

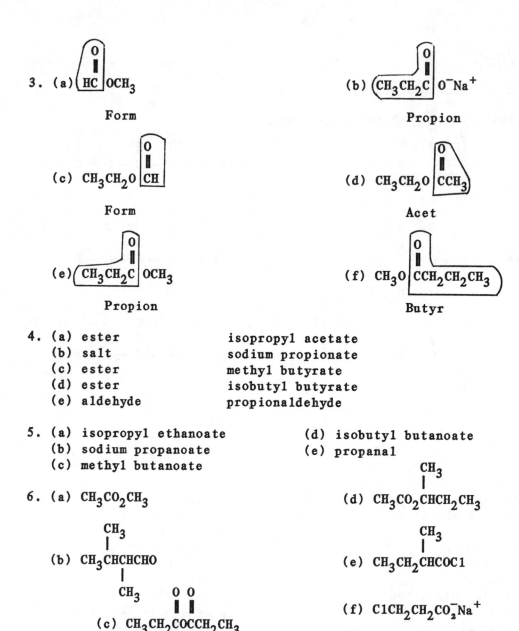

3. (a) HC—OCH₃ (with O double bond)
   Form

   (b) CH₃CH₂C—O⁻Na⁺ (with O double bond)
   Propion

   (c) CH₃CH₂O—CH (with O double bond)
   Form

   (d) CH₃CH₂O—CCH₃ (with O double bond)
   Acet

   (e) CH₃CH₂C—OCH₃ (with O double bond)
   Propion

   (f) CH₃O—CCH₂CH₂CH₃ (with O double bond)
   Butyr

4. (a) ester   isopropyl acetate
   (b) salt   sodium propionate
   (c) ester   methyl butyrate
   (d) ester   isobutyl butyrate
   (e) aldehyde   propionaldehyde

5. (a) isopropyl ethanoate   (d) isobutyl butanoate
   (b) sodium propanoate   (e) propanal
   (c) methyl butanoate

6. (a) $CH_3CO_2CH_3$

   (b) $CH_3CHCHCHO$ with CH_3 on second carbon and CH_3 below

   (c) $CH_3CH_2COCCH_2CH_3$ with two O (double bonds)

   (d) $CH_3CO_2CHCH_2CH_3$ with CH_3 substituent

   (e) $CH_3CH_2CHCOCl$ with CH_3 substituent

   (f) $ClCH_2CH_2CO_2^-Na^+$

II. Drill on the Reactions of Carboxylic Acids with Strong Base

1. $CH_3CH_2CH_2CO_2^-K^+$
2. $CH_3(CH_2)_6CO_2^-Na^+$
3. $CH_3 — O — CH_2CO_2^-Na^+$
4. $HOCH_2CH_2CO_2^-K^+$
5. $Na^{+-}O_2CCH_2CH_2CO_2^-Na^+$

III. Drill on the Reactions of Carboxylic Acid Salts with Strong Acids

1. HCO_2H

2. $HO_2CCH_2CH_2CH_3$

3. $CH_3(CH_2)_8CO_2H$

4. 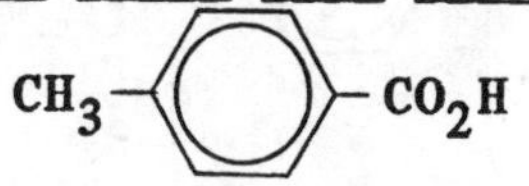 $CH_3-\langle\bigcirc\rangle-CO_2H$

5. $CH_3CH_2OCH_2CH_2CO_2H$

IV. Drill on Writing the Structures of Esters That Can Form From Given Acids and Alcohols

1. $CH_3CH_2CH_2CO_2CH_3$

2. $HCO_2CH_2CH_2CH_3$

3. $CH_3CH_2CO_2CH_2\overset{\overset{\displaystyle CH_3}{|}}{C}HCH_3$

4. $CH_3CO_2\overset{\overset{\displaystyle CH_3}{|}}{\underset{\underset{\displaystyle CH_3}{|}}{C}}CH_3$

5. $CH_3O_2CCH_2CH_2CH_2CO_2CH_3$

V. Drill on Writing the Products of the Hydrolysis of Esters

1. $CH_3CH_2CO_2H$ + $HOCH_3$

2. CH_3OH + HO_2CCH_3

3. 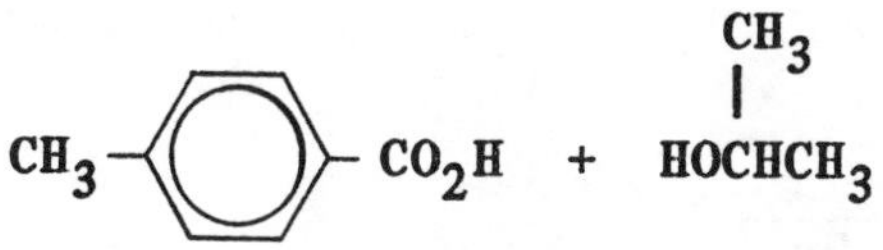 $CH_3-\langle\bigcirc\rangle-CO_2H$ + $HO\overset{\overset{\displaystyle CH_3}{|}}{C}HCH_3$

4. $CH_3CH_2\overset{\overset{\displaystyle CH_3}{|}}{C}HOH$ + $HO_2C\overset{\overset{\displaystyle }{}}{\underset{\underset{\displaystyle CH_3}{|}}{C}}HCH_3$

5. CH_3CH_2OH + $HO_2CCH_2CH_2CH_2CO_2H$ + $HOCH_3$

VI. Drill in Writing the Products of the Saponification of Esters

1. $CH_3CH_2CO_2Na$ + $HOCH_3$

2. CH_3OH + CH_3CO_2Na

3. CH_3—⬡—CO_2Na + $HOCH(CH_3)_2$

4. $CH_3CH_2\overset{\overset{\displaystyle CH_3}{|}}{C}HOH$ + $CH_3\overset{\overset{\displaystyle CH_3}{|}}{C}HCO_2Na$

5. CH_3CH_2OH + $NaO_2CCH_2CH_2CH_2CO_2Na$ + $HOCH_3$

ANSWERS TO SELF-TESTING QUESTIONS

Completion

1. (a) $CH_3CO_2^-Na^+$ (+ H_2O)

   (b) $CH_3CO_2CH_3$ (+ H_2O)

   (c) $CH_3CH_2CH_2CH_2CO_2H$ (+ $NaCl$)

   (d) $CH_3CH_2O\overset{\overset{\displaystyle O}{\|}}{C}$—⬡—$\overset{\overset{\displaystyle O}{\|}}{C}OCH_2CH_3$ (+ $2H_2O$)

   (e) CH_3OH + CH_3CO_2H (f) $CH_3\overset{\overset{\displaystyle O}{\|}}{C}CO_2H$

   (g) $2CH_3OH$ + $HO_2CCH_2CH_2CO_2H$

   (h) $CH_3OCH_2CH_2CO_2H$ + $HOCH_3$

   (i) $CH_3CH_2CH_2\overset{\overset{\displaystyle O}{\|}}{C}-O-CH_2CH_3$ (+ HCl)

   (j) $CH_3S-SCH_2CO_2H$ (+ $NaCl$)

   (k) CH_3OH + $CH_3CO_2^-Na^+$

2. (a) b, d, and i   (b) k

Multiple-Choice

1.	b	2.	d
3.	b	4.	c
5.	a	6.	b
7.	a and d	8.	c
9.	e	10.	b
11.	b	12.	b

17 Amines and Amides

OBJECTIVES

The chemistry of the amino group and the amide function is essential to our later study of proteins and nucleic acids. After you have studied Chapter 17 in the text and worked its Practice Exercises and Review Exercises, you should be able to do the following.

1. Identify the amine and amide functions in given structures, whether open-chain or heterocyclic.
2. Write the names of simple amines and amides.
3. Write the structures of simple amines and amides from their names.
4. Describe hydrogen bonding as it occurs among amines and amides and discuss its effects on physical properties.
5. Write equations for the reactions of amines with strong, aqueous acids.
6. Write equations for the reactions of protonated amines with strong, aqueous bases.
7. Write the structure of amides that can be formed (directly or indirectly) from given carboxylic acids and NH_3, RNH_2, or R_2NH.
8. Given the structure of an amide, write the products that form when it is hydrolyzed.
9. Define the terms in the Glossary.

GLOSSARY

Alkaloid. A physiologically active, heterocyclic amine isolated from plants.

Amide. Any organic compound whose molecules have a carbonyl-nitrogen

unit, $-\overset{\overset{\textstyle O}{\|}}{C}-\overset{|}{N}-$.

Amide Bond. The single bond that holds the carbonyl group to the nitrogen atom in an amide.

Amine. Any organic compound whose molecules have a trivalent nitrogen atom, as in $R-NH_2$, $R-NH-R$, or R_3N.

Amine Salt. Any organic compound whose molecules have a positively charged, tetravalent, protonated nitrogen atom, as in RNH_3^+, $R_2NH_2^+$, or R_3NH^+.

Base Ionization Constant (K_b). For the equilibrium (where B is some

base) $B + H_2O \rightleftharpoons BH^+ + OH^-$ $\qquad K_b = \dfrac{[BH^+][OH^-]}{[B]}$

You should prepare the 5 x 8" "reactions" cards for the amino group, the protonated amino group, and the amide group. Now is the time to bring up to date the lists of functional groups that react with aqueous acids, aqueous bases, and water.

KEY MOLECULAR "MAP SIGNS"

Key Molecular "Map Signs" in Organic Molecules	What to Expect When This Functional Group Is Present
$-\overset{\|}{\underset{\|}{N}}:$ Amino group	Influence on physical properties: · If present as $R-NH_2$ or R_2NH, the amino group can both donate and accept hydrogen bonds to and from water, amines, or alcohols.

• If present as R_3N, the amino group can only accept hydrogen bonds; it is as soluble in water as alcohols are.

Influence on chemical properties:

• The presence of an amino group makes the molecule a proton acceptor, a Brønsted base.

• The easy room-temperature changes:

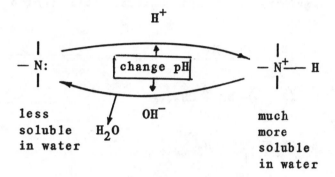

makes the amino group one of the major solubility switches in nature's biochemical molecules.

Amines (with at least one H on N)

• Forms amides with acids; best done by a reaction of the amine with the acid chloride or anhydride.

$$\overset{O}{\underset{}{\overset{\|}{-}}\,C - \overset{|}{N} -}$$

Amide

Influence on physical properties:

Amides have a very polar group (particularly if at least one hydrogen is attached to the nitrogen); virtually all amides are solids at room temperature.

Influence on chemical properties:

• Amides are not basic; they are not proton acceptors in the way that makes amines basic.

• Can be hydrolyzed by aqueous acids or bases, if heated, to give the carboxylic acid and the amine (or ammonia).

DRILL EXERCISES

I. DRILL IN WRITING PRODUCTS OF THE REACTIONS OF AMINES WITH STRONG ACIDS

Example 17.1 in the text describes how these products can be written. For additional practice, write the structures of the organic cations that form when each of the following amines reacts with something like hydrochloric acid (which is really a reaction with H_3O^+).

1. $CH_3CH_2CH_2NH_2$ _____

2. 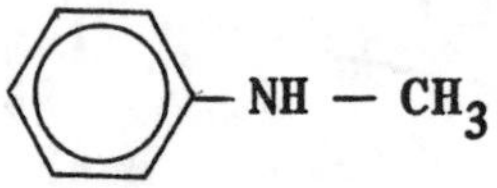 ⬡— NH — CH_3 _____

3. $CH_3CH_2NCH_2CH_3$
$\qquad\qquad |$
$\qquad\quad CH_3$ _____

4. 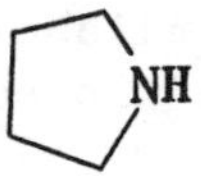
$\qquad$⬠NH _____

5. $CH_3NHCH_2CH_2NHCH_3$ (and 2HCl)
_____

II. DRILL IN WRITING THE PRODUCTS OF THE REACTIONS OF PROTONATED AMINES WITH STRONG, AQUEOUS BASE

Example 17.2 in the text explains how to work this kind of problem. For further drill, write the products of the deprotonation of the following cations.

1. $CH_3CH_2NH_3^+$ _____

2. $CH_3CH_2\overset{+}{N}HCH_3$
$\qquad\qquad |$
$\qquad\quad CH_3$ _____

3. 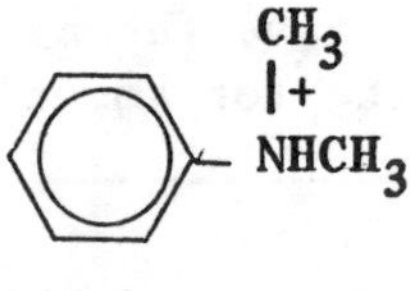 ⬡—
$\qquad\qquad CH_3$
$\qquad\qquad |+$
$\qquad\quad NHCH_3$ _____

$$\overset{+}{N}H_3CH_2CH_2\overset{\overset{\displaystyle O}{\|}}{C}CH_3$$

4. $\overset{+}{N}H_3CH_2CH_2\overset{O}{\overset{\|}{C}}CH_3$ _____

5. $\overset{+}{N}H_3\overset{|}{C}HCO_2^-$ _____
  $\overset{|}{C}H_3$

III. DRILL IN WRITING THE STRUCTURES OF THE AMIDES THAT CAN BE MADE (DIRECTLY OR INDIRECTLY) FROM GIVEN CARBOXYLIC ACIDS AND AMINES

See Example 17.3 for a discussion of how to work this kind of problem. The following will give extra opportunities to drill yourself. Write the structure of the amides that can be made from the given acids and amines (or ammonia).

1. ⬡—CO_2H + NH_3 _____

2. $CH_3CH_2CO_2H$ + NH_3 _____

3. $CH_3\overset{\overset{\displaystyle CH_3}{|}}{C}HCH_2CO_2H$ + NH_3 _____

4. ⬡—CO_2H + CH_3NH_2 _____

5. CH_3CO_2H + $CH_3CH_2NH_2$ _____

IV. DRILL IN WRITING THE PRODUCTS OF THE HYDROLYSIS OF AMIDES

Example 17.4 in the text discusses how this kind of problem can be worked. For more practice, do the following. Write the structures of the products that can form by the hydrolysis of the following amides. Show the acids as acids, not as their salts. Similarly, show the amines as amines, not in protonated forms.

1. $CH_3CH_2\overset{O}{\overset{\|}{C}}NH_2$ _____

$$
\begin{array}{c} O \\ \parallel \end{array}
$$

2. $CH_3CH_2CH_2CNHCH_3$ _____

$$
\begin{array}{c} O \\ \parallel \end{array}
$$

3. $CH_3CH_2NHCCH_2CH_3$ _____

$$
\begin{array}{cc} O & CH_3 \\ \parallel & \mid \end{array}
$$

4. $CH_3CHNHCCH_2CHCH_3$
$\quad\ \mid$
$\quad\ CH_3$ _____

5. 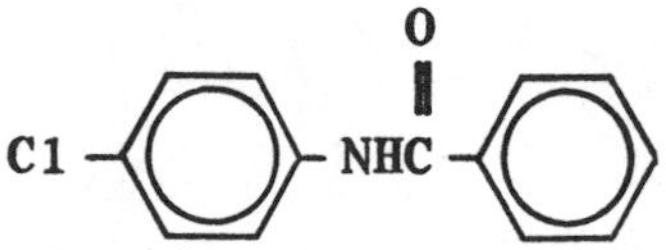 Cl ⬡ $-NHC$ ⬡ (with O double bonded to C) _____

V. <u>EXERCISES</u> <u>IN</u> <u>HYDROGEN</u> <u>BONDS</u>

Amines can accept and donate H-bonds like alcohols, but the H-bonds in amines are weaker. (Amines with three groups on nitrogen and no N — H bond left can only accept hydrogen bonds.) The questions in this exercise refer to the following structures.

A. $CH_3CH_2CH_2 - \ddot{O} - H$

$$
\begin{array}{c} H \\ \mid \end{array}
$$

B. $CH_3CH_2CH_2 - \underset{\cdot\cdot}{N} - H$

$$
\begin{array}{c} H \\ \mid \end{array}
$$

C. $CH_3 - \underset{\cdot\cdot}{N} - H$

$$
\begin{array}{c} CH_3 \\ \mid \end{array}
$$

D. $CH_3 - N: \\ \quad\ \mid \\ \quad\ CH_3$

E. $CH_3CH_2CH_2CH_2 - Br$

F. 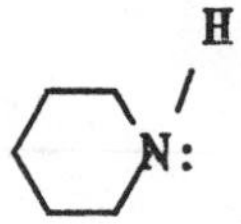 ⬡$N:$ with H above

$$
\begin{array}{c} H \\ \mid \end{array}
$$

G. 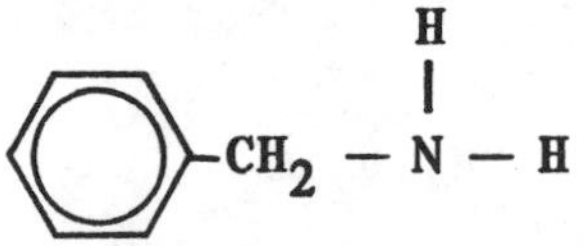 ⬡$-CH_2 - N - H$

$$
\begin{array}{c} H \\ \mid \end{array}
$$

H. ⬡$-\underset{\cdot\cdot}{N} - H$

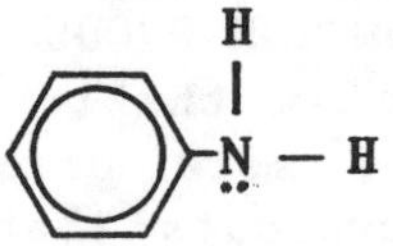

I. 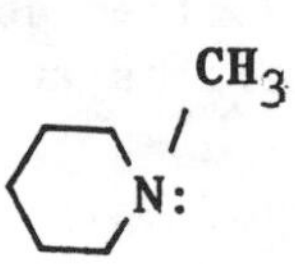 ⬡$N:$ with CH_3 above

1. Which can donate H-bonds? _____

2. Which can accept H-bonds? _____

3. Which can **only** accept H-bonds? _____

4. Which is totally insoluble in water? _____

5. Which has a higher boiling point, A or B? _____

6. Which is (are) the aromatic amine(s)? _____

A BRIEF SURVEY OF THE PRINCIPAL FUNCTIONAL GROUPS ATTACKED BY WATER, OXIDIZING AGENTS, REDUCING AGENTS OR NEUTRALIZING AGENTS

[Note: We summarize here only those reactions we have studied. In terms of the whole field of organic chemistry, the list is of course very incomplete. Omitted are the reactions of acid chlorides and anhydrides, including the phosphoric anhydrides.]

1. Groups that are split apart by water (hydrolysis) in the presence of acids or appropriate enzymes.

   Acetals are hydrolyzed to aldehydes and alcohols.

$$
\begin{array}{c}
\quad OR' \\
\quad / \\
RCH \qquad + \quad H_2O \xrightarrow{\ H^+\ } \quad RCHO \ + \ 2HOR' \\
\quad \backslash \\
\quad OR'
\end{array}
$$

   Esters are hydrolyzed to acids and alcohols.

$$
\begin{array}{c}
O \\
\parallel \\
RC - OR' \ + \ H_2O \xrightarrow{\ H^+\ } RCO_2H \ + \ HOR'
\end{array}
$$

   Amides are hydrolyzed to acids and amines (or ammonia).

$$
\begin{array}{c}
O \quad | \\
\parallel \ \ | \\
RC - N - \ + \ H_2O \xrightarrow{\ H^+\ } RCO_2H \ + H - N -
\end{array}
$$

2. Groups that are affected by oxidizing agents.

Alkenes are attacked by ozone and permanganate.

Alcohol groups are converted to carbonyl groups.

$$-\overset{|}{\underset{\underset{H}{|}}{C}} - \overset{}{\underset{\underset{H}{\backslash}}{O}} \; + \; (O) \; \longrightarrow \; \overset{\backslash}{\underset{/}{C}} = O \; + \; H_2O$$

Mercaptans are changed to disulfides.

$$2R - S - H \; + \; (O) \; \longrightarrow \; R - S - S - R \; + \; H_2O$$

Aldehydes are converted to carboxyl groups.

$$RCHO \; + \; (O) \; \longrightarrow \; RCO_2H$$

3. Groups that are affected by reducing agents (e.g., H_2 or donors of $H:^-$).

Carbon-carbon double bonds become saturated.

$$\overset{\backslash}{\underset{/}{C}} = \overset{/}{\underset{\backslash}{C}} \; + \; H_2 \; \xrightarrow[\text{P, heat}]{\text{Ni}} \; H - \overset{|}{\underset{|}{C}} - \overset{|}{\underset{|}{C}} - H$$

Disulfides are changed to mercaptans.

$$R - S - S - R \; + \; 2(H) \; \longrightarrow \; 2RSH$$

Aldehydes and **ketones** are changed to 1° and 2° alcohols.

$$\overset{\backslash}{\underset{/}{C}} = O \; + \; H_2 \; \xrightarrow[\text{P, heat}]{\text{Ni}} \; \overset{\backslash}{\underset{/}{C}}HOH$$

4. Groups that can neutralize strong acids at room temperature.

Amines: $RNH_2 \; + \; H_3O^+ \; \longrightarrow \; R\overset{+}{N}H_3 \; + \; H_2O$

Acid Salts: $\overset{\overset{\displaystyle O}{\|}}{RCO^-} + H_3O^+ \longrightarrow \overset{\overset{\displaystyle O}{\|}}{RCOH} + H_2O$

5. Groups that can neutralize strong bases at room temperature.

Amine Salts: $R\overset{+}{N}H_3 + {}^-OH \longrightarrow RNH_2 + H_2O$

Acids: $RCO_2H + {}^-OH \longrightarrow RCO_2^- + H_2O$

Phenols: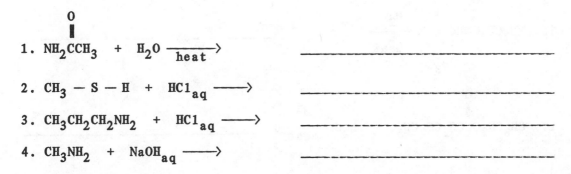

6. Other reactions involving water.
   Water is a reactant when:
   carbon-carbon double bonds can add water to form alcohols.
   Water is a product when:
   carbon-carbon double bonds are introduced by the dehydration of an alcohol;
   acetals are formed from hemiacetals and alcohols;
   esters are formed from acids and alcohols;
   amides are formed from acids and amines.

SELF-TESTING QUESTIONS

COMPLETION

Complete the following equations by writing the structures of the principal organic products. If no reaction occurs, write "none." Several of the following involve a review of reactions studied in earlier chapters.

1. $NH_2\overset{\overset{\displaystyle O}{\|}}{C}CH_3 + H_2O \xrightarrow[\text{heat}]{}$ _____

2. $CH_3 - S - H + HCl_{aq} \longrightarrow$ _____

3. $CH_3CH_2CH_2NH_2 + HCl_{aq} \longrightarrow$ _____

4. $CH_3NH_2 + NaOH_{aq} \longrightarrow$ _____

5. $CH_3CH_2CH_2 - SH$ + (O) $\longrightarrow$ _____

6.  $- S - S -$ + 2(H) $\longrightarrow$

_____

7. 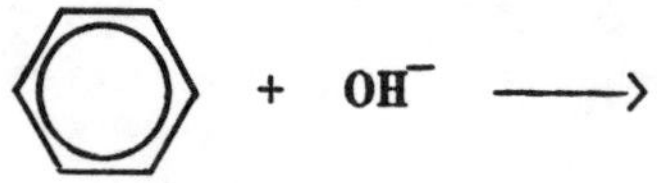 + OH^- $\longrightarrow$ _____

8. $CH_3CH_2NHCH_2CH_2$ + HCl_{aq} $\longrightarrow$ _____

9. $CH_3CH_2CH_2\overset{+}{N}H_3$ + OH^- $\longrightarrow$ _____

10. 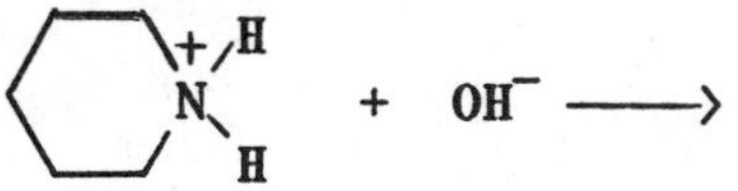 + OH^- $\longrightarrow$ _____

11. $CH_3 - O - CH_2CH_2 - SH$ + (O) $\longrightarrow$

_____

12. $CH_2 = CHCH_2\overset{+}{N}H_3$ + OH^- $\longrightarrow$ _____

13. $CH_3CH_2 - O - CH = CHCH_2NH_2$ + H_2 $\xrightarrow[\text{heat, p}]{\text{Ni}}$ _____

14. 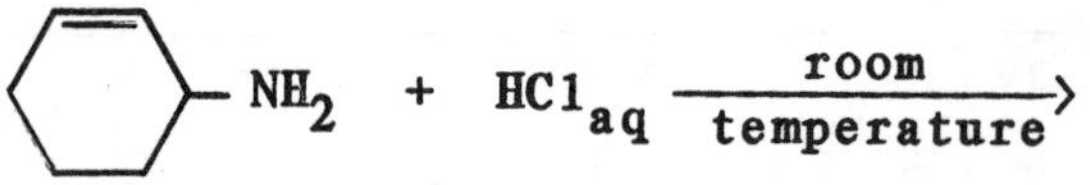 $- NH_2$ + HCl_{aq} $\xrightarrow[\text{temperature}]{\text{room}}$ _____

15. 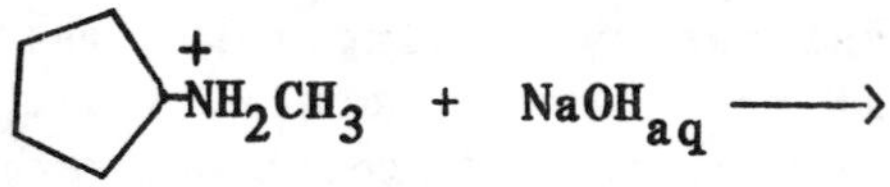 $\overset{+}{N}H_2CH_3$ + $NaOH_{aq}$ $\longrightarrow$ _____

16. $NH_2CH_2\overset{\overset{O}{\|}}{C}NHCH\overset{\overset{O}{\|}}{C}OH$ $\xrightarrow[\text{heat}]{H_2O}$ _____ + _____
    with CH_3 branch

17. $CH_3CH_2O\overset{\overset{O}{\|}}{C}CH_2CH_2CH_3$ + H_2O $\xrightarrow[\text{heat}]{H^+}$ _____

18. CH_3CO_2H + $NaOH(aq)$ ———————> _____

19. $CH_3CH_2CH_2CO_2^-$ + H_3O^+ ———————> _____

20. $CH_3\underset{\underset{OCH_3}{|}}{CHOCH_3}$ + H_2O $\xrightarrow[\text{heat}]{H^+}$ _____

MULTIPLE-CHOICE

1. Organic functional groups that are hydrolyzed by water (usually in the presence of an acid catalyst and heat) are
   (a) ethers   (d) disulfides
   (b) acetals   (e) esters
   (c) amides   (f) carboxylic acids

2. Organic functional groups that are rather easily oxidized are
   (a) alkanes   (d) ketones
   (b) aromatic hydrocarbons   (e) aldehydes
   (c) mercaptans   (f) amides

3. Organic functional groups that are good proton acceptors are
   (a) amino groups   (d) carboxylate ions
   (b) amides   (e) armoatic hydrocarbons
   (c) alkanes   (f) mercaptans

4. Organic functional groups that are good proton donors are
   (a) amino groups   (d) substituted ammonium ions
   (b) amides   (e) carboxylic acids
   (c) alkanes   (f) alcohols

5. Which compound is the most acidic?

   (a) CH_3CH_2OH (c) $CH_3\overset{\overset{\textstyle O}{\|}}{C} - NH_2$

   (b) $CH_3CH = O$ (d) CH_3CO_2H

6. Which compound is the most basic?
   (a) CH_3CH_2OH (c) $CH_3CH_2NH_3^+Cl^-$

   (b) $CH_3CH_2NH_2$ (d) $CH_3\overset{\overset{\textstyle O}{\|}}{C} - NH_2$

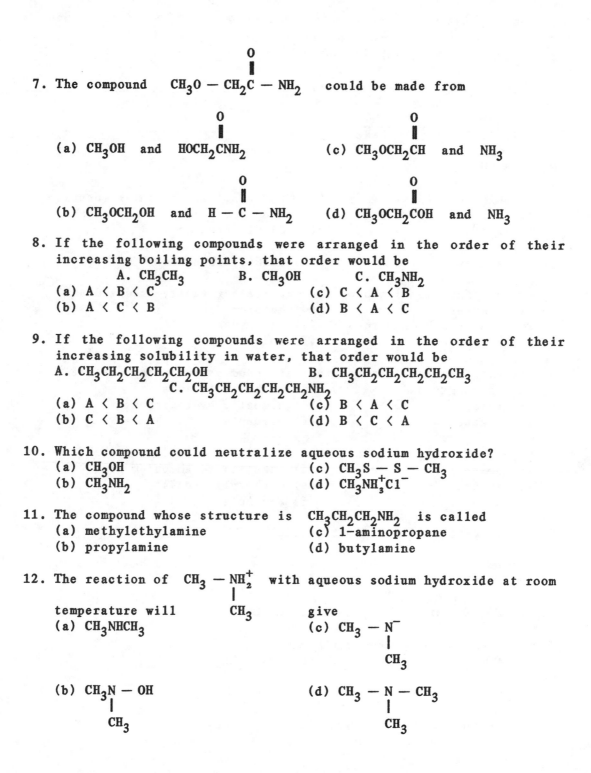

7. The compound $CH_3O - CH_2\overset{\overset{\displaystyle O}{\|}}{C} - NH_2$ could be made from

(a) CH_3OH and $HOCH_2\overset{\overset{\displaystyle O}{\|}}{C}NH_2$

(c) $CH_3OCH_2\overset{\overset{\displaystyle O}{\|}}{C}H$ and NH_3

(b) CH_3OCH_2OH and $H - \overset{\overset{\displaystyle O}{\|}}{C} - NH_2$

(d) $CH_3OCH_2\overset{\overset{\displaystyle O}{\|}}{C}OH$ and NH_3

8. If the following compounds were arranged in the order of their increasing boiling points, that order would be
    A. CH_3CH_3 B. CH_3OH C. CH_3NH_2
    (a) A < B < C   (c) C < A < B
    (b) A < C < B   (d) B < A < C

9. If the following compounds were arranged in the order of their increasing solubility in water, that order would be
    A. $CH_3CH_2CH_2CH_2CH_2OH$ B. $CH_3CH_2CH_2CH_2CH_2CH_3$
          C. $CH_3CH_2CH_2CH_2CH_2NH_2$
    (a) A < B < C   (c) B < A < C
    (b) C < B < A   (d) B < C < A

10. Which compound could neutralize aqueous sodium hydroxide?
    (a) CH_3OH (c) $CH_3S - S - CH_3$
    (b) CH_3NH_2 (d) $CH_3\overset{+}{N}H_3Cl^-$

11. The compound whose structure is $CH_3CH_2CH_2NH_2$ is called
    (a) methylethylamine   (c) 1-aminopropane
    (b) propylamine   (d) butylamine

12. The reaction of $CH_3 - \overset{\overset{\displaystyle |}{\underset{\displaystyle CH_3}{|}}}{\overset{+}{N}H_2}$ with aqueous sodium hydroxide at room temperature will   give
    (a) CH_3NHCH_3

    (c) $CH_3 - \overset{\overset{\displaystyle |}{\underset{\displaystyle CH_3}{|}}}{N^-}$

    (b) $CH_3\overset{\overset{\displaystyle |}{\underset{\displaystyle CH_3}{|}}}{N} - OH$

    (d) $CH_3 - \overset{\overset{\displaystyle |}{\underset{\displaystyle CH_3}{|}}}{N} - CH_3$

13. The presence of the $-NH_2$ group in an organic compound makes the compound
   (a) a good proton donor   (c) more soluble in water
   (b) less soluble in water   (d) a good proton acceptor

14. What is the best representation for hydrogen bonding in methylamine?

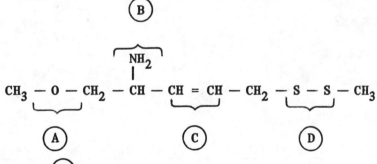

Questions fifteen through eighteen refer to this compound:

15. The group at (A) would

   (a) react with $NaOH_{aq}$
   (b) react with HCl_{aq}
   (c) be reduced by catalytic hydrogenation
   (d) be oxidizable by mild reagents
   (e) none of these

16. The group at (B) would

   (a) neutralize aqueous HCl

(b) accept H-bonds from water
(c) react with H_2
(d) none of these

17. The group at ⓒ would

(a) add hydrogen chloride
(b) donate hydrogen bonds
(c) be attacked by OH^-
(d) none of these

18. The group at Ⓓ would

(a) be easily reduced
(b) react with HCl_{aq}

(c) neutralize $NaOH_{aq}$
(d) react with water

ANSWERS

ANSWERS TO DRILL EXERCISES

I. <u>Drill in Writing Products of the Reactions of Amines with Strong Acids</u>

1. $CH_3CH_2CH_2NH_3^+$

2. 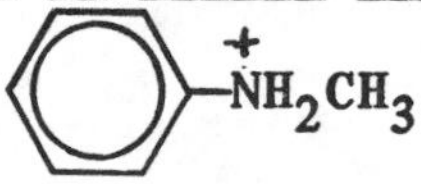 $\overset{+}{N}H_2CH_3$

3. $CH_3CH_2\overset{+}{N}HCH_2CH_3$
   $\quad\quad\quad\quad\underset{|}{CH_3}$

4.  $\overset{+}{N}H_2$

5. $CH_3\overset{+}{N}H_2CH_2CH_2\overset{+}{N}H_2CH_3$

II. <u>Drill in Writing the Products of the Reactions of Protonated Amines with Strong, Aqueous Base</u>

1. $CH_3CH_2NH_2$

2. $CH_3CH_2NCH_3$
   $\quad\quad\quad\underset{|}{CH_3}$

3. $-NCH_3$
   with $\overset{CH_3}{\underset{|}{}}$ above N

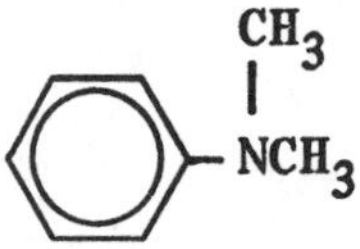

4. $NH_2CH_2CH_2\overset{O}{\overset{\|}{C}}CH_3$

5. $NH_2\underset{\underset{CH_3}{|}}{CH}CO_2^-$

III. <u>Drill in Writing the Structures of the Amides That Can be Made</u>
     <u>(Directly or Indirectly) from Given Carboxylic Acids and Amines</u>

1.
$$\text{C}_6\text{H}_5-\overset{\displaystyle O}{\overset{\|}{C}}\text{NH}_2$$

4.
$$\text{C}_6\text{H}_5-\overset{\displaystyle O}{\overset{\|}{C}}\text{NHCH}_3$$

2. $\text{CH}_3\text{CH}_2\overset{\displaystyle O}{\overset{\|}{C}}\text{NH}_2$

5. $\text{CH}_3\overset{\displaystyle O}{\overset{\|}{C}}\text{NHCH}_2\text{CH}_3$

3. $\text{CH}_3\overset{\displaystyle CH_3}{\overset{|}{C}}\text{HCH}_2\overset{\displaystyle O}{\overset{\|}{C}}\text{NH}_2$

IV. <u>Drill in Writing the Products of the Hydrolysis of Amides</u>

1. $\text{CH}_3\text{CH}_2\text{CO}_2\text{H}$ + NH_3

4. $\text{CH}_3\overset{\displaystyle}{\text{CHNH}_2}$ + $\text{HO}_2\text{CCH}_2\overset{\displaystyle CH_3}{\overset{|}{C}}\text{HCH}_3$
   $\quad\;\;\overset{|}{\text{CH}_3}$

2. $\text{CH}_3\text{CH}_2\text{CH}_2\text{CO}_2\text{H}$ + NH_2CH_3

3. $\text{CH}_3\text{CH}_2\text{NH}_2$ + $\text{HO}_2\text{CCH}_2\text{CH}_3$

5. $\text{Cl}-\text{C}_6\text{H}_4-\text{NH}_2$ + $\text{HO}_2\text{C}-\text{C}_6\text{H}_5$

V. <u>Exercises in Hydrogen Bonds</u>
1. A, B, C, F, G, and H
2. A, B, C, D, F, G, H, and I
3. D and I
4. E
5. A   (The amine, B, boils at a lower temperature because it
            forms weaker H-bonds than does the alcohol. The formula
            weights of both are essentially equal.)
6. H   (To be an aromatic amine, the amino group must be joined
            directly to the aromatic ring. While G is an aromatic
            compound, it is not an aromatic amine; instead, it is an
            aliphatic amine.)

ANSWERS TO SELF-TESTING QUESTIONS
Completion
  1. NH_3 + HO_2CCH_3 (The hydrolysis of an amide.)
  2. none (The only reaction of mercaptans, as far as we are concerned,
          is their oxidation to disulfides.)

3. $CH_3CH_2CH_2NH_3Cl$

4. none   (Amines do not react with aqueous bases.)

5. $CH_3CH_2CH_2 - S - S - CH_2CH_2CH_3$ + H_2O

6. 2 ⬡-SH

7. none   (Benzene has no reaction with bases.)

8. $CH_3CH_2 - \overset{\overset{\displaystyle H}{|+}}{\underset{\underset{\displaystyle H}{|}}{N}} - CH_2CH_3Cl^-$

9. $CH_3CH_2CH_2NH_2$ + H_2O 10. ⬡$\overset{\displaystyle H}{N}$ + H_2O

11. $CH_3 - O - CH_2CH_2 - S - S - CH_2CH_2 - O - CH_3$ + H_2O (The ether group is unaffected; it remains intact during the reaction.)

12. $CH_2 = CHCH_2NH_2$ + H_2O (The double bond is not affected by an aqueous hydroxide ion.)

13. $CH_3CH_2 - O - CH_2CH_2CH_2NH_2$ (Only the double bond is hydrogenated; neither the ether group nor the amino group is changed by hydrogenation.)

14. ⬡$-\overset{+}{N}H_3Cl^-$ (Aqueous hydrochloric acid does not affect the double bond at room temperature.)

15. ⬠$-NH - CH_3$ (+ H_2O)

16. $NH_2CH_2CO_2H$ + $NH_2\underset{\underset{\displaystyle CH_3}{|}}{CH}CO_2H$ (Only the amide group is affected.)

17. CH_3CH_2OH + $HO_2CCH_2CH_2CH_3$ (The hydrolysis of an ester.)

18. $CH_3CO_2^-Na^+$ (+ H_2O)

19. $CH_3CH_2CH_2CO_2H$ (+ H_2O)

20. CH_3CHO + $2HOCH_3$ (The hydrolysis of an acetal.)

Multiple-Choice

1. b, c, e	2. c, e	3. a, d
4. d, e	5. d	6. b
7. d	8. b	9. d
10. d	11. b, c	12. a
13. c, d	14. d	15. e
16. a, b	17. a	18. a

18 Optical Isomerism

OBJECTIVES

Life is as dependent on molecular geometry as it is on other molecular features. Nearly all biochemical substances consist of molecules that exhibit some type of chirality, or handedness. This is particularly true of all enzymes. Here we learn that the chemical properties of a substance whose molecules are chiral can depend, often dramatically, upon its kind of chirality. The chapter develops this point and gives some illustrations. You, in turn, should learn the definitions of the terms of optical isomerism given in the Glossary in such a way that you are able to illustrate them by examples.

After studying this chapter and working the Practice and Review Exercises in it, you should be able to do all of the following.

1. Examine the structures of a set of isomers and pick out which are related as structural isomers and which are related as stereo—isomers.
2. Examine a molecular structure and pick out any chiral carbons.
3. Examine the chiral carbons in a structure and pick out which are identical and which are different.
4. If all of the chiral carbons in a structure are different, calculate the number of optical isomers possible.
5. Examine a structure and predict whether it can exist in two forms related as enantiomers.
6. Examine the structures of a set of stereoisomers and pick out

which are related as enantiomers, which are related as diaster-
eomers, and which are meso compounds. (From Special Topic 1.)

7. Write structures that illustrate the composition of a racemic
   mixture. (See Special Topic 1.)

8. Calculate the specific rotation of a substance from its observed
   rotation, its concentration, and the path length of light passing
   through it.

9. Calculate the concentration of an optically active substance from
   data on its observed optical rotation, its specific rotation, and
   the path length of light passing through it.

10. Define all of the terms in the Glossary and give illustrations
    where applicable.

GLOSSARY

Achiral. Not possessing chirality; that quality of a molecule (or
other object) that allows it to be superimposed on its mirror
image.

Chiral. Having handedness in a molecular structure. (See also
Chirality.)

Chiral Carbon. A carbon that holds four different atoms or groups.

Chirality. The quality of handedness that a molecular structure has
that prevents this structure from being superimposable on its
mirror image.

Dextrorotatory. That property of an optically active substance by
which it can cause the plane of plane-polarized light to rotate
clockwise.

Enantiomer. One of a pair of stereoisomers that are related as an
object is related to its mirror image but that cannot be
superimposed one on the other.

Levorotatory. The property of an optically active substance that
causes a counterclockwise rotation of the plane of plane-
polarized light.

Optical Activity. The ability of a substance to rotate the plane of
polarization of plane-polarized light.

Optical Isomer. One of a set of compounds whose molecules differ only in their chiralities.

Optical Rotation. The degrees of rotation of the plane of plane-polarized light caused by an optically active solution; the observed rotation of such a solution.

Plane-Polarized Light. Light whose electrical field vibrations are all in the same plane.

Polarimeter. An instrument for detecting and measuring optical activity.

Racemic Mixture. A 1:1 mixture of enantiomers which is therefore optically inactive.

Specific Rotation [α]. The optical rotation of a solution per unit of concentration per unit of path length.

$$[\alpha] = \frac{(100)(\alpha)}{(c)(\ell)}$$

where α = observed rotation; c = concentration in g/100 mL; and ℓ = path length.

Stereoisomer. One of a set of isomers whose molecules have the same atom-to-atom sequences but different geometric arrangements; a geometric (cis-trans) or optical isomer.

Structural Isomer. One of a set of isomers whose molecules differ in their atom-to-atom sequence.

Substrate. The substance on which an enzyme performs its catalytic work.

Superimposition. An operation to see if one molecular model can be made to blend simultaneously at exactly every point with another model.

SELF-TESTING QUESTIONS

COMPLETION

1. These two structures:

$$\begin{array}{ccc}
Cl \quad\quad Cl & & H \quad\quad Cl \\
\backslash \quad / & & \backslash \quad / \\
C = C & \text{and} & C = C \\
/ \quad\quad \backslash & & / \quad\quad \backslash \\
H \quad\quad H & & Cl \quad\quad H
\end{array}$$

represent what kind of isomerism: _____

2. These two structures:

$$\begin{array}{cc}
H \quad\quad Cl & Cl \quad\quad H \\
\backslash \quad / & \backslash \quad / \\
C = C & C = C \\
/ \quad\quad \backslash & / \quad\quad \backslash \\
H \quad\quad Cl & H \quad\quad Cl
\end{array}$$

represent _____.

3. These two structures: $CH_3 - O - CH_2 - CO_2H$ and $HO - CH_2CO_2CH_3$
represent what kind of isomerism? _____

4. Complete the following figure to
represent the two enantiomers of

$$\begin{array}{c}
Br \\
/ \\
CH_3CH \\
\backslash \\
Cl
\end{array}$$

5. Add two Cls and two Hs to each of
these structures to show two di-
astereomers. (Cf. Special Topic 1.)

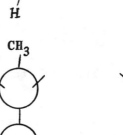

6. Let the surface depicted below represent the surface of an enzyme
with the structural features A, B, and C that are involved in the

enzyme's work as a catalyst. Examine structures 1, 2, and 3. If the reaction of this enzyme requires that A meet A, B meet B, and C meet C simultaneously as the molecule nestles to the surface of the enzyme, then which molecule(s) can interact with the enzyme?

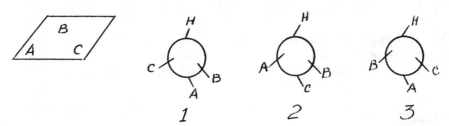

MULTIPLE-CHOICE

1. A substance that is optically active can
   (a) rotate polarized light
   (b) rotate the plane of light
   (c) polarize light
   (d) rotate the plane of plane-polarized light

2. If a substance has a specific rotation described as $[\alpha]_D^{20} = -15.6^{\circ}$ the substance is
   (a) levorotatory   (c) optically active
   (b) dextrorotatory   (d) superimposable

3. A bottle labeled "(+)-glucose" contains a substance that is
   (a) achiral   (c) dextrotatory
   (b) optically active   (d) positively charged

4. If a solution of 6 g/100 mL of compound X in a polarimeter tube 20 centimeters long has an optical rotation of 1.2°, then its specific rotation is
   (a) 23° (b) 10° (c) 100° (d) 1°

5. If the solution of question 4 were put in a tube 1 dm long, the observed rotation would be
   (a) 5° (b) 10° (c) 50° (d) 0.6°

6. The molecule:

$$CH_3CHCH-CHCHCH_3$$

with CH_3 groups above the two CH carbons, and OH and Br below, has how many chiral carbons?

(a) 1   (b) 2   (c) 3   (d) 4

7. The molecule in question 6 could exist as
   (a) one enantiomer
   (b) two enantiomers
   (c) two pairs of enantiomers
   (d) one pair of enantiomers

ANSWERS

ANSWERS TO SELF-TESTING QUESTIONS

Completion

1. stereoisomerism (or cis-trans isomerism or geometrical isomerism)
2. structural isomerism
3. structural isomerism

4.

5.

one pair

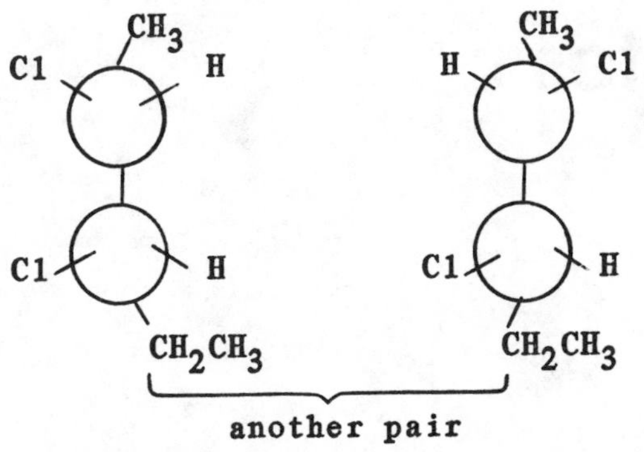

another pair

6. Structure 1 (Imagine that you rotate it counterclockwise 120°
                about the bond from the central atom to the H. The
                other two structures are enantiomers of structure 1.)

Multiple-Choice
1. d
2. a, c,
3. b, c
4. b (20 cm = 2 dm)

5. d
6. b
7. c

19 Carbohydrates

OBJECTIVES

After studying this chapter and working the Practice and Review Exercises in it, you should be able to do the following.

1. Describe carbohydrates by their structural features.
2. Name the three classes of carbohydrates we have studied.
3. Interpret such terms as "aldose," "hexose," and "aldohexose."
4. Name the three nutritionally important monosaccharides and give at least one source of each.
5. Name the three nutritionally important disaccharides, give a source of each, and name the products each gives when hydrolyzed.
6. Name three polysaccharides made entirely from glucose units and state where each is found in nature.
7. Write the structures for the α-, β-, and the open forms of glucose. (This would be a minimum goal as far as monosaccharide structures are concerned.)
8. Look at the cyclic structure of a given monosaccharide and point out its hemiacetal system.
9. Look at the cyclic structure of a given disaccharide and point out its acetal-oxygen bridge; tell if the disaccharide will give a positive result in Benedict's or Tollens' Test.
10. Explain what is meant by "deoxy-."
11. Write the structure of α- and β- maltose. (This is a minimum goal as far as disaccharide structures are concerned.)
12. Write the structure of the repeating unit in amylose (as a minimum

goal for illustrating what polysaccharides are like).

13. Describe an instance in which two polysaccharides differ only in the orientation (i.e., geometry) of their oxygen bridges.
14. Describe what one does and sees in the starch-iodine test.
15. Name the principal components of starch.
16. Describe the structural relations between amylopectin and amylose. (Do this in words if not by writing the structures.)
17. Write the plane projection structure of D-glucose and explain why it is in the D-family.
18. Tell from the plane projection structure of any aldose or ketose whether it is in the D- or L- family.
19. Define all of the terms in the Glossary.

GLOSSARY

Absolute Configuration. The actual arrangement in space about each chiral center in a molecule.

Aldohexose. A monosaccharide whose molecules have six carbon atoms and an aldehyde group.

Aldose. A monosaccharide whose molecules have an aldehyde group.

Amylopectin. A polymer found in starch in which linear amylose chains have joined to each other by α(1→6) glycosidic bonds to give a branched polymer of α-glucose.

Amylose. A linear polymer of glucose in which glucose units are joined by α(1→4) glycosidic bonds.

Biochemistry. The study of the structures and properties of substances found in living systems.

Blood Sugar. The carbohydrates – mostly glucose – that are present in blood.

Carbohydrate. Any naturally occurring substance whose molecules are polyhydroxyaldehydes or polyhydroxyketones or can be hydrolyzed to such compounds.

Cellulose. A linear polymer of glucose in which glucose units are joined by β(1→4) glycosidic linkages. The chief constitutent of cotton.

Dextrin. A polymer of α-glucose that forms from the incomplete hydrolysis of starch.

D-Family; L-Family. The names of the two optically active families to which substances can belong when they are considered solely according to one kind of molecular chirality (molecular handedness) or the other.

Disaccharide. A carbohydrate that can be hydrolyzed into two monosaccharides.

Fructose. A ketohexose present in honey and one product of the hydrolysis of sucrose; levulose.

Galactose. An aldohexose that forms, together with glucose, when lactose (milk sugar) is hydrolyzed.

Glucose. An aldohexose whose molecules serve as building blocks for glycogen, starch, cellulose, dextrin, maltose, sucrose, and lactose; the chief carbohydrate in blood; blood sugar; dextrose.

Glycogen. The starchlike polymer of α-glucose that serves as an animal's means of storing glucose units.

Iodine Test. The test for starch by which a drop of iodine produces an intensely purple color if starch is present.

Ketohexose. A monosaccharide whose molecules contain six carbon atoms and have a keto group.

Ketose. A monosaccharide whose molecules have a ketone group.

Lactose. A disaccharide that can be hydrolyzed to glucose and galactose; milk sugar.

Maltose. A disaccharide that can be hydrolyzed to two glucose molecules; malt sugar.

Monosaccharide. A carbohydrate that cannot be hydrolyzed.

Mutarotation. The gradual change in the specific rotation of a substance in solution but without a permanent, irreversible chemical change occurring.

Polysaccharide. A carbohydrate whose molecules are polymers of monosaccharides.

Reducing Carbohydrate. A carbohydrate that gives a positive Benedict's Test.

Simple Sugar. Any monosaccharide.

Starch. A naturally occurring mixture, obtained from plants, of
amylose and amylopectin.

Sucrose. A disaccharide that can be hydrolyzed to glucose and
fructose.

DRILL EXERCISES

EXERCISES IN CARBOHYDRATE STRUCTURES AND SYMBOLS

In the simplified structures of the cyclic forms of carbohydrates, the bonds or lines that are parts of the rings (hexagons or pentagons) form a flat surface. You should imagine that this surface comes out of the page, perpendicular to it. Bonds or lines that point downward from corners of these rings actually point below the plane of the ring. Those pointing up from a corner project above the plane. These relations hold even if we rotate the ring around an imaginary axis going through the center of the ring and perpendicular to the plane of the ring.

1. Just to make sure that the condensed structural symbols for the monosaccharides are understood, convert this symbol into its full structural formula with the atomic symbols given for all carbon, hydrogen, and oxygen atoms and with all bonds represented by lines.

2. Study the following Fischer projection structures and answer the questions. (Not all of these plane projectons are in strict accord with all of the rules for making them, but the rearward projections of vertical lines and the forward projections of horizontal lines are intended to be as specified by these rules.) Remember that in imaginary operations with Fischer projections, they may be slid around on the plane, but they may not be taken out of the plane and tipped in any way.

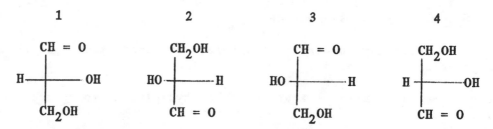

    1   2   3   4

(a) Which, if any, are D-glyceraldehyde? _____

(b) Which, if any, are L-glyceraldehyde? _____

3. Given the Fischer projection structure for D-galactose, write the Fischer projection structures for L-galactose.

$$
\begin{array}{c}
\text{CH} = \text{O} \\
\text{H} \rule{0.5cm}{0.4pt} \text{OH} \\
\text{HO} \rule{0.5cm}{0.4pt} \text{H} \\
\text{HO} \rule{0.5cm}{0.4pt} \text{H} \\
\text{H} \rule{0.5cm}{0.4pt} \text{OH} \\
\text{CH}_2\text{OH}
\end{array}
$$

    D-galactose   L-galactose

                                                   _____

4. Modify the structure of D-galactose given above to show the structure of D-2-deoxygalactose.

_____

SELF-TESTING QUESTIONS

<u>COMPLETION</u>

1. A carbohydrate whose molecules will react with water to produce two sugar units is a _____.

2. The structural feature involved in the link between two glucose units in maltose is the _____.

3. The hydrolysis of sucrose produces _____ and _____.

4. The partial hydrolysis of starch gives a mixture called _____.

5. The iodine test is used to detect the presence of _____.

6. Carbohydrates all have high melting points, which means that their molecules are strongly attracted to each other in the crystals. The force of attraction responsible for this property is the _____.

7. The names of the two very broad families of optically active compounds, which are organized solely on the bases of "handedness," are the _____ family and the _____ family.

8. The hydrolysis of lactose produces _____ and _____.

9. The linear polymer of α-glucose is _____.

10. Starch is a mixture of _____ and _____.

11. The technical name for a potential aldehyde group is the _____.

12. The hydrolysis of maltose produces _____.

13. Because maltose and lactose have the _____ group, they are reducing disaccharides.

14. The storage form of glucose molecules in animals is _____,
which structurally is similar to _____, one of the
storage forms of glucose units in plants.

15. If a polysaccharide can be hydrolyzed into nothing but glucose, it
might be any one of the following (make a complete list):

MULTIPLE-CHOICE

1. Which is (are) the structure(s) of α-glucose?

(a)   (b)   (c)   (d)

2. If the letter G is used to represent a glucose unit, then the
symbol:
        etc. $- O - G - O - G - O - G - O - G - O - G -$ etc.
could be a way of showing the basic structural feature of
(a) amylopectin   (b) amylose   (c) glycogen   (d) galactose

3. An example of a reducing carbohydrate is
(a) sucrose   (b) maltose   (c) cellulose   (d) galactose

4. An aldohexose would have
(a) a potential aldehyde group
(b) five hydroxyl groups in its open form
(c) six carbons
(d) one $- CH_2OH$ group

5. The 50:50 mixture of glucose and fructose
(a) is called dextromaltose
(b) is obtainable by the hydrolysis of sucrose
(c) is called invert sugar
(d) is a disaccharide

6. If the molecules of a substance have the structure

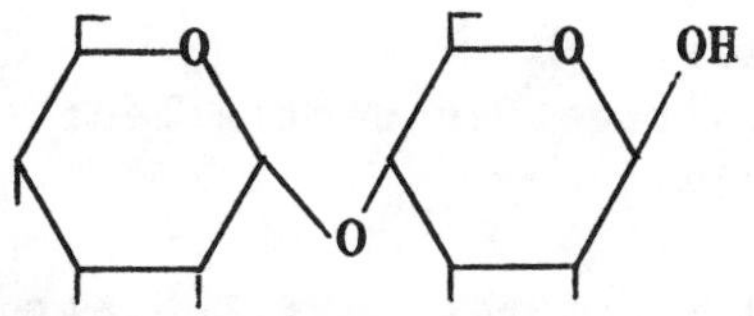

(a) it is a disaccharide
(b) it is in the D-family
(c) it can mutarotate
(d) it has a β-acetal oxygen bridge

7. If the molecules of a substance have the structure:

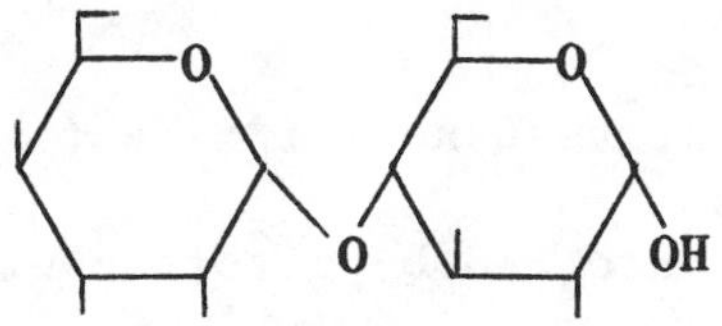

(a) it can be hydrolyzed into two glucose units
(b) it can be hydrolyzed into one glucose unit plus another monosaccharide
(c) it will not give a Benedict's Test
(d) it will not mutarotate

8. If the molecules of a substance have the structure:

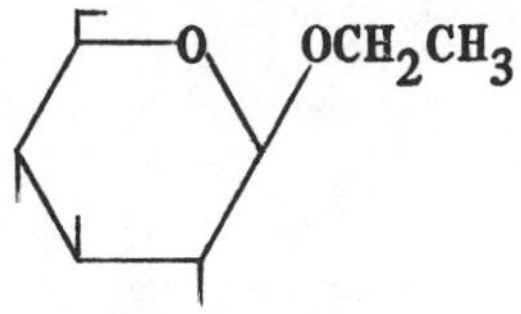

(a) it is an α-glucoside
(b) it is a β-glucoside
(c) it will mutarotate
(d) it will give a positive Benedict's Test

9. If the molecules of a substance have the structure:

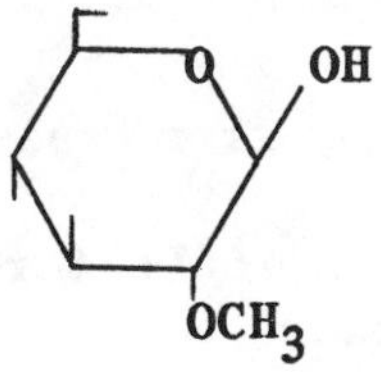

(a) it is an α-glucoside
(b) it is a β-glucoside
(c) it will give a positive Benedict's Test
(d) it is a disaccharide

10. If the molecules of a substance have the structure:

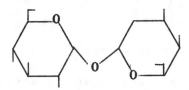

(a) it is a disaccharide
(b) it will give a positive Benedict's Test
(c) it will mutarotate
(d) it will hydrolyze into two D-galactose units

ANSWERS

ANSWERS TO DRILL EXERCISES

Exercises in Carbohydrate Structures and Symbols

1.

```
              H
              |
        H  —  C  — OH
              |
              C — — O
        HO    |        OH
          \  /  H     /
           C         C
          /    OH  H  \
        H     |   |/   H
              C — — C
              |     |
              H     OH
```

2. (a) 1 and 2 (Note that 2 results from spinning 1   180° in the plane of the paper.)

   (b) 3 and 4 (3 is the mirror image of 1; 4 results from turning 3 180° in the plane of the paper.)

3.

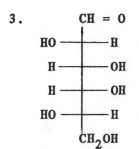

```
          CH = O
HO ———————— H
H  ———————— OH
H  ———————— OH
HO ———————— H
          CH₂OH
```

4.

```
          CH = O
H  ———————— H
HO ———————— H
HO ———————— H
H  ———————— OH
          CH₂OH
```

ANSWERS TO SELF-TESTING QUESTIONS
Completion
1. disaccharide
2. α-acetal bridge (or α 1→4 glycosidic link)
3. fructose, glucose
4. dextrin
5. starch
6. hydrogen bond
7. D- and L- (families)
8. glucose, galactose
9. amylose
10. amylose, amylopectin
11. hemiacetal group
12. glucose
13. hemiacetal (or potential aldehyde)
14. glycogen, amylopectin
15. starch, amylose, amylopectin, glycogen, cellulose, dextrin

Multiple-Choice
1. a, b, and d
2. b
3. b and d
4. a, b, c, and d
5. b and c
6. a, b, and c
7. b
8. b   (Only when the hemiacetal group – or the hemiketal group – is present will the substance mutarotate and give a positive Benedict's Test.)
9. c   (To be a glucoside, the CH_3 group would have to be part of an acetal system at carbon number one.)
10. a   (It has no hemiacetal system and therefore cannot mutrarotate or give a positive Benedict's Test.)

20 Lipids

OBJECTIVES

Although all of these objectives are important, one through five constitute minimum goals in our preparation for later chapters. Objectives 11 and 12 are necessary for an understanding of how the membranes of animal cells are organized and how they can hold a cell together while letting substances in and out.

After you have studied this chapter and worked the exercises in it, you should be able to do the following.

1. Write the structure of a molecule that includes at least one carbon-carbon double bond and is typically found in a triacyl-glycerol.
2. Using that structure, write the products of its reaction with
   (a) water – catalyzed by an enzyme (as in digestion);
   (b) aqueous sodium hydroxide – saponification; and
   (c) hydrogen – hydrogenation with a catalyst, heat, and pressure.
3. Write the names and structures for a least three saturated and three unsaturated fatty acids.
4. Describe the principal structural differences between animal fats and vegetable oils.
5. Explain what polyunsaturated means when used to describe vegetable oils.
6. Look at the structures of two triacylglycerols that have essentially identical formula weights and tell which has the higher

degree of unsaturation.

7. Name two kinds of glycerol-based phospholipids and tell where they may be found in the body.
8. Do the same for two kinds of sphingosine-based lipids.
9. Name several steroids and briefly describe their purposes.
10. Briefly explain why steroids are classified as lipids.
11. Describe the composition and the structure of a biological membrane.
12. Explain how active transport is necessary for certain cellular functions.
13. Define each of the terms in the Glossary.

GLOSSARY

<u>Active</u> <u>Transport</u>. The movement of a substance through a biological membrane against a concentration gradient and caused by energy-consuming chemical changes that involve parts of the membrane.

<u>Amphipathic</u> <u>Molecule</u>. A molecule with both hydrophilic and hydrophobic groups.

<u>Fatty</u> <u>Acid</u>. Any carboxylic acid that can be obtained by the hydrolysis of animal fats or vegetable oils.

<u>Glycolipid</u>. A lipid whose molecules include a glucose unit, a galactose unit or some other carbohydrate unit.

<u>Gradient</u>. The presence of a change in value of some physical quantity with distance, as in a <u>concentration</u> gradient in which the concentration of a solute is different in different parts of the system.

<u>Hydrophilic</u> <u>Group</u>. Any part of a molecular structure that attracts water molecules; a polar or ionic group such as $-OH$, $-CO_2^-$, $-NH_3^+$, or $-NH_2$.

<u>Hydrophobic</u> <u>Group</u>. Any part of a molecular structure that has no attraction for water molecules; a nonpolar group such as any alkyl group.

<u>Lipid</u>. A plant or animal product that tends to dissolve in such nonpolar solvents as ether, carbon tetrachloride, and benzene.

Lipid Bilayer. The sheetlike array of two layers of lipid molecules, interspersed with molecules of cholesterol and proteins, that make up the membranes of cells in animals.

Nonsaponifiable Lipid. Any lipid, such as the steroids, that cannot be hydrolyzed or similarly broken down by aqueous alkali.

Phosphoglyceride. A phospholipid such as a plasmalogen or a lecithin whose molecules include a glycerol unit.

Phospholipid. Lipids such as the phosphoglycerides, the plasmalogens, and the sphingomyelins whose molecules include phosphate ester units.

Plasmalogens. Glycerol-based phospholipids whose molecules also include an unsaturated fatty alcohol unit.

Receptor Molecule. A molecule of a protein built into a cell membrane that can accept a molecule of a hormone or a neurotransmitter.

Saponifiable Lipid. Any lipid with ester groups.

Sphingolipid. A lipid that, when hydrolyzed, gives sphingosine instead of glycerol, plus fatty acids, phosphoric acid, and a small alcohol or a monosaccharide; sphingomyelins and cerebrosides.

Steroids. Nonsaponifiable lipids such as cholesterol and several sex hormones whose molecules have the four fused rings of the steroid nucleus.

Triacylglycerol. A lipid that can be hydrolyzed to glycerol and fatty acids; a triglyceride; sometimes, simply called a glyceride.

Triglyceride. (See Triacylglycerol.)

Wax. A lipid whose molecules are esters of long-chain monohydric alcohols and long-chain fatty acids.

SELF-TESTING QUESTIONS

COMPLETION

1. Write the structures of all of the products that would form from

the complete digestion of the following lipid.

$$
\begin{array}{l}
\quad\quad\quad\quad\quad\quad O \\
\quad\quad\quad\quad\quad\quad \| \\
CH_2 - O - C(CH_2)_7CH = CHCH_2CH = CH(CH_2)_4CH_3 \\
| \\
\quad\quad\quad\quad\quad O \\
\quad\quad\quad\quad\quad \| \\
CH - O - C(CH_2)_8CH_3 \\
| \\
\quad\quad\quad\quad\quad\quad O \\
\quad\quad\quad\quad\quad\quad \| \\
CH_2 - O - C(CH_2)_7CH = CH(CH_2)_7CH_3
\end{array}
$$

2. If the lipid of question 1 had been saponified by sodium hydroxide instead of digested, the fatty acids would have been produced in the form of their _____.

3. Write the structure of the product of the hydrogenation of all the alkene groups in the lipid in question 1.

4. A molecule of vegetable oil will normally have more _____ _____ than a molecule of animal fat.

5. The two functional groups in a simple lipid are _____ and _____.

6. An example of a nonsaponifiable lipid is any member of the family of _____.

7. Lipids with the general structure

$$R - O - \overset{\overset{\displaystyle O}{\|}}{C} - R'$$

(where R and R' are both long-chain) are in the family of _____.

8. The acid, other than fatty acids, liberated when complex lipids are hydrolyzed is _____.

9. A vegetable oil is said to be more _____ than an animal fat (referring to double bonds).

10. The complete hydrolysis (digestion) of this phosphatidycholine would give what products?   (Write their structures and names.)

$$
\begin{array}{l}
CH_2 - O - \overset{\overset{\displaystyle O}{\|}}{C}(CH_2)_7 CH = CH(CH_2)_7 CH_3 \\[2ex]
\;| \\[1ex]
CH - O - \overset{\overset{\displaystyle O}{\|}}{C}(CH_2)_7 CH = CHCH_2 CH = CH(CH_2)_4 CH_3 \\[2ex]
\;| \\[1ex]
CH_2 - O - \overset{\overset{\displaystyle O}{\|}}{P} - O - CH_2 CH_2 \overset{+}{N}(CH_3)_3 \\[1ex]
\qquad\quad\; | \\
\qquad\quad\; O^-
\end{array}
$$

11. Another name for the structure in question 10 is _____.

12. Because the hydrocarbon chains in the structure in question 10 are water-avoiding, they are said to be _____. In the region of the phosphate group, however, the molecule is _____. The molecule in question 10, taken as a whole, is described as _____.

13. Because the hydrocarbon chains of the structure in question 10 have double bonds, the lipid would be described as poly- _____ _____.

MULTIPLE-CHOICE

1. To reduce the degree of polyunsaturation of a triacylglycerol, a manufacturer might
   (a) hydrate it   (c) hydrolyze it
   (b) hydrogenate it   (d) dehydrogenate it

2. If a manufacturer hydrogenated a vegetable oil, the substance might
   (a) turn rancid   (c) become a solid at room temperature
   (b) become a detergent   (d) become a diglyceride

3. In glycerides, aqueous sodium hydroxide will attack
   (a) ester linkages   (c) alkene groups
   (b) ether linkages   (d) alkenelike portions

4. Fatty acids obtained from natural lipids are generally
   (a) of even carbon number   (c) insoluble in water
   (b) monocarboxylic   (d) long chain

5. The hydrolysis of a naturally occurring triacylglycerol will give

   (a) glycerol + RCO_2H + $R'CO_2H$ + $R''CO_2H$

   (b) glycerol + RCO_2^- + $R'CO_2^-$ + $R''CO_2^-$

   (c) glycerol + $3RCO_2H$

   (d) glycerol + $3RCO_2^-$

6. The name of a C_{18} acid with three alkene groups is

   (a) stearic acid   (c) linoleic acid
   (b) oleic acid   (d) linolenic acid

7. Phospholipids are esters of
   (a) sphingosine only   (c) cholesterol only
   (b) either glycerol or sphingosine (d) glycerol only

8. Phosphoglycerides are
   (a) esters of phosphatidic acids   (c) esters of sphingosine
   (b) cerebrosides   (d) glycolipids

9. Cholesterol is classified as a lipid because
   (a) it is an ester
   (b) it dissolves in fat solvents
   (c) it is present in gallstones
   (d) it can be saponified

10. The cell membranes of animals are made of
    (a) lipids   (c) proteins
    (b) polysaccharides   (d) cholesterol

11. Molecules of saponifiable lipids have
    (a) hydrophobic groups   (c) amide groups
    (b) ester groups   (d) alkene groups

12. The cell membranes of animals are organized
    (a) with hydrophobic groups projecting outward away from the
        membrane
    (b) as lipid bilayers with imbedded proteins
    (c) with hydrophilic groups projecting inward
    (d) with cellulose molecules lending structural support

ANSWERS

ANSWERS TO SELF-TESTING QUESTIONS

Completion

1. $HOCH_2CHCH_2OH$ + $CH_3(CH_2)_4CH = CHCH_2CH = CH(CH_2)_7CO_2H$
            |
            OH

                     + $CH_3(CH_2)_8CO_2H$

                     + $CH_3(CH_2)_7CH = CH(CH_2)_7CO_2H$

2. sodium salts

3.

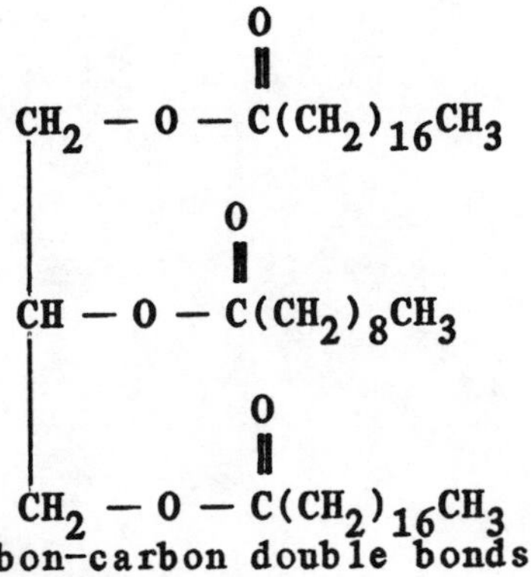

$$CH_2 - O - \overset{\overset{\textstyle O}{\|}}{C}(CH_2)_{16}CH_3$$

$$CH - O - \overset{\overset{\textstyle O}{\|}}{C}(CH_2)_8CH_3$$

$$CH_2 - O - \overset{\overset{\textstyle O}{\|}}{C}(CH_2)_{16}CH_3$$

4. carbon-carbon double bonds (or alkene groups or unsaturation)
5. esters, alkenes
6. steroids
7. waxes
8. phosphoric acid
9. unsaturated
10.

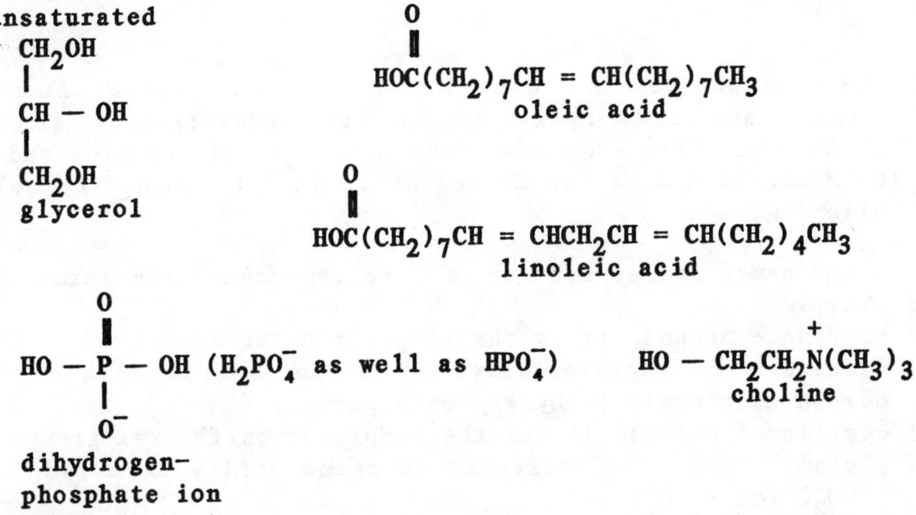

CH_2OH

$CH - OH$

CH_2OH
glycerol

$$HO\overset{\overset{\textstyle O}{\|}}{C}(CH_2)_7CH = CH(CH_2)_7CH_3$$
oleic acid

$$HO\overset{\overset{\textstyle O}{\|}}{C}(CH_2)_7CH = CHCH_2CH = CH(CH_2)_4CH_3$$
linoleic acid

$$HO - \overset{\overset{\textstyle O}{\|}}{\underset{\underset{\textstyle O^-}{|}}{P}} - OH \quad (H_2PO_4^- \text{ as well as } HPO_4^-)$$
dihydrogen-
phosphate ion

$$HO - CH_2CH_2\overset{+}{N}(CH_3)_3$$
choline

11. lecithin
12. hydrophobic; hydrophilic; amphipathic
13. unsaturated

Multiple-Choice

1. b
2. c
3. a
4. a, b, c, and d
5. a
6. d

7. b
8. a
9. b
10. a, c, and d
11. a, b, and d
12. b

21 Proteins

OBJECTIVES

These objectives are designed to promote general knowledge about proteins and are important for our understanding of later chapters. When you have completed your study of this chapter and have worked the Practice and Review Exercises in it, you should be able to do the following.

1. Write the names and structures of five representative amino acids, for example
   (a) glycine – because it is the simplest amino acid;
   (b) alanine – as representative of an amino acid with a hydro-carbon side chain (a hydrophobic group);
   (c) cysteine – because it has the important sulfhydryl group;
   (d) glutamic acid – to represent an amino acid with a side chain $-CO_2H$ (or $-CO_2^-$) group;
   (e) lysine – to represent amino acids with a side chain $-NH_2$ (or $-NH_3^+$) group.
2. Based on the amino acids you have learned, write the structure of any di-, tri-, tetra-, or pentapepetide and identify the peptide bonds.
3. Translate a structure such as Gly·Ala·Glu into a condensed structural formula.
4. Write the structures that show how disulfide bonds occur in proteins.
5. Write structures that illustrate how hydrogen bonds and salt

bridges participate in the structure of proteins.

6. Name the four levels of structure in proteins and briefly describe each in terms of the kinds of forces that stabilize it and the kinds of geometric forms it takes.
7. Explain how hydrophobic and hydrophilic side chains influence the shape of a protein.
8. Explain the relation between a polypeptide and a protein.
9. Explain how the solubility of a protein can be changed by changing the pH of its medium.
10. Given the structure of a small polypeptide, write the structures of the products of its digestion.
11. Describe what happens structurally when a protein is denatured.
12. Describe, in general terms, how denaturation affects the properties of a protein.
13. List at least four denaturing agents.
14. Name five fibrous proteins.
15. Name two globular proteins.
16. Define each of the terms in the Glossary.

GLOSSARY

Albumin. One of a family of globular proteins that tend to dissolve in water, and that in blood contribute to the blood's colloidal osmotic pressure and aid in the transport of metal ions, fatty acids, cholesterol, triacylglycerols, and other water-insoluble substances.

Amino Acid. Any organic compound whose molecules have both an amino group and a carboxyl group.

Amino Acid Residue. A structural unit in a polypeptide,

$$- NH - CH - CO - \quad ,$$
$$\underset{\underset{G}{|}}{}$$

furnished by an amino acid, where G is a side chain group.

Collagen. The fibrous protein of connective tissue that changes to gelatin in boiling water.

Denaturation. The loss of the natural shape and form of a protein molecule together with its ability to function biologically, but not necessarily accompanied by the rupture of any of its peptide bonds.

Dipeptide. A compound whose molecules have two α-amino acid residues joined by a peptide (amide) bond.

Dipolar Ion. A molecule that carries one plus charge and one minus charge, such as an α-amino acid.

Elastin. The fibrous protein of tendons and arteries.

Fibrin. The fibrous protein of a blood clot that forms from fibrinogen during clotting.

Fibrous Proteins. Water-insoluble proteins found in fibrous tissues.

Globular Proteins. Proteins that are soluble in water or in water that contains certain dissolved salts.

Globulins. Globular proteins in the blood that include γ-globulin, an agent in the body's defense against infectious diseases.

α-Helix. One kind of secondary structure of a polypeptide in which its molecules are coiled.

Hemoglobin (HHb). The oxygen-carrying protein in red blood cells.

Isoelectric. A condition of a molecule in which it has an equal number of positive and negative sites.

Isoelectric Point (pI). The pH of a solution in which a specified amino acid or a protein is in an isoelectric condition; the pH at which there is no net migration of the amino acid or protein in an electric field.

Keratin. The fibrous protein of hair, fur, fingernails, and hooves.

Myosins. Proteins in contractile muscle.

Peptide Bond. The amide linkage in a protein; a carbonyl-to-nitrogen bond.

pI. (See Isoelectric Point.)

β-Pleated Sheet. A secondary structure for a polypeptide in which the molecules are aligned side by side in a sheetlike array with the sheet partially pleated.

Polypeptide. A polymer with repeating α-aminoacyl units joined by peptide (amide) bonds.

Primary Structure. The sequence of aminoacyl residues held together by bonds in a polypeptide.

Prosthetic Group. A nonprotein molecule joined to a polypeptide to make a biologically active protein.

Protein. A naturally occurring polymeric substance made up wholly or mostly of polypeptide molecules.

Quaternary Structure. An aggregation of two or more polypeptide strands each with its own primary, secondary, and tertiary structure.

Salt Bridge. A force of attraction between (+) and (−) sites on polypeptide molecules.

Secondary Structure. A shape, such as the α-helix or a unit in a β-pleated sheet, that all or a large part of a polypeptide molecule adopts under the influence of hydrogen bonds or salt bridges after its peptide bonds have been made.

Side Chain. An organic group that can be appended to a main chain or to a ring.

Tertiary Structure. The shape of a polypeptide molecule that arises from further folding or coiling of secondary structures.

Triple Helix. The quaternary structure of tropocollagen in which three polypeptide chains are twisted together.

SELF-TESTING QUESTIONS

COMPLETION
1. In their dipolar ionic forms, all amino acids have the same

basic unit that (without side chains) has the structure:

_____

2. Objective 1 asks you to know the structures of five representative amino acids. The best way to learn them is to learn their side chains, because these are always affixed to the basic unit you wrote in question 1. On the lines provided, write the structures of these side chains.

_____

glycine side chain

_____

alanine side chain

_____

cysteine side chain

_____

glutamic acid side chain

_____

lysine side chain

3. Write the structure of the dipeptide that could form from two glycine units.

_____

4. In terms of the three-letter symbols for amino acids, the dipeptide in question 3 would be represented as _____ .

5. Write the structure of a tripeptide that could be made from alanine, glutamic acid, and cysteine. Let alanine be the N-terminal residue and cysteine the C-terminal residue.

6. In terms of the three-letter symbols, the tripeptide in question 5 has the formula _____ and it contains _____ peptide bonds.

7. Write the structure of the tetrapeptide: Ala·Gly·Lys·Cys

8. Write the structure of cysteine in its dipolar ionic form.

    _____

9. If the following polypeptide were subjected to mild reducing conditions, what would form? Write the structure(s) using the three-letter symbols.

    Gly·Ala·Cys·Lys·Glu
            |
            S
            |
            S
            |
    Glu·Ala·Cys·Lys·Glu   _____

10. In a discussion of how a polypeptide coils or otherwise assumes some geometric shape, we are talking about what structural level?

    _____

11. Three non-covalent forces that can determine the shape that will be adopted by a polypeptide are _____
    _____ .

12. When we speak of an α-helix itself undergoing folding or twisting, we are talking about what level of protein structure?

    _____

13. Which level of protein structure is not necessarily attacked by denaturating agents? _____

14. The side chains of which of the five amino acids in Objective 1 are the most susceptible to being altered by a change in the pH of the medium? (Name the amino acids.) _____
    _____

15. When we speak of two or more polypeptides becoming associated in some way in a gigantic protein system, we are talking about which level of protein structure? _____

MULTIPLE-CHOICE

1. The partial hydrolysis of a protein produces a number of amino

acids together with several dipeptides. Which of the following fragments would **not** be present?

(a) $NH_2CH_2CH_2CO_2H$

(b)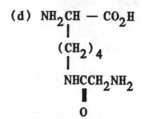

(c) $HO_2CCH_2NHCCH_3$ (with O double bond on the C)

(d) $NH_2CH - CO_2H$
   $\quad\quad |$
   $\quad (CH_2)_4$
   $\quad\quad |$
   $\quad NHCCH_2NH_2$
   $\quad\quad ||$
   $\quad\quad O$

2. If the dipolar ionic form of alanine neutralized H^+, it would be changed into

(a) $\overset{+}{NH_3}CHCO_2^-$
   $\quad\quad |$
   $\quad\quad CH_3$

(b) NH_2CHCO_2H
   $\quad\quad |$
   $\quad\quad CH_3$

(c) $NH_2CHCO_2^-$
   $\quad\quad |$
   $\quad\quad CH_3$

(d) $\overset{+}{NH_3}CHCO_2H$
   $\quad\quad |$
   $\quad\quad CH_3$

3. The peptide bonds in a protein can be broken by
   (a) hydrolysis
   (b) buffer action
   (c) denaturation
   (d) hydration

4. Two proteins might be joined together by the action of a mild oxidizing agent if among their amino acid units there was present
   (a) glutamic acid
   (b) cysteine
   (c) glycine
   (d) lysine

5. Ions of heavy metals (e.g., Hg^{2+} or Pb^{2+}) denature proteins by combining with
   (a) — SH groups
   (b) — CH_3 groups
   (c) peptide bonds
   (d) the protein backbones

6. In very strongly acidic solutions, the molecules of alanine would mostly be in what form?
   (a) NH_2CHCO_2H
   $\quad\quad |$
   $\quad\quad CH_3$

   (c) $NH_2CHCO_2^-$
   $\quad\quad |$
   $\quad\quad CH_3$

(b) $\overset{+}{N}H_3CHCO_2H$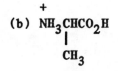
$\quad\quad\quad |$
$\quad\quad\quad CH_3$

(d) $\overset{+}{N}H_3CHCO_2^-$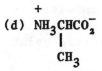
$\quad\quad\quad |$
$\quad\quad\quad CH_3$

7. In their dipolar ionic forms, amino acids are
   (a) electrically neutral   (c) weak bases
   (b) weak acids   (d) solids at room temperature

8. An amino acid with an isoelectric point of 6.8 would consist of molecules with side chains that have
   (a) an extra — NH_2 group   (c) a hydrophobic group
   (b) an extra — CO_2H group   (d) an extra — SO_3H group

9. All the common, naturally occurring amino acids (except glycine) are members of the
   (a) D—family   (b) L—family   (c) lipid family   (d) peptide family

10. The level of protein structure that has amide bonds is the
    (a) primary   (b) secondary   (c) tertiary   (d) quaternary

11. If two molecules of alanine were joined as a dipeptide
    (a) two isomeric dipeptides are possible
    (b) one dipeptide is possible
    (c) the structural symbol would be Ala·Ala
    (d) one peptide bond would be present

12. When proteins are digested
    $\quad\quad\quad \overset{O}{\overset{\|}{}}$
    (a) — C — NH — units become — CO_2H + NH_2 — units
    (b) — SH groups become — S — S — groups
    (c) — S — S — groups become — SH groups
    (d) — CO_2H and NH_2 units join to become $-\overset{O}{\overset{\|}{C}} - NH -$ units

13. If somehow the protein 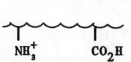 were changed   to

    the protein would be

(a) less soluble in water
(b) more soluble in water
(c) at its isoelectric point
(d) digested

14. In the pentapeptide Gly·Ala·Lys·Cys·Glu, the N-terminal unit is
    (a) glycine   (b) glutamic acid   (c) lysine   (d) — NH_2

15. If the conventions for writing the structures of proteins with the three-letter symbols of amino acids are properly obeyed, the symbol for this tripeptide would be

$$HOCCH_2NHCCHNHCCHNH_2$$

with O (double bond) above each C, and CH_3 and CH_2SH as side chains

    (a) Gly·Ala·Cys   (c) Ala·Gly·Cys
    (b) Cys·Ala·Gly   (d) Ala·Cys·Gly

16. The principal non-covalent force in protein structure is the
    (a) salt bridge   (c) helix
    (b) disulfide link   (d) hydrogen bond

17. An important secondary structural feature in proteins is the
    (a) salt bridge   (c) disulfide link
    (b) hydrogen bond   (d) α-helix

18. Nonprotein molecules that often are associated with proteins are called
    (a) prosthetic groups   (c) side chains
    (b) zwitterions   (d) 3^0 structures

19. The principal protein in the walls of blood vessels is
    (a) keratin   (b) myosin   (c) elastin   (d) fibrin

20. An important protein in contractile muscle is
    (a) keratin   (b) myosin   (c) elastin   (d) fibrin

21. A protein in hair is
    (a) keratin   (b) myosin   (c) elastin   (d) fibrin

22. One important hemoprotein is
    (a) myoglobin   (b) λ-globulin   (c) casein   (d) nucleoprotein

ANSWERS

<u>ANSWERS</u> <u>TO</u> <u>SELF-TESTING</u> <u>QUESTIONS</u>
<u>Completion</u>

1. $\overset{+}{N}H_3 - CH - \overset{\overset{\displaystyle O}{\|}}{C} - O^-$

2. glycine   alanine   cysteine   glutamic acid   lysine
    $- H$ $- CH_3$ $- CH_2SH$ $- CH_2CH_2CO_2H$ $- CH_2CH_2CH_2CH_2NH_2$

3. $NH_2 - CH_2\overset{\overset{\displaystyle O}{\|}}{C} - NH - CH_2\overset{\overset{\displaystyle O}{\|}}{C} - OH$

   (Better yet, the dipolar ionic form: $^+NH_3 - CH_2\overset{\overset{\displaystyle O}{\|}}{C} - NH - CH_2\overset{\overset{\displaystyle O}{\|}}{C} - O^-$)

4. Gly·Gly

5. $NH_2 - \underset{\underset{\displaystyle CH_3}{|}}{CH} - \overset{\overset{\displaystyle O}{\|}}{C} - NH - \underset{\underset{\displaystyle CH_2}{|}}{CH} - \overset{\overset{\displaystyle O}{\|}}{C} - NH - \underset{\underset{\displaystyle CH_2SH}{|}}{CH} - \overset{\overset{\displaystyle O}{\|}}{C} - OH$

   with the middle residue chain continuing:
   CH_2
   CH_2
   CO_2H

   6. Ala·Glu·Cys, two

7. $NH_2 - \underset{\underset{\displaystyle CH_3}{|}}{CH} - \overset{\overset{\displaystyle O}{\|}}{C} - NH - \underset{\underset{\displaystyle H}{|}}{CH} - \overset{\overset{\displaystyle O}{\|}}{C} - NH - \underset{\underset{\displaystyle CH_2}{|}}{CH} - \overset{\overset{\displaystyle O}{\|}}{C} - NH - \underset{\underset{\displaystyle CH_2SH}{|}}{CH} - \overset{\overset{\displaystyle O}{\|}}{C} - OH$

   with the third residue chain continuing:
   CH_2
   CH_2
   CH_2
   CH_2
   NH_2

8. $\overset{+}{N}H_3 - CH - \overset{\overset{\textstyle O}{\textstyle \|}}{C} - O^-$
       |
       CH_2SH

9. Gly·Ala·Cys·Lys·Glu
Gly·Ala·Cys·Lys·Glu   (Two identical molecules would form.)
10. the secondary structure
11. salt bridges, hydrogen bonds, and the water-avoiding (or at-
tracting) responses of hydrophobic (or hydrophilic) groups
12. the tertiary structure
13. the primary structure
14. lysine and glutamic acid
15. the quaternary structure

Multiple-Choice

1. a, c, and d   (Choice b is a dipeptide, Ala·Gly, and could be one
of the several dipeptides formed. Choice a is not an
alpha-amino acid. While choice c has an amide bond,
the unit on the right is not an amino acid but rather
an acetyl unit. Choice d involves lysine but not in
a form found in proteins; the amide bond involves the
side chain and not the alpha-amino group.)

2. d   3. a
4. b   5. a
6. b   7. a, b, c, and d
8. c   9. b
10. a   11. b, c, and d
12. a   13. b (It no longer is isoelectric.)
14. a
15. b   (The N-terminal unit is Cys, and the conventions tell us to
write the N-terminal unit on the left.)
16. d
17. d   (All of the others stabilize secondary structures.)
18. a   19. c
20. b   21. a
22. a

22 Nucleic Acids

OBJECTIVES

Heredity is not only involved with substances having certain functions; it is also concerned with molecules having particular structures that carry out these functions. You are not expected to memorize the structures of the heterocyclic bases, their corresponding nucleotides, or sections of nucleic acid molecules. You should, however, be able to name the hydrolysis products of the nucleic acids. You should also be prepared to give some of the condensed structural representations for these products using the symbols for the phosphate-pentose-phosphate-pentose chain and the letters — A, T, G, C, and U — that represent the side chains.

After you have completed your study of this chapter, you should be able to do the following.

1. Give the name and the abbreviation for the chemical of an individual gene.
2. Give the general name for the monomer unit of a nucleic acid.
3. Give the names of the compounds produced when these monomeric units are hydrolyzed.
4. Describe in words the two main structural differences between DNA and RNA.
5. Using simple symbols, describe the features of a DNA strand.
6. Describe what in that structure is the genetic code.
7. Describe in words the contribution of Crick and Watson to the

chemistry of heredity.

8. Describe in words the structure and shape of paired strands of DNA and the forces that stabilize it.

9. Explain why, regardless of species, A and T are always found in a ratio of 1:1, and why G and C are found in the same ratio.

10. Name and outline the process by which a gene copies itself.

11. Describe the functions of the four types of RNA and where they work.

12. Using words and simple drawings, explain how a gene specifies a unique amino acid sequence on a polypeptide.

13. Using letter symbols — A, T, C, G, and U — explain what a codon and its anticodon are and name the chemicals that bear them.

14. In general terms, describe some of the things that can be done by recombinant DNA technology.

15. Give four examples of diseases related to genetic disorders.

16. In general terms, explain how a virus works in a host cell.

17. Define each of the terms in the Glossary.

GLOSSARY

Anticodon. A sequence of three adjacent side chain bases on a mole- cule of tRNA that is complementary to a codon and that fits to its codon on an mRNA chain during polypeptide synthesis.

Base, Heterocyclic. A heterocyclic amine obtained from the hydrolysis of nucleic acids: adenine, thymine, guanine, cytosine, or uracil.

Base Pairing. In nucleic acid chemistry, the association by means of hydrogen bonds of two heterocyclic, side-chain bases — adenine with thymine (or uracil) and guanine with cytosine.

Chromatin. Filaments that consist of DNA and polypeptides in a cell nucleus.

Chromosome. Small threadlike bodies in a cell nucleus that carry genes in a linear array and that are microscopically visible during cell division.

Codon. A sequence of three adjacent side-chain bases in a molecule of mRNA that codes for a specific amino acid residue when the mRNA participates in polypeptide synthesis.

Deoxyribonucleic Acid (DNA). The chemical of a gene; one of a large
     number of polymers of deoxyribonucleotides and whose sequences of
     side-chain bases constitute the genetic messages of genes.

Deoxyribonucleotides. The monomers of deoxyribonucleic acids (DNA)
     that consist of deoxyribose-phosphate esters with each deoxyribose
     unit carrying a side-chain base (one of four heterocyclic amines,
     adenine, thymine, guanine, or cytosine).

Double Helix DNA Model. A spiral arrangement of two intertwining DNA
     molecules held together by hydrogen bonds between side-chain
     bases.

Exon. A segment of a DNA strand that eventually becomes expressed as
     a corresponding sequence of aminoacyl residues in a polypeptide.

Gene. A unit of heredity carried on a cell's chromosomes and consist-
     ing of DNA.

Genetic Code. The set of correlations that specify which codons on
     mRNA chains are responsible for which aminoacyl residues when the
     latter are steered into place during the mRNA-directed synthesis
     of polypeptides.

Heterogeneous Nuclear RNA (hnRNA). RNA made directly at the guidance
     of DNA and from which messenger RNA (mRNA) is made; primary
     transcript RNA.

Inducer. A substance whose molecules remove repressor molecules from
     operator genes and so open the way for structural genes to direct
     the overall syntheses of particular polypeptides.

Intron. A segment of a DNA strand that separates exons and that does
     not become expressed as a segment of a polypeptide.

Messenger RNA (mRNA). RNA that carries the genetic code as a specific
     series of codons for a specific polypeptide from the cell's
     nucleus to the cytoplasm.

Nucleic Acid. A polymer of nucleotides in which the repeating units
     are pentose phosphate esters, each pentose unit bearing a side-
     chain base (one of four heterocyclic amines); polymeric compounds
     that are involved in the storage, transmission, and expression of

genetic messages.

Nucleotide. A monomer of a nucleic acid that consists of a pentose phosphate ester in which the pentose unit carries one of five heterocyclic amines as a side-chain base.

Plasmid. A circular molecule of supercoiled DNA in a bacterial cell.

Radiomimetic Substance. A substance whose chemical effect in a cell mimics the effect of ionizing radiation.

Recombinant DNA. DNA made by combining the natural DNA of plasmids in bacteria or the natural DNA in yeasts with DNA from external sources, such as the DNA for human insulin, and made as a step in a process that uses altered bacteria or yeasts to make specific proteins (e.g., interferons, human growth hormone, or insulin).

Replication. The reproductive duplication of a DNA double helix.

Replisome. A small cellular body on the nuclear matrix through which DNA strands loop in and out during replication.

Repressor. A substance whose molecules can bind to a gene and prevent the gene from directing the synthesis of a polypeptide.

Ribonucleic Acids (RNA). Polymers of ribonucleotides that participate in the transcription and the translation of the genetic messages into polypeptides. (See also Heterogeneous Nuclear RNA, Messenger RNA, Ribosomal RNA, and Transfer RNA.)

Ribonucleotides. The monomers for ribonucleic acids that consist of ribose phosphate esters with each ribose unit carrying a side-chain base (one of four heterocyclic amines: adenine, uracil, guanine, and cytosine).

Ribosomal RNA (rRNA). RNA that is incorporated into cytoplasmic bodies called ribosomes.

Ribosome. A granular complex of rRNA that becomes attached to a mRNA strand and that supplies some of the enzymes for mRNA-directed polypeptide synthesis.

Transcription. The synthesis of messenger RNA under the direction of

DNA.

Transfer RNA (tRNA). RNA that serves to carry an aminoacyl group to a specific acceptor site of a mRNA molecule at a ribosome where the aminoacyl group is placed into a growing polypeptide chain.

Translation. The synthesis of a polypeptide under the direction of messenger RNA.

Virus. One of a large number of substances that consist of nucleic acid (usually RNA) surrounded (usually) by a protein overcoat and that can enter host cells, multiply, and destroy the host.

SELF-TESTING QUESTIONS

COMPLETION

1. Nucleic acids are polymers whose monomers have the general name of _____. These monomers can be hydrolyzed to give one or the other of two sugars named _____ and _____, an inorganic _____ ion and a set of heterocyclic _____.

2. Nucleic acids that are made with ribose have the full name of _____ which is usually abbreviated _____. The four bases usually obtained from this kind of nucleic acid have the names and one-letter symbols of

_____ ____   _____ ____

_____ ____   _____ ____

3. Nucleic acids made with deoxyribose have the full name of _____ which is usually abbreviated _____. The four bases obtained from this kind of nucleic acid have the names and one-letter symbols of

_____ ____   _____ ____

_____ ____   _____ ____

4. The "backbones" of all nucleic acids contain a chain of diesters of _____ and the particular kind of sugar molecule. The bases are attached to the backbone, one base at

each _____ unit.

5. The bases have functional groups and molecular geometries that permit them to form base-pairs with each other by means of _____ bonds. Base A always pairs with either _____ or with _____; G always pairs with _____.

6. Two strands of DNA form a twisting _____ according to evidence cited by _____ and _____. Base A of one strand pairs to base _____ opposite it on the other strand; and base C pairs to base _____.

7. The helices are further coiled into super helices that are woven in and out of tiny packages that are mostly made of protein, that occur at fixed points in the nuclear matrix, and that are called

_____.

8. An individual hereditary unit called a _____ consists of a particular series of triplets of nucleotides in one of the kinds of nucleic acids, _____.

9. In higher organisms, at least many complete genes come in interrupted sequences of triplets. The interrupting segments in DNA molecules are called _____, and the segments that make up a full gene are called _____.

10. The process whereby a gene becomes reproduced exactly, just prior to cell division, is called _____. The faithfulness of the copying depends on _____

_____.

11. When DNA is used to direct the synthesis of a polypeptide, the DNA first directs the synthesis of a specific form of RNA called _____, abbreviated _____. This product is then processed to delete the triplets in its chain that correspond to the _____ segments of DNA. Then the remaining triplets are "knitted" back together to give another form of RNA called _____, abbreviated _____.

12. Each triplet in mRNA is called a _____, and each is able to specify a particular _____ residue in a completed _____.

13. The overall process of using DNA to direct the formation of mRNA is called _____.

14. Following this process is another complicated process called _____, and its final product is a specific _____.

15. To accomplish this last process, the cell needs two other kinds of RNA. One kind is used to make ribosomes and is called _____, abbreviated _____. The other kind is called _____, abbreviated _____, and its function is to carry _____ to polypeptide assembly sites on _____.

16. One particular triplet of bases on a tRNA molecule can match a complementary triplet on mRNA. This tRNA triplet is called _____.

17. In some organisms at least, genes are in a switched-off status because a _____ molecule has become bound to a segment of the gene. A molecule that can combine with this and remove it is called _____. Several drugs called _____ kill bacteria by interfering with bacterial gene-directed polypeptide synthesis.

18. Radiations such as X rays or gamma rays cause the most damage to a cell when they strike _____. Those that cause cancer are called _____. If a birth defect is the result, the agent is called a _____.

19. Substances that can invade particular "host" cells and take over the genetic machinery are called _____.

20. In a recently developed technology called _____ DNA technology, particles called _____ in a bacterium are modified and given new genetic material. Then the bacteria will manufacture some _____ that corresponds to its additional genes.

21. Four diseases attributed to defective genes are
    (1) Associated with the overproduction of thick mucus in the lungs: _____
    (2) Associated with an impairment in the metabolism of the amino

acid phenylalanine: _____
(3) Associated with a blood disorder: _____
(4) Associated with poor pigmentation of the eyes and the skin: _____

MULTIPLE-CHOICE

1. What is transmitted from parents to offspring is a complete set of
   (a) DNA molecules
   (b) polypeptides
   (c) enzymes
   (d) hormones

2. The site of polypeptide synthesis in a cell is
   (a) a chromosome
   (b) a gene
   (c) a ribosome
   (d) a nuclear membrane

3. If DNA were fully hydrolyzed, the products would be
   (a) ribose
   (b) deoxyribose
   (c) phosphoric acid
   (d) a few heterocyclic amines

4. Molecules of tRNA differ from molecules of mRNA in being
   (a) longer
   (b) triple helices
   (c) shorter
   (d) inside ribosomes

5. The overall process of transcription proceeds in which order?
   (a) exon to intron to polypeptide
   (b) mRNA to polypeptide
   (c) DNA to hnRNA to mRNA
   (d) DNA to rRNA to tRNA to mRNA

6. DNA segments that appear to be uninvolved in polypeptide synthesis are called
   (a) introns   (b) exons   (c) triplets   (d) anticodons

7. Human insulin can be made by bacteria or yeasts by a method called
   (a) recombinant bacteria
   (b) recombinant RNA
   (c) recombinant plasmids
   (d) recombinant DNA

8. Gene-directed polypeptide synthesis proceeds in which order of events?
   (a) gene to mRNA to hnRNA to tRNA
   (b) gene to hnRNA to mRNA to polypeptide
   (c) gene to rRNA to tRNA to hnRNA
   (d) gene to seplicated gene to polypeptide

9. According to the Crick–Watson theory, the genetic message carried by a gene is related most particularly to
    (a) the kinds of heterocyclic amines projecting from the phosphate–pentose chain in the gene
    (b) the sequence in which the heterocyclic amines are lined up along the "backbone" of the gene
    (c) the absence of one of the — OH groups normally found in RNA
    (d) the sequence of amino acids in gene molecule

10. The functional groups in adenine are geometrically arranged to enable adenine to pair by hydrogen bonding with
    (a) adenine   (b) guanine   (c) thymine   (d) uracil

11. The molecular basis of a mutation is most closely linked to a
    (a) defect in the transcription of a genetic message to mRNA
    (b) change in the sequence or identity of heterocyclic amines on a DNA molecule
    (c) defect in the arrival sequence of tRNA molecules at mRNA codon sites
    (d) defect in the rRNA of ribosomes

12. If a codon triplet were U–C–G, the anticodon would be
    (a) A–C–G   (b) C–G–A   (c) T–G–C   (d) A–G–C

ANSWERS

ANSWERS TO SELF-TESTING QUESTIONS

Completion
1. nucleotides; ribose and deoxyribose; phosphate; bases (or amines)
2. ribonucleic acids; RNA
   adenine   A   guanine   G
   uracil   U   cytosine   C
3. deoxyribonucleic acids; DNA
   adenine   A   guanine   G
   thymine   T   cytosine   C
4. phosphoric acid; sugar (or pentose)
5. hydrogen; T or U; C
6. double helix; Crick and Watson; T; G
7. replisomes
8. gene; DNA
9. introns; exons

10. replication; base pairing of A to T and of G to C
11. heterogeneous nuclear RNA; hnRNA; intron; messenger RNA; mRNA
12. codon; amino acid; polypeptide
13. transcription
14. translation; polypeptide
15. ribosomal RNA; rRNA; transfer RNA; tRNA; aminoacyl units; a ribosome (or at an mRNA site at a ribosome)
16. an anticodon
17. repressor; an inducer; antibiotics
18. DNA in a cell nucleus; carcinogens; teratogens
19. viruses
20. recombinant; plasmids; polypeptide (or protein)
21. (1) cystic fibrosis
    (2) PKU (phenylketonuria)
    (3) sickle cell anemia
    (4) albinism

Multiple-Choice

1. a
2. c
3. b, c, d
4. c
5. c
6. a
7. d
8. b
9. b
10. c, d
11. b
12. b

23 Nutrition

OBJECTIVES

Health, happiness, and life are all greatly influenced by what
we eat and drink. When most people learn why various things must be
in the diet in their proper proportions, they usually make an effort
to insure that they have the right kind of diet. The objectives that
follow, which you should be able to do once you have studied the
chapter and have worked its Exercises and Review Questions, are
designed to emphasize the knowledge that will help you have the best
physical well-being possible.

1. Explain why the recommended daily allowances of the National
   Academy of Sciences are higher than the minimum daily
   requirements.
2. Explain why the best nutrition is obtained from a variety of
   foods.
3. Explain why food energy should come from a mix of lipids and
   carbohydrates.
4. Explain why several amino acids, but not all, are called
   "essential."
5. Compare meat, cereal, and fruit proteins in their digestibility.
6. Give both the positive and negative consequences of milling
   grains.
7. Compare the proteins of meats, dairy products, cereals, and nuts
   in their biological values.
8. Describe the major factor affecting the biological values of

proteins.

9. Explain why one could not eat enough maize (corn) or cassava per day to satisfy one's needs for amino acids.
10. Describe the problems vegetarians must solve in order to have good nutrition.
11. Name the essential fatty acids and tell why they are important.
12. In general ways, explain how vitamins are different from other nutrients.
13. Name the vitamins and, where known, identify the human deficiency diseases or syndromes associated with each.
14. Give one good source for each vitamin.
15. Name and give the chemical formulas of the six minerals needed at levels above 100 mg/day.
16. Name and give the principal functions of the ten chief trace elements.
17. Define the terms in the Glossary.

GLOSSARY

Adequate Protein. A protein that, when digested, makes available all of the essential amino acids in suitable proportions to satisfy both the amino acid and total nitrogen requirements of good nutrition without providing excessive calories.

Biological Value. In nutrition, the percentage of the nitrogen of ingested protein that is absorbed from the digestive tract and retained by the body when the total protein intake is less than normally required.

Biotin. A water-soluble vitamin needed to make enzymes used in fatty acid synthesis.

Choline. A compound needed to make complex lipids and acetylcholine; classified as a vitamin.

Coefficient of Digestibility. The proportion of an ingested protein's nitrogen that enters circulation rather than elimination (in feces); the difference between the nitrogen ingested and the nitrogen in the feces divided by the nitrogen ingested.

Dietetics. The application of the findings of the science of nutrition to the feeding of individual humans, whether well or ill.

Essential Amino Acid. An α-amino acid that the body cannot make from other amino acids and that must be supplied by the diet.

Essential Fatty Acid. A fatty acid that must be supplied by the diet.

Folacin. A vitamin supplied by folic acid or pteroylglutamic acid and that is needed to prevent megaloblastic anemia.

Food. A material that supplies one or more nutrients without contributing materials that, either in kind or quantity, would be harmful to most healthy people.

Limiting Amino Acid. The essential amino acid most poorly provided by a dietary protein.

Minerals. Ions that must be provided in the diet at levels of 100 mg/day or more; Ca^{2+}, Mg^{2+}, Na^+, K^+, Cl^-, and phosphate.

Niacin. A water-soluble vitamin needed to prevent pellagra and essential to the coenzymes in NAD^+ and $NADP^+$; nicotinic acid or nicotinamide.

Nitrogen Balance. A condition of the body in which it excretes as much nitrogen as it receives in the diet.

Nutrition. The science of the substances of the diet that are necessary for growth, operation, energy, and repair of bodily tissues.

Pantothenic Acid. A water-soluble vitamin needed to make coenzyme A.

Recommended Dietary Allowance (RDA). The level of intake of a particular nutrient as determined by the Food and Nutrition Board of the National Research Council of the National Academy of Sciences to meet the known nutritional needs of most healthy individuals.

Riboflavin. A B vitamin needed to give protection against the breakdown of tissue around the mouth, the nose, and the tongue, as well as to aid in wound healing.

Thiamin. A B vitamin needed to prevent beri beri.

Trace Element. Any element that the body needs each day in an amount of no more than 20 mg.

<u>Vitamin</u>. An organic substance that must be in the diet; whose absence causes a deficiency disease; which is present in foods in trace concentrations; and that isn't a carbohydrate, lipid, protein, or amino acid.

<u>Vitamin A</u>. Retinol; a fat-soluble vitamin in yellow-colored foods and needed to prevent night blindness and certain conditions of the mucous membranes.

<u>Vitamin B_6</u>. Pyridoxine, pyridoxal, or pyridoxamine; a vitamin needed to prevent hypochromic microcytic anemia and used in enzymes of amino acid catabolism.

<u>Vitamin B_{12}</u>. Cobalamin; a vitamin needed to prevent pernicious anemia.

<u>Vitamin C</u>. Ascorbic acid; a vitamin needed to prevent scurvy.

<u>Vitamin D</u>. Cholecalciferol (D_3) or ergocalciferol (D_2); a fat-soluble vitamin needed to prevent rickets and to ensure the formation of healthy bones and teeth.

<u>Vitamin Deficiency Diseases</u>. Diseases caused not by bacteria or viruses but by the absence of specific vitamins, such as pernicious anemia (B_{12}), hypochromic microcytic anemia (B_6), pellagra (niacin), the breakdown of certain tissues (riboflavin), megaloblastic anemia (folacin), beri beri (thiamin), scurvy (C), hemorrhagic disease (K), rickets (D), and night blindness (A).

<u>Vitamin E</u>. A mixture of tocopherols; a fat-soluble vitamin apparently needed for protection against edema and anemia (in infants) and possibly against dystrophy, paralysis, and heart attacks.

<u>Vitamin K</u>. The antihemorrhagic vitamin that serves as a cofactor in the formation of a blood clot.

SELF-TESTING QUESTIONS

COMPLETION

1. With respect to the terms "nutrient" and "food," we may say that bread is a _____ and the wheat protein in bread is a _____ .

2. The symbol RDA stands for _____.

3. A quantitative measure of the digestibility of a given protein is its _____, and the equation that defines it is _____.

4. The extent to which a given protein has a high biological value is determined by the extent to which it supplies _____ _____.

5. A protein with a very low amount of lysine would have a low _____ _____.

6. The essential amino acid most poorly supplied by a given protein is called the _____ of that protein.

7. A protein with all of the essential amino acids in approximately the right proportions for humans is called an _____.

8. An individual whose intake of nitrogen in the diet equals the quantity of nitrogen excreted is said to have a _____ _____.

9. In order to grow and develop an infant should have a _____ _____ nitrogen balance.

10. Two essential fatty acids are _____ acid and _____ acid.

11. List four criteria that must be satisfied by a substance if it is to be considered a vitamin.

    (a) _____

    (b) _____

    (c) _____

    (d) _____

12. The two broad classes of vitamins are the _____ vitamins and the _____ vitamins.

13. The four vitamins that are most hydrocarbon-like are vitamins _____, _____, _____, and _____.

14. Because the molecules of some of the hydrocarbon-like vitamins have carbon-carbon _____ bonds, these vitamins can be slowly destroyed by contact with air.

15. Lack of vitamin D leads to _____, a disorder of _____.

16. Lack of vitamin _____ may lead to impaired vision in dim light.

17. Lack of vitamin _____ may lead to problems in controlling hemorrhaging.

18. In the absence of ascorbic acid, vitamin _____, the disease known as _____ would develop.

19. Because thiamin is a part of an enzyme essential to the catabolism of _____, the daily requirement for it is related to the daily intake of _____.

20. Beri-beri is a disease that occurs when _____ is lacking in the diet.

21. Pantothenic acid is needed to make coenzyme _____.

22. The principal minerals in the body are (6) _____, _____, _____, _____, _____, and _____. (Give their correct formulas).

23. To qualify as a trace element, the mass that the element contributes to the total mass of the body doesn't exceed _____.

24. The names of eleven trace elements known or believed to be needed by the body are (in any order)

    _____   _____   _____

    _____   _____   _____

    _____   _____   _____

    _____   _____

25. All trace metal elements occur as their _____, not as their atoms.

26. The central ion in hemoglobin has the name _____ and the formula _____.

MULTIPLE-CHOICE

1. A zero-carbohydrate diet will cause the body to make its own
   - (a) linoleic acid
   - (b) sucrose
   - (c) glucose
   - (d) nonessential amino acids

2. The proteins with the highest digestibility coefficients are generally those of
   - (a) vegetables   (b) meat   (c) fruit   (d) cereals

3. One of the essential amino acids that often is a limiting amino acid is
   - (a) lysine   (b) alanine   (c) glycine   (d) hydroxyproline

4. If 102 g of wheat protein has to be eaten to result in 80.4 g being actually digested and absorbed, the digestibility coefficient of wheat protein is
   - (a) 80.4   (b) 82   (c) 1.27   (d) 0.79

5. The vegetarian diet supplies sufficient and adequate protein each day without also furnishing large amount of calories provided that the diet emphasizes
   - (a) brown rice
   - (b) beans
   - (c) two different vegetables about equally
   - (d) cassava

6. Even though phosphate is essential in the diet, it is not classified as a vitamin because it is not
   - (a) organic
   - (b) required for normal growth
   - (c) present in foods
   - (d) a carbohydrate

7. The vitamin supplied by yellow-colored vegetables is vitamin
   - (a) A   (b) B   (c) C   (d) D

8. A vitamin that can be stored in adipose tissue is
   - (a) vitamin K
   - (b) vitamin B_6
   - (c) vitamin D
   - (d) thiamine

9. The trace element that serves as a cofactor for the action of insulin is

(a) Cr^{3+} (b) K^+ (c) F^- (d) Mn^{2+}

10. The trace element needed to make heme is

(a) Cu^{2+} (b) Na^+ (c) Fe^{2+} (d) K^+

ANSWERS

ANSWERS TO SELF-TESTING QUESTIONS

Completion
1. food, nutrient
2. recommended dietary allowances
3. coefficient of digestibility,

$$\text{coefficient of digestibility} = \frac{[\text{N in food eaten} - \text{N in feces}]}{\text{N in food eaten}}$$

4. essential amino acids
5. biological value
6. limiting amino acid
7. adequate protein
8. nitrogen balance
9. positive
10. linoleic, arachidonic
11. (a) an organic compound that cannot be made in the body
    (b) its absence causes a deficiency disease
    (c) its presence is required for growth and health
    (d) found only in trace concentrations in foods and is not a carbohydrate, a lipid, or a protein
12. fat-soluble, water-soluble
13. A, D, E, and K
14. double
15. rickets, bone metabolism
16. A
17. K
18. C, scurvy
19. carbohydrate, calories
20. thiamin
21. A
22. Ca^{2+}, Mg^{2+}, Na^+, K^+, Cl^-, and inorganic phosphate ion (P)
23. 20 mg

24. iron, cobalt, zinc, chromium, molybdenum, copper, manganese, nickel, tin, vanadium, and silicon
25. ions
26. iron(II) ion (or the ferrous ion); Fe^{2+}

Multiple-Choice

 1. c
 2. b
 3. a
 4. d
 5. c
 6. a
 7. a
 8. c
 9. a
10. c

24 Enzymes, Hormones, and Neurotransmitters

OBJECTIVES

Objectives 1 through 8 are mostly concerned with the basic vocabulary of enzymes; 9 and 10 are about how enzymes work in general. The heart of this chapter is the study of the ways in which the body controls which reactions go, which shut down, which accelerate, and which go slower. The different chemical and physiological functions of enzymes, hormones, and neurotransmitters are "must" topics of the chapter.

After you have studied this chapter and worked its Exercises, you should be able to do the following.

1. Describe the composition of enzymes and the kinds of cofactors in general terms.
2. Describe what enzymes do.
3. Explain how enzymes possess "specificity."
4. Explain why enzymes are sensitive to denaturing conditions and pH.
5. Describe in general terms how certain B vitamins are vital to some enzymes.
6. Recognize from the name of a substance whether it is an enzyme.
7. Tell what an enzyme's substrate (or type of reaction) is from its name.
8. Define and give an example of an isoenzyme and describe how an analysis of serum isoenzymes can be used in medical diagnosis.
9. Describe the "lock and key" mechanism of enzyme action.

10. Describe the "induced fit" model for enzyme action.
11. Explain (using drawings to illustrate your explanation) how an allosteric activation of an enzyme with two active sites results in a sigmoid rate curve.
12. Explain how an effector influences enzyme activity.
13. Describe what happens in a zymogen–enzyme conversion and give an example.
14. Explain how feedback inhibition works and what homeostasis means.
15. Describe how allosteric inhibition works.
16. Explain how some poisons work.
17. Name three kinds of poisons that act as irreversible enzyme inhibitors.
18. Explain in general terms what kind of substance is used to measure the concentration of an enzyme in some body fluid.
19. Name the enzymes whose serum levels are measured in (a) viral hepatitis and (b) myocardial infarction.
20. Name the two kinds of primary chemical messengers.
21. Without necessarily reproducing molecular structures, describe what cyclic nucleotides are.
22. List the steps in the overall process that occurs when cyclic AMP is involved after a signal releases a hormone or a neurotransmitter.
23. Describe how a hormone finds its target cells and recognizes them.
24. In general terms, explain how neurotransmitters work and why they must eventually be deactivated.
25. Describe in general terms what the following do.
    (a) acetylcholine and cholinesterase
    (b) norepinephrine
    (c) monoamine oxidases
    (d) antidepressant drugs
    (e) dopamine
    (f) drugs for schizophrenia, like Thorazine   and Haldol
    (g) GABA
    (h) mild tranquilizers, like Valium   and Librium
    (i) enkaphalins and endorphins
    (k) substance P
26. Define each of the terms in the Glossary.

GLOSSARY

Active Site. That region of an enzyme molecule most directly responsible for the catalytic effect of the enzyme.

Agonist. A compound whose molecules can bind to a receptor on a cell membrane and cause a response by the cell.

Allosteric Activation. The activation of an enzyme's catalytic site by the binding of some molecule at a position elsewhere on the enzyme.

Allosteric Inhibition. The inhibition of the activity of an enzyme caused by the binding of an inhibitor molecule at some site other than the enzyme's catalytic site.

Antagonist. A compound that can bind to a membrane receptor but not cause any response by the cell.

Antimetabolite. A substance that inhibits the growth of bacteria.

Apoenzyme. The wholly polypeptide part of an enzyme.

Binding Site. That part of an enzyme molecule that holds the substrate molecule and positions it over the active site.

Coenzyme. An organic compound needed to make a complete enzyme from an apoenzyme.

Cofactor. A nonprotein compound or ion that is an essential part of an enzyme.

Competitive Inhibition. The inhibition of an enzyme by the binding of a molecule that can compete with the substrate for the occupation of the catalytic site.

Effector. A chemical other than a substrate that can allosterically activate an enzyme.

Endocrine Gland. An organ that makes one or more hormones.

Enzyme. A catalyst in a living system.

Enzyme-Substrate Complex. The temporary combination that an enzyme must form with its substrate before catalysis can occur.

Esterase. An enzyme that catalyzes the hydrolysis of an ester.

Feedback Inhibition. The competitive inhibition of an enzyme by a product of its own action.

Homeostasis. The response of an organism to a stimulus such that the organism is restored to its pre-stimulated state.

Hormone. A primary chemical messenger made by an endocrine gland and carried by the bloodstream to a target organ where a particular chemical response is initiated.

Induced Fit Theory. Certain enzymes are induced by their substrate molecules to modify their shapes to accommodate the substrate.

Inhibitor. A substance that interacts with an enzyme to prevent its acting as a catalyst.

Isoenzymes. Enzymes that have identical catalytic functions but which are made of slightly different polypeptides.

Kinase. An enzyme that catalyzes the transfer of a phosphate group.

Lipase. An enzyme that catalyzes the hydrolysis of lipids.

Lock-and-Key Theory. The specificity of an enzyme for its substrate is caused by the need for the substrate molecule to fit to the enzyme's surface much as a key fits to and turns only one tumbler lock.

Monoamine Oxidase. An enzyme that catalyzes the inactivation of neurotransmitters or other amino compounds of the nervous system.

Neurotransmitter. A substance released by one nerve cell to carry a signal to the next nerve cell.

Oxidase. An enzyme that catalyzes an oxidation.

Oxidoreductase. An enzyme that catalyzes the formation of an oxidation-reduction equilibrium.

Peptidase. A digestive enzyme that catalyzes the hydrolysis of a peptide.

Poison. A substance that reacts in some way in the body to cause changes in metabolism that threaten health or life.

Proenzyme.   An inactive form of an enzyme; a zymogen.

Reductase.   An enzyme that catalyzes a reduction.

Target Cell.   A cell at which a hormone molecule finds a site where it can become attached and then cause some action that is associated with the hormone.

Transferase.   An enzyme that catalyzes the transfer of some group.

Zymogen.   A polypeptide that is changed into an enzyme by the loss of a few amino acid residues or by some other change in its structure; a proenzyme.

SELF-TESTING QUESTIONS

COMPLETION

1. An organic compound that is needed, in some cases, to complete an enzyme is called a _____.

2. The specific site on a large enzyme molecule where a substrate experiences the catalytic activity of the enzyme is called the _____.

3. An enzyme might lose its catalytic ability through the action of heat or a change in pH because an enzyme is made up mostly of a _____.

4. The temporary union of an enzyme with the compound on which it acts is called the _____.

5. The theory that accounts for the unusual specificity of an enzyme is the _____ theory.

6. Coenzymes can be made in the body, but in many cases it is essential that the body be supplied with _____ via the diet because they are a necessary part of many coenzyme molecules.

7. Some metabolic sequences are shut down when molecules of their final product combine with and inactivate an _____ for one of the early steps in the sequence.

8. An enzyme that catalyzes the hydrolysis of an ester link would be called an _____; and an enzyme that helps in transferring a phosphate group is called a _____.

9. The enzyme sucrase catalyzes the hydrolysis of _____.

10. Some enzymes will remove a pair of electrons from a substrate; because of this particular kind of action, they are called _____.

11. What are the names of the vitamins needed to make each of these coenzymes?

   NAD$^+$ _____

   FAD _____

   thiamine pyrophosphate _____

12. Each of the following statements describes a way in which an enzyme might be activated or otherwise controlled. On the blank line write the name of the kind of control described.
   (a) An enzyme's catalytic site is activated by the binding of a nonsubstrate molecule elsewhere on the enzyme.

   _____

   (b) an enzyme with two (or more) catalytic sites has the second site activated by the binding of a substrate to the first site.

   _____

   (c) The activity of an enzyme for its substrate is suppressed by the binding at its active site of a nonsubstrate molecule of similar shape to the substrate molecule.

   _____

   (d) The activity of an enzyme is suppressed by its binding at its active site a molecule of one of the products of the action of the enzyme.

   _____

   (e) The activity of an enzyme is suppressed by its binding at a place away from its active site some other molecule.

(f) An enzyme forms when a small fragment is cut off from a polypeptide chain letting the active site emerge or unfold.

13. Feedback inhibition is an example of a general kind of ability for self-regulation called _____.

14. Compounds that inhibit normal metabolism in disease-causing bacteria have the general name of _____.

15. By using the _____ for an enzyme as an analytical reagent, the concentration of the enzyme in some body fluid can be measured.

16. In diseases or injuries of the liver, one of the liver enzymes with the symbol _____ escapes into the _____ along with a lower level of another liver enzyme with the symbol _____. The ratio of these two is typically _____
    (higher or lower)
    in victims of viral hepatitis than in healthy individuals.

17. In a myocardial infarction, three enzymes with the short symbols of _____, _____ and _____ leak from heart tissue into general circulation. One of these can be found in the form of three isoenzymes which have the synmbols _____ if present in skeletal muscle, _____ when found in heart muscle, and _____ when present in brain tissue. A procedure called _____ can be used to separate these isoenzymes from each other in a sample of blood.

18. In an infarct, the enzyme lactate dehydrogenase, symbolized as _____, appears in the blood, but there are _____ isoenzymes for it. Normally, the first two of these isoenzymes appear in relative concentrations in which the level of the first is less than that of the second. This relative relationship is inverted in an infarct, and the phenomenon is called an _____ _____.

19. The general names for the two kinds of primary chemical messengers are _____ and _____. Endocrine glands make the _____ and _____ make the other kind of chemical messenger.

20. Many primary chemical messengers function at _____ cells by activating a membrane-bound enzyme called _____ cyclase. This enzyme then catalyzes the conversion of _____ to _____. This then activates an _____ inside the cell. To halt the process and switch the signal off, another enzyme called _____ hydrolyzes _____.

21. Hormones exert their influence in a variety of ways. Adrenaline, for example, is an _____ activator. Insulin and human growth hormone affect the _____ of the membranes of their target cells. Some sex hormones function by activating _____ which then direct the synthesis of _____.

22. Chemical communication from one neuron to another occurs across a very narrow, fluid-filled space called a _____, and the general name for substances that move across this space carrying a "message" is _____.

23. The neuron from which this messenger moves is called the _____ neuron, and the neuron to which the messenger migrates is called the _____ neuron.

24. A complex forms between a receptor in the postsynaptic neuron and the _____ that works to activate _____ _____. This enzyme then catalyzes the formation of _____ in the membrane of the postsynaptic neuron.

25. To switch a nerve signal off, enzyme-catalyzed reactions have to degrade or otherwise deactivate molecules of _____.

26. Cholinesterase catalyzes the hydrolysis of the neurotransmitter called _____. Nerve gases work by _____ _____ and the botulinus bacillus works by _____ _____. Local anesthetics like nupercaine and procaine block pain signals by _____ _____.

27. Several antidepressants work by interfering with the normal use of a neurotransmitter named _____. The

deactivation of excess or unused amounts of this substance is catalyzed by enzymes called the _____, symbolized as _____. If this enzyme is inhibited by one kind of antidepressant, then the signal carried by the neurotransmitter is _____.
   (kept up or blocked)

28. The neurotransmitter called _____ is believed to be involved with schizophrenia. Drugs that bind to receptors for this neurotransmitter inhibit its _____ to these receptors. Symptoms similar to those of schizophrenia can be induced by the abuse of drugs called _____ that work by triggering the _____ of this neuro-transmitter.   (release or decomposition)

29. The full name of GABA is _____. If the work of this neurotransmitter is enhanced, the result is that nerve signals are _____.
            (accelerated or inhibited)

30. Two families of neurotransmitters that include powerful pain-killers are named the _____ and the _____.

MULTIPLE-CHOICE

1. The site on an enzyme that is able to accept the corresponding substrate is called the
   (a) allosteric site   (c) catalytic site
   (b) binding site   (d) effector site

2. A substance with the name protease is
   (a) a prostaglandin   (c) an enzyme
   (b) a coenzyme   (d) a hormone

3. An enzyme for the reaction: $A + B \rightleftharpoons C + D$
   (a) increases the quantity of C that forms
   (b) increases the quantity of A that remains unreacted
   (c) shifts the equilibrium to favor the right side
   (d) accelerates the establishment of the equilibrium

4. The catalytic activity of an enzyme will generally
   (a) increase with increasing temperature

(b) increase with decreasing temperature
(c) be unaffected by temperature
(d) be greatest in a particular temperature range

5. Several coenzymes
   (a) activate apoenzymes
   (b) are made from various B—vitamins
   (c) catalyze hydrolysis reactions
   (d) are esters of carboxylic acids

6. The symbol NAD^+ stands for
   (a) an enzyme   (b) a vitamin   (c) a coenzyme   (d) an isoenzyme

7. When an enzyme that contains FMN interacts with one having NADH, then which can form?

   (a) $FMNH + NAD^+$ (c) $NADP^+ + FMNH_2$

   (b) $NAD^+ + FMNH_2$ (d) $NAD^+ + H^+ + FMN$

8. Isoenzymes usually differ in
   (a) the substrates they will accept
   (b) their coenzymes
   (c) their apoenzyme portions
   (d) their cofactors

9. The symbol CK stands for
   (a) creatine kinase   (c) cofactor kinase
   (b) carbon—potassium   (d) an isoenzyme

10. A theory that is used to explain enzyme specificity is called the
    (a) theory of allosteric activation
    (b) the sigmoid rate curve theory
    (c) induced fit theory
    (d) lock—and—key theory

Questions 11 through 13 refer to this sequence of reactions:

$$A \xrightarrow{E_a} B \xrightarrow{E_b} C \xrightarrow{E_c} D \xrightarrow{E_d} E$$

11. If the enzyme, E_a, for the conversion of A to B is activated by molecules of A, and if the graph or plot of the rate of the conversion versus the concentration of E_a has a sigmoid shape, then

we probably have the operation of
(a) feedback inhibition   (c) activation by an effector
(b) gene activation   (d) allosteric activation

12. If molecules of final product, E, deactivate the enzyme E_a for the
    first step by binding to the active site of that enzyme, we are
    probably seeing the operation of
    (a) competitive inhibition by nonproduct
    (b) feedback inhibition
    (c) allosteric inhibition
    (d) inhibition by effector

13. If molecules of a substance, A', that resemble the molecules of A
    but that cannot be changed to B, deactivate the enzyme E_a by
    binding to the active site of that enzyme, then we are probably
    observing
    (a) competitive inhibition by nonproduct
    (b) allosteric inhibition by nonproduct
    (c) feedback inhibition by nonproduct
    (d) allosteric action by nonproduct

14. A polypeptide that changes into an enzyme when a small fragment is
    broken off by some activating reaction is
    (a) a prostaglandin   (c) a coenzyme
    (b) an apoenzyme   (d) a proenzyme

15. The general principle that makes possible the measurement of just
    one enzyme out of several that are present in blood serum is that
    of
    (a) enzyme specificity   (c) enzyme inhibition
    (b) electrophoresis   (d) allosteric activation

16. Symptoms of a myocardial infarction include
    (a) LD_1–LD_2 flip
    (b) a rise in the serum level of the CK(MB) isoenzyme
    (c) a drop in the serum level of GOT
    (d) a change in the serum level of glucose oxidase

17. Primary chemical messengers are
    (a) enzymes and hormones
    (b) hormones and neurotransmitters
    (c) neurotransmitters and cyclic nucleotides
    (d) cyclic nucleotides and axons

18. The ability of a hormone to recognize its own target cells depends
    on a "lock-and-key" fit of the hormone molecule to
    (a) a receptor protein   (c) cyclic AMP
    (b) a neurotransmitter   (d) adenyl cyclase

19. An example of a hormone that activates an enzyme is
    (a) adrenalin   (c) testosterone
    (b) insulin   (d) human growth hormone

20. In a postsynaptic neuron, adenylate cyclase is activated by
    (a) a neurotransmitter
    (b) a receptor
    (c) a neurotransmitter-receptor complex
    (d) phosphodiesterase

21. A substance that serves as both a neurotransmitter and a hormone
    is
    (a) acetylcholine   (c) epinephrine
    (b) cholinesterase   (d) norepinephrine

22. Norepinephrine that is reabsorbed by a presynaptic neuron is
    degraded by
    (a) GABA   (b) MAO   (c) LD   (d) CK(MB)

23. A drug that binds to a postsynaptic neuron's receptor protein
    (a) inhibits the transmission of a nerve signal
    (b) accelerates the activation of adenylate cyclase
    (c) prolongs the receipt of signals from the presynaptic neuron
    (d) inactivates MAO enzymes

24. Amphetamines stimulate presynaptic neurons to release
    (a) L-DOPA   (b) dopamine   (c) GABA   (d) MAO

25. A pain-killer produced in the brain itself is
    (a) substance P   (b) GABA   (c) enkephalin   (d) morphine

ANSWERS

ANSWERS TO SELF-TESTING QUESTIONS

Completion
1. coenzyme   (Cofactor is not correct because it would include

trace elements too. Vitamin as the answer is close, but coenzyme is better here because vitamins usually have to be changed into coenzymes first. Also, remember that just a few vitamins work this way.)

2. catalytic site
3. protein (it is more specific than apoenzyme.)
4. enzyme-substrate complex
5. lock-and-key
6. vitamins
7. enzyme
8. esterase; kinase
9. sucrose
10. oxidases
11. nicotinamide (or nicotinic acid, but it is the amide that is used); riboflavin; thiamine
12. (a) activation by an effector
    (b) allosteric activation
    (c) competitive inhibition by a nonsubstrate
    (d) feedback inhibition
    (e) allosteric inhibition
    (f) zymogen activation
13. homeostasis
14. antimetabolite
15. substrate
16. GPT; bloodstream; GOT; higher
17. CK, LD, and GOT; CK(MM); CK(MB); CK(BB); electrophoresis
18. LD; 5; LD_1-LD_2 flip
19. hormones and neurotransmitters; hormones; neurons (nerve cells)
20. target; adenylate cyclase; ATP to cyclic AMP; enzyme; phosphodiesterase; cyclic AMP
21. enzyme; permeabilities; genes; enzymes
22. synapse; neurotransmitter
23. presynaptic; postsynaptic
24. neurotransmitter; adenylate cyclase; cyclic AMP
25. neurotransmitter
26. acetylcholine; inhibiting cholinesterase; blocking the synthesis of acetylcholine; blocking the receptor protein for acetylcholine
27. norepinephrine; monoamine oxidases; MAO; kept up
28. dopamine; binding; amphetamines; release
29. gamma-aminobutyric acid; inhibited
30. endorphins and enkephalins

Multiple-Choice

1. b	13. a
2. c	14. d
3. d	15. a
4. d	16. a, b
5. b	17. b
6. c	18. a
7. b	19. a
8. c	20. c
9. a	21. d
10. c, d	22. b
11. d	23. a
12. b	24. b
	25. c

25 Extracellular Fluids of the Body

OBJECTIVES

Objectives 1 through 4 are concerned with the chemistry of digestion and the digestive juices. The first objective should be considered a minimum goal for this area.

In the professional health areas, the chemistry of respiration and the chemistry of the blood are areas of vital importance because respiratory problems arise in a number of emergency situations. To deal swiftly and correctly with the emergency, health care personnel must obtain and quickly evaluate several measurements, including the blood pH, its bicarbonate level, and the partial pressures of the blood gases. Early in your career, you will most likely have to learn about the chemistry of respiration and the chemistry of blood in order to increase your professional capabilities. Objectives 5 through 17 deal with these areas (although a study of the chemistry of blood is not complete without a study of the function of the kidneys, the subject of objectives 19 and 20). If you use this occasion for a thorough study, your value as a professional will improve that much more quickly.

After you have studied this chapter and worked the exercises in it, you should be able to do the following.

1. Name the end products of the digestion of carbohydrates, proteins, and lipids.

340

2. List the major digestive reactions of the mouth, the stomach, and the duodenum.
3. Name the digestive juices and their principal enzymes and zymogens.
4. Describe the functions of bile salts.
5. Describe the functions of the principal components of the blood.
6. Give the names and formulas and general uses of the two Group IA cations that occur in the body.
7. Describe the levels of the Group IA cations in plasma and intracellular fluids.
8. Describe in general terms the problems associated with hypo- and hypernatremia as well as hypo- and hyperkalemia.
9. Give the names, formulas, and general uses of the two Group IIA cations present in the body, and describe their levels in blood and intracellular fluids.
10. Describe in general terms the problems associated with hypo- and hypercalcemia as well as hypo- and hypermagnesemia.
11. Describe the chemical composition of bone giving the name and the formula of its chief mineral and the name of its principal non-mineral substance.
12. State the range of values for the chloride ion level in blood and inside cells, and describe (in general terms) its function.
13. Describe what problems are associated with hypo- and hyperchloremia.
14. Explain how fluids and nutrients exchange at capillary loops.
15. Name two situations in which proteins leave the blood and describe the consequences.
16. Give three ways edema may arise.
17. Describe the main features of the composition of hemoglobin.
18. Explain what 2,3-diphosphoglycerate (DPG) does.
19. Referring to the allosteric effect and the hemoglobin-oxygen dissociation curve, explain how oxygen binds cooperatively to hemoglobin.
20. Discuss how oxygen affinity varies with blood pH.
21. Discuss how oxygen affinity varies with the pCO_2 of blood.
22. Describe how partial pressure gradients aid in gas exchange.
23. Describe how localized changes in pH aid in gas exchange.
24. Describe the functions of the isohydric shift and the chloride shift.
25. Explain how waste CO_2 is transported in blood.
26. Explain how myoglobin in certain tissue aids in gas exchange.
27. Outline how values of blood pH, pCO_2, and $[HCO_3^-]$ change from the normal in clinical situations involving metabolic or respiratory

acidosis and alkalosis.
28. Describe the role of the kidneys in preventing acidosis.
29. Describe what vasopressin, aldosterone, and renin do.
30. Define the terms in the Glossary.

GLOSSARY

Acidosis. A condition in which the pH of the blood is below normal. Metabolic acidosis is brought on by a defect in some metabolic pathway. Respiratory acidosis is caused by a defect in the respiratory centers or in the mechanisms of breathing.

Albumin. One of a family of globular proteins that tend to dissolve in water, and that in blood contribute to the blood's colloidal osmotic pressure and aid in the transport of metal ions, fatty acids, cholesterol, triacylglycerols, and other water-insoluble substances.

Aldosterone. A steroid hormone, made in the adrenal cortex, secreted into the bloodstream when the sodium ion level is low, and that signals the kidneys to leave sodium ions in the bloodstream.

Alkalosis. A condition in which the pH of the blood is above normal. Metabolic alkalosis is caused by a defect in metabolism. Respiratory alkalosis is caused by a defect in the respiratory centers of the brain or in the apparatus of breathing.

Aminopeptidase. An enzyme that catalyzes the hydrolysis of N-terminal amino acid residues from small polypeptides.

α-Amylase. An enzyme that catalyzes the hydrolysis of amylose.

Bile. A secretion of the gall bladder that empties into the upper intestine and furnishes bile salts; a route of excretion for cholesterol and bile pigments.

Bile Salts. Steroid-based detergents in bile that emulsify fats and oils during digestion.

Carbaminohemoglobin. Hemoglobin that carries chemically bound carbon dioxide.

Carboxypeptidase. A digestive enzyme that catalyzes the hydrolysis of C-terminal amino acid residues from small polypeptides.

Cardiovascular Compartment.   The entire network of blood vessels and the heart.

Chloride Shift. An interchange of chloride ions and bicarbonate ions between a red blood cell and the surrounding blood serum.

Chyme. The mixture of partially digested food that forms in the stomach and that is released through the pyloric valve into the duodenum.

Chymotrypsin. A digestive enzyme that catalyzes the hydrolysis of peptide bonds in large polypeptides.

Digestive Juice. A secretion into the digestive tract that conists of a dilute aqueous solution of digestive enzymes (or their zymogens) and inorganic ions.

2,3-Diphosphoglycerate (DPG). An organic ion that nestles within the hemoglobin molecule in deoxygenated blood but is expelled from the hemoglobin molecule during oxygenation.

Edema. The swelling of tissue caused by the retention of water.

Electrolytes, Blood. The ionic substances dissolved in the blood.

Enterokinase. An enzyme of intestinal juice that changes trypsinogen into trypsin.

Erythrocyte. A red blood cell.

Extracellular Fluids. Body fluids that are outside of cells.

Fibrin. The fibrous protein of a blood clot that forms from fibrinogen during clotting.

Fibrinogen. A protein in blood that is changed to fibrin during clotting.

Gastric Juice. The digestive juice secreted into the stomach and that contains pepsinogen, hydrochloric acid, and gastric lipase.

Globulins. Globular proteins in thé blood that include γ-globulin, an agent in the body's defense againt infectious diseases.

Hemoblogin (**HHb**). The oxygen-carrying protein in red blood cells.

Hyperkalemia. An elevated level of potassium ion in blood – above 5.0 meq/L.

Hypernatremia. An elevated level of sodium ion in blood – above 145 meq/L.

Hypokalemia. A low level of potassium ion in blood – below 3.5 meq/L.

Hyponatremia. A low level of sodium ion in blood – below 134 meq/L.

Internal Environment. Everything enclosed within an organism.

Interstitial Fluids. Fluids in tissues but not inside cells.

Intestinal Juice. The digestive juice that empties into the duodenum from the intestinal mucosa and whose enzymes also work within the intestinal mucosa as molecules migrate through.

Isohydric Shift. In actively metabolizing tissue, the use of a hydrogen ion released from newly formed carbonic acid to react with and liberate oxygen from oxyhemoglobin; in the lungs, the use of hydrogen ion released when hemoglobin oxygenates to combine with bicarbonate ion and liberate carbon dioxide for exhaling.

Lactase. A digestive enzyme that catalyzes the hydrolysis of lactose.

Lipase. An enzyme that catalyzes the hydrolysis of lipids.

Maltase. A digestive enzyme that catalyzes the hydrolysis of maltose.

Mucin. A viscous glycoprotein released in the mouth and the stomach that coats and lubricates food particles and protects the stomach from the acid and pepsin of gastric juice.

Nuclease. An enzyme that catalyzes the hydrolysis of nucleic acids.

Oxygen Affinity. The percentage to which all of the hemoglobin molecules in the blood are saturated with oxygen molecules.

Pancreatic Juice. The digestive juice that empties into the duodenum from the pancreas.

Pepsin. A digestive, proteolytic enzyme in gastric juice that forms by the action of acid on pepsinogen.

Pepsinogen. The zymogen of pepsin.

Respiratory Gases. Oxygen and carbon dioxide.

Saliva. The digestive juice secreted in the mouth whose enzyme, amylase, catalyzes the partial digestion of starch.

Shock, Traumatic. A medical emergency in which relatively large volumes of blood fluid leave the vascular compartment and enter the interstitial spaces.

Sucrase. A digestive enzyme that catalyzes the hydrolysis of sucrose.

Trypsin. A digestive enzyme, made from the trypsinogen of pancreatic juice by the action of enterokinase, that catalyzes the hydrolysis of proteins.

Vascular Compartment. The entire network of blood vessels and their contents.

Vasopressin. A hypophysis hormone that acts at the kidneys to help regulate the concentrations of solutes in the blood by instructing the kidneys to retain water (if the blood is too concentrated) or to excrete water (if the blood is too dilute).

Zymogen. A polypeptide that is changed into an enzyme by the loss of a few amino acid residues or by some other change in its structure; a proenzyme.

SELF-TESTING QUESTIONS

COMPLETION

Questions 1 through 13 are concerned with the chemistry of digestion.
  1. The first digestive juice to have much effect on proteins in the diet is _____.

2. Some digestive enzymes occur in their respective digestive juices initially in inactive forms called _____.

3. The proteolytic enzyme active in the stomach is called _____ _____ and its zymogen is _____.

4. The partially digested material in the stomach that moves into the upper intestinal tract is called _____.

5. Starch in the diet is acted upon first in the _____ by enzyme _____ found in _____.

6. All of the reactions of digestion are classified as _____ reactions.

7. For the most effective digestion of fats and oils in the diet, we depend on the emulsifying action of _____, compounds that are best described as _____ and that are released into the _____ of the digestive tract from an organ called the _____.

8. A special enzyme in intestinal juice called _____ helps convert trypsinogen to trypsin.

9. Besides helping to digest proteins in the diet, trypsin also catalyzes the conversion of _____ to chymotrypsin and of _____ to carboxypeptidase.

10. The arrival of acidic chyme in the duodenum causes the release of enzyme-rich fluids called _____ and

11. Another fluid stimulated by the arrival of chyme in the upper intestinal tract is called _____.

12. The principal fat-digesting enzyme released in pancreatic juice is called _____.

13. Complete the following table by writing the names of the end products of the complete digestion of each family of foods given in the first column.

FOOD	END PRODUCTS OF COMPLETE DIGESTION

Proteins   _Amino acids_____

Starch   _glucose_____

A mixture of lactose,   _glucose fructose galactose__
maltose, and sucrose

Triacylglycerols   _fatty acids_____

14. The chief cation in the blood has the formula _NA^+_, and its
    normal concentration is from _____ to _____ (include the
    unit). When its level exceeds the higher of these two values, the
    condition is called _____, and when it falls
    below the lower value, the condition is _____.

15. The chief cation inside cells has the formula _____, and its
    concentration is generally _____ (include the unit).

16. When the potassium ion level of the blood drops below _____
    _____ (include the unit), the condition is called _____
    _____, and when its level rises above _____
    (include the unit), the condition is _____.

17. Three health conditions that can cause severe hyperkalemia are
    ____Burns_____, ____HEART ATTACKS_____, and
    _____.

18. A general and unusual decrease in body fluids can cause _____
                                                       (hypo or
    ___hypo___ -natremia as well as ____hyper____ -kalemia.
      hyper)   (hypo or hyper)

19. The formulas of the two Group IIA cations in the body are _____
    and _____.

20. The second most abundant cation inside cells has the formula
    _____.

21. The chief cation in bones has the formula _____.

22. Besides being necessary in bones, calcium ions also participate in

the _____Contraction_____ of muscles.

23. The overuse of milk of magnesia can cause _____ -
                                              (~~hypo~~ or hyper)
    magnesemia.

24. A deficiency of vitamin _D___ can cause hypocalcemia.

25. The overuse of calcium-based antacids can cause _____.

26. Both Ca^{2+} and Mg^{2+} can activate certain __enzymes_____.

27. From the standpoint of osmosis and dialysis, which is the more
    concentrated mixture, blood or interstitial fluid? _blood_____

28. In which direction will water naturally have a net tendency to
    migrate: from a blood capillary into the interstitial compartment
    or from the interstitial compartment into the bloodstream?
    _____Intest_____

29. Besides the forces generated by osmosis and dialysis, what force
    is present in the circulatory system that is especially important
    on the arterial side of a capillary loop? _____

30. Materials are taken away from tissues by both veins and the
    _____ ducts.

31. If fluids accumulate in some tissue (e.g., the lower limbs) a con-
    dition of _____ exists.

32. Within the red blood cells, called _____ are mole-
    cules of _____, which carry oxygen from the
    lungs to tissues needing oxygen.

33. Each molecule of _____ in red blood cells has
    _____ subunits and each subunit can carry _____ molecule(s) of
    oxygen.

34. The reaction whereby oxygen is picked up may be symbolized as
    follows: HHb + O_2 $\rightleftharpoons$ _____ + _____

35. To help force the reaction from left to right, the system acts to

neutralize the _____ ion.

36. The ion largely responsible for doing this (question 35) has the formula _____, and the product of the neutralizing reaction subsequently breaks up into _____ and water.

37. If 95% of all the hemoglobin in a sample of blood is saturated with oxygen, then the _____ of the sample is 95%.

38. The first molecule of oxygen to bind to deoxygenated hemoglobin has an _____ effect that aids in bringing oxygen to the remaining oxygen-binding sites.

39. The symbol HbO_2^- is our symbol for _____, and that symbol is incorrect to the extent it does not show that each molecule of this substance carries _____ molecules of oxygen.

40. At active tissues needing oxygen, ions with the symbol _____ are generated which help HbO_2^- release oxygen. In other words, at a lower pH hemoglobin has a lower _____.

41. The equation for the formation of carbaminohemoglobin is:

$$\rightleftharpoons$$
_____

42. The equation of question 41 is one way that newly formed molecules of carbon dioxide are taken up by the bloodstream. The other chemical change to carbon dioxide that occurs when it enters a red blood cell has the equation (an equilibrium):

$$\rightleftharpoons \qquad \rightleftharpoons$$
_____

43. The equations (or equilibria) of questions 41 and 42 occur at actively metabolizing tissue, and their operation shifts what equilibrium involving oxygen release

$$\rightleftharpoons$$
_____

As you have written this equilibrium, how does it shift at actively metabolizing cells? _____
(left or right)

44. To recapitulate, the equilibrium of question 43 shifts to the
_____ at cells needing _____. They need it be-
cause they have done some chemical work that produces water and
_____ which reacts somewhat with the water to
give _____ according to the equilibrium of
question _____. It also reacts somewhat with hemoglobin
according to the equilibrium of question _____. These last two
reactions generate _____ ions that aid in
forcing the equilibrium of question 43 to the _____ and thereby
help in releasing _____.

45. The use of the hydrogen ions generated in the equilibria of
questions 41 and 42 to shift the equilibrium of 43 is called the
_____.

46. Bicarbonate ions travel back to the lungs in the serum, not inside
_____.

47. When bicarbonate ions leave the erythrocyte, _____
ions move in to replace them. This switch is called the
_____.

48. The combined action of H^+ and CO_2 generated at actively metabolizing cells serves to _____ the oxygen affinity
                         (raise or lower)
of _____.

49. A hemoprotein at muscles that has a higher oxygen affinity than
hemoglobin is called _____.

50. An acidosis brought on by an error in metabolism is called _____
_____.

51. An acidosis caused by a breakdown in the respiratory centers or by
some deterioration of the lungs is called _____.

52. Hyperventilation is a method used by a system to expel excess
_____, the loss of which should help
_____ (raise or lower)   the pH of the blood.

53. Prolonged vomiting may cause metabolic _____.

54. Shallow breathing, or _____, is the way the system tried to retain _____ which has the effect of retaining _____ acid in the carbonate buffer. This helps _____ the pH of the blood. Thus,
                         (raise or lower)
shallow breathing may be a response to metabolic
_____.
      (acidosis or alkalosis)

55. If shallow breathing occurs because the respiratory centers are not working, the individual cannot efficiently expel _____ _____ and may experience respiratory _____.

56. The hyperventilation of a patient in hysterics causes an over-removal of _____, a loss of the _____ buffer, and a _____ in the pH of the blood. These responses result in respiratory _____. Re-breathing exhaled air helps suppress this because it supplies more _____ to the lungs and thence to the bloodstream.

57. If air too enriched in oxygen is given to a patient for too long a period, that individual will have trouble getting _____ _____ out of actively metabolizing cells.

58. What organ(s), in effect, removes acid from the blood? _____ _____.

59. If the blood pressure drops greatly, the kidneys may not be able to _____ the blood because the blood flow through them is reduced.

60. The kidneys respond to a fall in blood pressure by secreting an enzyme called _____ into the blood which acts on a circulating proenzyme called _____. This pro-enzyme is changed to _____ which helps to generate _____ the most powerful _____ _____ known.

61. The endocrine gland called the _____ is somehow sensitive to the _____pressure of the blood. If this pressure goes up, it means that the concentration of

dissolved and dispersed substances in the blood is too _____
.   (high
_____. If this happens, the gland secretes the hormone
or low)
_____, whose target organs are
_____.

62. Secretion of this hormone (in question 61) eventually results in
the formation of a _____ volume of urine.
                        (lesser or greater)

63. The hormone that helps the bloodstream to conserve its sodium ions
is called _____.

MULTIPLE-CHOICE

1. The end products of the digestion of milk sugar are
   (a) glucose and galactose   (c) glucose and fructose
   (b) only glucose   (d) glucose and maltose

2. If gastric juice were completely devoid of its hydrochloric acid,
   this would impair the digestion in the stomach of
   (a) lipids   (c) proteins
   (b) carbohydrates   (d) chyme

3. Removal of the gall bladder would reduce the efficiency of the
   digestion of
   (a) lipids   (c) proteins
   (b) carbohydrates   (d) fatty acids

4. The enzyme that catalyzes the conversion of trypsinogen to trypsin
   is
   (a) amylase   (b) pepsin   (c) mucin   (d) enterokinase

5. The enzyme in saliva is
   (a) mucin   (b) pepsin   (c) α-amylase   (d) amylose

6. Bile contains
   (a) a lipase   (c) a carbohydrase
   (b) a protease   (d) no enzymes

7. Without bile salts, which substances would not be as easily absorbed from the intestinal tract into the bloodstream?

(a) fat-soluble vitamins   (c) glucose
(b) water-soluble vitamins   (d) amino acids

8. The most abundant cation in the blood is
   (a) Na   (b) K   (c) K^+ (d) Na^+

9. Any injury that causes a large number of cells to break open can
   lead to
   (a) hypernatremia   (c) hypermagnesemia
   (b) hyperkalemia   (d) hyperchloremia

10. If the intake of K^+ ion is high, the body spontaneously works to
    lose
    (a) Cl^- (b) Ca^{2+} (c) Na^+ (d) Mg^{2+}

11. The blood can become hypocalcemic in
    (a) vitamin D deficiency
    (b) a misfunctioning thyroid gland
    (c) overdoses of antacids based on $Ca(OH)_2$
    (d) blood tranfusions

12. The blood can become hypermagnesemic
    (a) in vitamin B_{12} deficiency
    (b) when it becomes hypercalcemic
    (c) if milk of magnesia is overused as a laxative
    (d) if the thyroid gland's activity becomes impaired

13. A normal value for the level of Cl^- in the blood is
    (a) 2 meq/L   (c) 35.5 mg/L
    (b) 100 meq/L   (d) 100 eq/L

14. If the level of Cl^- ion in blood drops, the body tends to retain

    (a) HCO_3^- (b) H^+ (c) Na^+ (d) H_2O

15. If the level of Cl^- in blood rises, one result can be
    (a) hypochloremia   (b) hyponatremia   (c) acidosis   (d) alkalosis

16. Of all of the anions in blood, about two-thirds are

    (a) HPO_4^{2-} (b) $H_2PO_4^-$ (c) SO_4^{2-} (d) Cl^-

17. Plasma and interstitial fluids are most unlike in their
    concentrations of

(a) electrolytes   (c) lipids
(b) proteins   (d) carbohydrates

18. Oxygen is transported in the bloodstream chiefly as
    (a) molecules of O_2 (c) oxyhemoglobin ions
    (b) hydronium ions   (d) methemoglobin ions

19. Hemoglobin in blood has a relatively high oxygen affinity in the
    tissue capillaries of those tissues where
    (a) the pH is dropping   (c) pCO_2 is rising
    (b) the pH is rising   (d) pCO_2 is low

20. Hemoglobin more easily accepts its second, third, and fourth
    molecules of oxygen than it does its first because the first has
    (a) an allosteric effect   (c) an isohydric effect
    (b) a Bohr effect   (d) a releasor effect

21. The equilibrium $HHB + O_2 \rightleftharpoons HbO_2^- + H^+$ will respond to a drop in
    pH by
    (a) shifting to the right   (c) remaining unchanged
    (b) shifting to the left   (d) absorbing more O_2

22. If the equilibria $CO_2 + H_2O \rightleftharpoons H_2CO_3 \rightleftharpoons H^+ + HCO_3^-$ shift to the
    right, then the equilibrium of question 21 will tend to
    (a) shift to the right   (c) remain unchanged
    (b) shift to the left   (d) absorb oxygen

23. The equilibrium $HHb + CO_2 \rightleftharpoons Hb - CO_2^- + H^+$ shifts to the right
    in a region where
    (a) the oxygen affinity of HHb is high
    (b) the pO_2 is relatively high
    (c) the pH is relatively high
    (d) the pCO_2 is relatively high

24. When the carbonate buffer system in blood acts, a hydrogen ion is
    replaced by a water molecule at the expense of a
    (a) hydroxide ion   (c) bicarbonate ion
    (b) hydronium ion   (d) calcium ion

25. Something that has a higher oxygen affinity under identical con-
    ditions than adult hemoglobin is
    (a) sickle cell hemoglobin   (c) fetal hemoglobin
    (b) myoglobin   (d) none of these

26. Healthy kidneys respond to acidosis by
    (a) putting H^+ into urine   (c) retaining Na^+
    (b) retaining HCO_3^- (d) removing ketone bodies

27. When respiratory centers are healthy, the body can respond to acidosis by
    (a) hyperventilation   (c) retaining CO_2
    (b) hypoventilation   (d)   retaining H^+

28. Hyperventilation aids in controlling acidosis by
    (a) removing CO_2 from blood
    (b) bringing in more O_2
    (c) promoting the chloride shift
    (d) increasing the pCO_2 of blood

29. In severe emphysema, acidosis may develop because
    (a) hyperventilation cannot be stopped
    (b) shallow breathing cannot be used
    (c) oxygen toxicity has become a problem
    (d) the removal of CO_2 from the blood at the lungs is impaired

30. If for any reason hemoglobin molecules leave the lungs not fully saturated with oxygen, the condition is called
    (a) oxygen affinity   (c) hypoxia
    (b) oxygen toxicity   (d) anoxia

31. If for any reason a tissue cannot get oxygen, the condition is called
    (a) oxygen affinity   (c) hypoxia
    (b) oxygen toxicity   (d) anoxia

32. The nitrogen wastes present in urine is (are)
    (a) urea   (c) uric acid (and the urate ion)
    (b) creatine   (d) ammonia

33. If the osmotic pressure of the blood rises by even as little as 2%,
    (a) the kidneys release renin
    (b) the hypophysis releases vasopressin
    (c) the adrenal cortex releases aldosterone
    (d) the liver releases fibrinogen

34. The hormone that helps to regulate the level of sodium ions in the

blood is
(a) angiotensin I   (c) renin
(b) vasopressin   (d) aldosterone

35. On the arterial side of a capillary loop, the blood pressure is
    (a) lower than on the venous side
    (b) equal to that on the venous side
    (c) lower than the osmotic pressure from the interstitial areas
    (d) higher than the osmotic pressure from the interstitial areas

ANSWERS

ANSWERS TO SELF-TESTING QUESTIONS

Completion
 1. gastric juice
 2. zymogens
 3. pepsin, pepsinogen
 4. chyme
 5. mouth, amylase, saliva
 6. hydrolysis (or hydrolytic)
 7. bile salts, detergents (or soaps or steroid-based detergents),
    upper intestine (or duodenum), gall bladder
 8. enterokinase
 9. chymotrypsinogen, procarboxypeptidase
10. intestinal juice and pancreatic juice
11. bile
12. pancreatic lipase
13. amino acids
    glucose
    glucose, fructose, galactose
    fatty acids and glycerol

14. Na^+; 135 to 145 meq/L; hypernatremia; hyponatremia

15. K^+; 125 meq/L

16. 3.5 meq/L; hypokalemia; 5.0 meq/L; hyperkalemia

17. burns, crushing injuries, and heart attacks

18. hyper-; hypo-

19. Mg^{2+}; Ca^{2+}

20. Mg^{2+}

21. Ca^{2+}
22. contraction
23. hyper-
24. D
25. hypercalcemia
26. enzymes
27. blood
28. from the interstitial compartment into the bloodstream
29. simple blood pressure (from the pumping action of the heart)
30. lymph
31. edema
32. erythrocytes, hemoglobin
33. hemoglobin, four, one
34. $HbO_2^- + H^+$
35. hydrogen
36. HCO_3^-, carbon dioxide
37. oxygen affinity
38. allosteric
39. oxyhemoglobin, four

40. H^+ (H_3O^+ is better), oxygen affinity

41. $HHb + CO_2 \rightleftharpoons Hb - CO_2^- + H^+$

42. $CO_2 + H_2O \rightleftharpoons H_2CO_3 \rightleftharpoons HCO_3^- + H^+$

43. $HHb + O_2 \rightleftharpoons HbO_2^- + H^+$
    to the left (or right to left)
44. left, oxygen; carbon dioxide, carbonic acid, 42; 41; hydrogen
    (H^+), left, oxygen
45. isohydric shift
46. red blood cells (or erythrocytes)
47. chloride (Cl^-); chloride shift
48. lower, hemoglobin
49. myoglobin
50. metabolic acidosis
51. respiratory acidosis
52. carbon dioxide, raise
53. alkalosis
54. hypoventilation, carbon dioxide, carbonic; lower; alkalosis
55. carbon doxide, acidosis
56. carbon dioxide, carbonate, rise; alkalosis; carbon dioxide
57. carbon dioxide
58. the kidneys
59. filter or cleanse

60. renin, angiotensinogen; angiotensin I, angiotensin II, vasoconstrictor
61. hypophysis (or pituitary), osmotic; high; vasopressin, the kidneys
62. lesser than usual
63. aldosterone

Multiple-Choice

1. a	13. b	25. b and c
2. c	14. a	26. a, b, c, and d
3. a	15. c	27. a
4. d	16. d	28. a
5. c	17. b	29. d
6. d	18. c	30. c
7. a	19. b and d	31. d
8. d	20. a	32. a, b, c, and d
9. b	21. b	33. b
10. c	22. b	34. d
11. a, b	23. d	35. d
12. c	24. c	

26 Biochemical Energetics

OBJECTIVES

After you have studied this chapter and worked the Exercises in it, you should be able to do the following.

1. Name the two products of digestion that are most frequently used for chemical energy in living systems.
2. Name the end products of the complete catabolism of carbohydrates and simple lipids.
3. Write the structures of the triphosphate and diphosphate networks.
4. Explain the basis for classifying organophosphates as high or low energy.
5. Name the principal triphosphate used as an immediate source of chemical energy in cells.
6. Outline (by a flow sheet) the principal pathways in biochemical energetics between products of digestion and the synthesis of ATP.
7. Give the symbols for the chief hydride-accepting coenzymes and write equations (in two ways) illustrating their activity.
8. In general terms, explain what kinds of gradients are forced into existence by the operation of the respiratory chain, and where they occur.
9. In general terms, describe the connections between these gradients and the synthesis of ATP (using the chemiosmotic theory).
10. Give the general purpose of the citric acid cycle (Kreb's Cycle).
11. Describe the place acetyl coenzyme A has in biochemical energetics.

12. Give the ATP yield per glucose molecule.
13. Define the terms in the Glossary.

GLOSSARY

Acetyl Coenzyme A. The molecule from which acetyl groups are transferred into the citric acid cycle or into the lipigenesis cycle.

$$CH_3C \overset{\overset{\displaystyle O}{\|}}{} - S - CoA$$

Adenosine Diphosphate (ADP). A high-energy diphosphate ester obtained
  from adenosine triphosphate (ATP) when part of the chemical energy
  in ATP is tapped for some purpose in a cell.

Adenosine Monophosphate (AMP). A low-energy phosphate ester that can
  be obtained by the hydrolysis of ATP or ADP; a monomer for the
  biosynthesis of nucleic acids.

Adenosine Triphosphate (ATP). A high-energy triphosphate ester used
  in living systems to provide chemical energy for metabolic needs.

Aerobic Sequence. An oxygen-consuming sequence of catabolism that
  starts with glucose or with glucose units in glycogen, and
  proceeds through glycolysis, the citric acid cycle, and the
  respiratory chain.

Anaerobic Sequence. The oxygen-independent catabolism of glucose or
  of glucose units in glycogen to lactate ion.

Chemiosmotic Theory. An explanation of how oxidative phosphorylation
  is related to a flow of protons in a proton gradient that is
  established by the respiratory chain, and that extends across the
  inner membrane of a mitochondrion.

Citric Acid Cycle. A series of reactions that dismantle acetyl units
  and send electrons (and protons) into the respiratory chain; a
  major source of metabolites for the respiratory chain.

Fatty Acid Cycle. The catabolism of a fatty acid by a series of repeating steps that produce acetyl units (in acetyl CoA).

Glycolysis. A series of chemical reactions that break down glucose or glucose units in glycogen until pyruvate remains (when the series is operated aerobically) or lactate forms (when the conditions are anaerobic).

High-Energy Phosphate. An organophosphate with a phosphate group transfer potential equal to or higher than that of ADP or ATP.

Oxidative Phosphorylation. The synthesis of high energy phosphates such as ATP from lower energy phosphates and inorganic phosphate by the reactions that involve the respiratory chain.

Phosphate Group Transfer Potential. The relative ability of an organophosphate to transfer a phosphate group to some acceptor.

Respiratory Chain. The reactions that transfer electrons from the intermediates made by other pathways to oxygen; the mechanism that creates a proton gradient across the inner membrane of a mitochondrion and that leads to ATP-synthesis; the enzymes that handle these reactions.

Respiratory Enzymes. The enzymes of the respiratory chain.

Substrate Phosphorylation. The direct transfer of a phosphate unit from an organophosphate to a receptor molecule.

SELF-TESTING QUESTIONS

COMPLETION

1. When glucose is burned in air, the elements present in the glucose emerge in molecules of _____ and _____.

2. When glucose is carried through both the anaerobic and aerobic sequences of metabolism in the body, its elements emerge in molecules of _____ and _____.

3. The energy produced by the combustion of glucose appears largely as _____.

4. Some of the energy produced by the breakdown of glucose in the

body appears as _____, but of greater
importance to the body, it also appears in the form of molecules
of _____.

5. The structure of ATP in its electrically neutral, un-ionized form
   is (complete the following):

   (adenosine)—

6. When adenosine triphosphate reacts, provides energy for some
   chemical change, and loses one phosphate, the remainder of its
   molecule is called _____ and has the symbol
   _____.

7. The symbol, P_i, stands for a mixture having (give the formulas)
   _____ and _____ as its two prin-
   cipal ions. The exact proportion of these ions in the mixture is
   largely a function of the _____ of the medium.

8. If structure II in the following reaction has a higher phosphate
   group transfer potential than structure I, where should the
   arrowhead be placed?

   $$R - O - PO_3^{2-} \;+\; R'OH \;\underline{\quad\quad}\; ROH \;+\; R' - O - PO_3^{2-}$$

   I   II

9. When proteins of a relaxed muscle interact with ATP, the muscle
   contracts as _____ and _____ are
   produced.

10. To put these two substances (question 9) back together again, the
    body taps energy from the _____, a series
    of reactions whereby electrons are taken from some metabolite of
    the _____ cycle and are delivered to _____
    _____ to produce water.

11. To fuel the _____ cycle, the body uses acetyl _____.

12. The name of the high energy phosphate in muscles that can quickly
    remake ATP from ADP is _____.

13. The chief location in the cell for the synthesis of ATP is the
_____ of a mitochondrion. When the respiratory
chain operates, the net effect is to change MH_2 and half a
molecule of oxygen to M and _____ and to generate a
gradient of _____ ions as well as a gradient of
_____ across the _____ of the mitochondrion.

14. As _____ ions move across the inner membrane of the mitochondrion at those places where such movement can occur, other
changes are initiated that lead to the synthesis of
_____ from _____ and P_i.

15. Our symbol for a donor of H:⁻ units to the respiratory chain is
_____, and the metabolic pathway that is the richest sup-
plier of them is called the _____.

16. The first receptor of H:⁻ in the respiratory chain is an enzyme
whose coenzyme has the short symbol of _____; the symbol for its
reduced form is _____. This reduced form can pass H:⁻ to
another enzyme with a coenzyme having the short symbol of _____
and whose reduced form is _____.

17. When $FMNH_2$ passes H:⁻ to the next enzyme in the chain having the
short symbol _____, only the electrons in H:⁻ go to this next
enzyme and the hydrogen nuclei, H^+, are put _____

_____.

In this way, the first "installment" of the two gradients, the
_____ gradient and the _____ gradient that extends from
relatively high concentrations of these species on the _____
(outside or
_____ of the mitochondrial membrane to lower concentrations on
inside)
_____ of this membrane.
(outside or inside)

18. Following the iron-sulfur enzyme and the enzyme containing coenzyme Q is a series of enzymes collectively called the
_____. The final acceptor of the electrons being trans-

ferred from one of these enzymes to the next is _____,
and water forms.

19. The respiratory chain has a "branch," and it involves one acceptor
of $H:^-$ from certain kinds of metabolites. The coenzyme for this
acceptor has the short symbol of _____, and its reduced form is
symbolized as _____. The latter can be reoxidized by
interacting with an enzyme of the main respiratory chain with the
coenzyme having the symbol _____.

20. When MH_2 is oxidized by an enzyme containing NAD^+, the whole
process eventually puts _____ H^+ ions from the inside to the
outside of the inner mitrochondrial membrane. This gradient can
drive the formation of a maximum of _____ molecules of ATP from
ADP and P_i. When a metabolite is oxidized by the enzyme at the
branch in the respiratory chain whose coenzyme is FAD, then the
whole process puts _____ H^+ ions into the gradient, which can
drive the formation of a maximum of _____ molecules of ATP.

21. The organic group that is joined to coenzyme A has the structure:
_____, and the symbol we use for the combination of this
with coenzyme A is _____.

22. The metabolic sequence that degrades acetyl coenzyme A is called
the _____. Certain intermediates in this
sequence are _____, and agents causing this kind
                    (oxidized or reduced)
of chemical change are enzymes of another sequence called the
_____.

MULTIPLE-CHOICE

1. Of the following substances, which has the most potential for gen-
erating molecules of ATP?
(a) glucose   (c) glycine
(b) $CH_3(CH_2)_{16}CO_2H$ (d) oxygen

2. Among the substances essential to the aerobic sequence of glucose
catabolism is (are)

(a) NAD^+ (b) O_2 (c) ATP   (d) ADP

3. The product of the operation of the respiratory chain that is most
needed by the body is

(a) ATP
(b) H_2O

(c) $[H:^- + H^+]$
(d) FAD

4. Among the accomplishments of the citric acid cycle is (are)
   (a) the synthesis of active acetyl
   (b) supplying $H:^-$ and H^+ to the respiratory chain
   (c) glycogenesis
   (d) glycolysis

5. To help get glucose inside cells of certain tissues, the blood
   should carry
   (a) epinephrine   (b) insulin   (c) cyclic AMP   (d) NAD

6. If a reaction is symbolized by

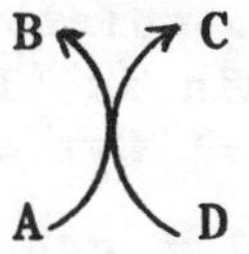

   $$\begin{matrix} B & & C \\ & \diagdown\diagup & \\ A & & D \end{matrix}$$

   it could be rewritten as

   (a) A + B $\longrightarrow$ C + D
   (b) B + C $\longrightarrow$ A + D
   (c) A + C $\longrightarrow$ B + D
   (d) A + D $\longrightarrow$ B + C

7. To complete the following reaction, we should include as a product

   $$ATP + H_2O \longrightarrow \underline{\hspace{4cm}} + PP_i$$

   (a) ADP

   (b) AMP

   (c) H_2O_2

   (d) $CO_2 + H_2O$

8. The chief use of acetyl CoA is to provide "fuel" for
   (a) the respiratory chain   (c) the anaerobic sequence
   (b) glycolysis   (d) homeostasis

9. The chief purpose of the respiratory chain is to
   (a) accept $H:^-$ (c) resupply ATP
   (b) use up acetyl CoA   (d) use O_2

10. The complete catabolism of glucose gives a maximum ATP yield per
    molecule of glucose of
    (a) 15   (b) 18   (c) 36   (d) 38

11. Between the respiratory chain and glycolysis in the catabolism of glucose occurs the
    (a) anaerobic sequence
    (b) oxidative phosphorylation
    (c) the citric acid cycle
    (d) homeostasis

12. The operation of the respiratory chain establishes two gradients within a mitochondrion. These involve specifically

    (a) OH^- and Cl^- (c) NAD^+ and $NADH_2$

    (b) H^+ and (+) charge   (d) electron pairs and ATP

13. One of the (incomplete) sequences of enzymes or coenzymes in the respiratory chain is (in the correct order)
    (a) FMN $\rightarrow$ FAD $\rightarrow$ Cyt c $\rightarrow$ Cyt a

    (b) FAD $\rightarrow$ NAD^+ $\rightarrow$ FeS $-$ P $\rightarrow$ Q

    (c) Q $\rightarrow$ Cyt c_1 $\rightarrow$ Cyt b $\rightarrow$ FeS $-$ P

    (d) NAD^+ $\rightarrow$ FMN $\rightarrow$ Q $\rightarrow$ Cyt a

14. The symbol FMN stands for
    (a) an enzyme
    (b) the reduced form of a flavin enzyme
    (c) the oxidized form of a flavoprotein
    (d) a coenzyme

15. The chemiosmotic theory was proposed by
    (a) Hans Krebe   (c) Charles MacMunn
    (b) Peter Mitchell   (d) David Keilin

16. The oxidation of MH_2 to M and H_2O that begins with the enzyme bearing the coenzyme NAD^+ in the respiratory chain can generate a maximum of how many molecules of ATP?
    (a) 2   (b) 3   (c) 4   (d) 6

17. One important source of acetyl CoA is

    (a) pyruvate + CoA$-$SH + NAD^+
    (b) citrate
    (c) the respiratory chain
    (d) oxidative phosphorylation

18. When glucose is catabolized to CO_2 and H_2O, oxygen atoms from O_2 in the air end up as parts of molecules of
    (a) CO_2 (b) H_2O (c) ATP   (d) P_i

19. Which of the following types of reactions occur in the citric acid cycle?
    (a) dehydration
    (b) hydrogenation
    (c) dehydrogenation
    (d) addition of water to a double bond

20. When acetyl CoA adds to the keto group of oxaloacetate, the product is
    (a) pyruvate   (b) ATP   (c) lactate   (d) citrate

ANSWERS

ANSWERS TO SELF-TESTING QUESTIONS

Completion
1. carbon dioxide, water
2. carbon dioxide, water
3. heat
4. heat, ATP
5.

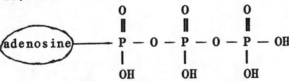

6. adenosine diphosphate, ADP
7. $H_2PO_4^-$, HPO_4^{2-};   pH
8. Place the arrow at the left end of the line
9. ADP, P_i
10. respiratory chain, citric acid, oxygen
11. citric acid cycle, CoA
12. creatine phosphate
13. inner membrane; H_2O; H^+; (+) charge; inner membrane
14. H^+; ATP; ADP
15. MH_2; citric acid cycle
16. NAD^+; NADH; FMN; $FMNH_2$
17. FeS — P; to the outside of the inner membrane of the mitochon-

drion; H^+ ion; (+) charge; outside; inner; inside
18. cytochromes; oxygen
19. FAD; $FADH_2$; Q (or coenzyme Q)

20. 6; 3; 4; 2

21. $CH_3 - \overset{\displaystyle O}{\overset{\|}{C}} - $; $CH_3 - \overset{\displaystyle O}{\overset{\|}{C}} - S - CoA$

22. citric acid cycle; oxidized; respiratory chain

Multiple-Choice

1. b	11. c
2. a, b, and d	12. b
3. a	13. d
4. b	14. d
5. b	15. b
6. d	16. b
7. b	17. a
8. a	18. b
9. c	19. a, c, and d
10. d	20. d

27 Metabolism of Carbohydrates

OBJECTIVES

This chapter continues our study of the molecular basis of energy for living and introduces the molecular basis of a very wide-spread disease – diabetes. After you have studied this chapter and worked the Review Exercises in it, you should be able to do the following.

1. Discuss the factors that influence the blood sugar level, using the technical terms for high and low values.
2. List some of the effects of hypoglycemia.
3. Describe the glucose tolerance test and its purpose.
4. Describe the influences that the hormones epinephrine, glucagon, insulin, and somatostatin have on glucose metabolism.
5. Describe in general terms what diabetes mellitus is.
6. Outline the Cori Cycle.
7. Write an overall equation for glycolysis – glucose to lactate.
8. Explain the importance of glycolysis.
9. Explain why anaerobic glycolysis ends at lactate, not pyruvate.
10. Give the general purpose of the pentose phosphate pathway of glucose catabolism.
11. Describe in general terms what the liver can do if the brain's chief nutrient is not available in the diet.
12. Define all of the terms in the Glossary.

GLOSSARY

Anaerobic Sequence. The oxygen–independent catabolism of glucose or of glucose units in glycogen to lactate ion.

Blood Sugar Level. The concentration of carbohydrate (mostly glucose) in the blood; usually stated in units of mg/dL.

Cori Cycle. The sequence of chemical events and transfers of substances in the body that describes the distribution, storage, and mobilization of blood sugar, including the reconversion of lactate to glycogen.

Diabetes Mellitus. A disease in which there is an insufficiency of effective insulin and an impairment of glucose tolerance.

Epinephrine. A hormone of the adrenal medulla that activates the enzymes needed to release glucose from glycogen.

Glucagon. A hormone, secreted by the α–cells of the pancreas in response to a decrease in the blood sugar level, that stimulates the liver to release glucose from its glycogen stores.

Gluconeogenesis. The synthesis of glucose from compounds with smaller molecules or ions.

Glucose Tolerance. The ability of the body to manage the intake of dietary glucose while keeping the blood sugar level from fluctuating widely.

Glucose Tolerance Test. A series of measurements of the blood sugar level after the ingestion of a considerable amount of glucose; used to obtain information about an individual's glucose tolerance.

Glucosuria. The presence of glucose in urine.

Glycogenesis. The synthesis of glycogen.

Glycogenolysis. The breakdown of glycogen to glucose.

Glycolysis. A series of chemical reactions that break down glucose or glucose units in glycogen until pyruvate remains (when the series is operated aerobically) or lactate forms (when the conditions are anaerobic).

Human Growth Hormone. One of the hormones that affects the blood sugar level; a stimulator of the relese of the hormone glucagon.

Hyperglycemia. An elevated level of sugar in the blood – above 95 mg/dL in whole blood.

Hypoglycemia. A low level of glucose in blood – below 65 mg/dL of whole blood.

Insulin. A protein hormone made by the pancreas, released in response to a rise in the blood sugar level, and used by certain tissues to help them take up glucose from circulation.

Normal Fasting Level. The normal concentration of something in the blood, such as blood sugar, after about four hours without food.

Oxygen Debt. The condition in a tissue when anaerobic glycolysis has operated and lactate has been excessively produced.

Pentose Phosphate Pathway. The synthesis of NADPH that uses chemical energy in glucose 6-phosphate and that involves pentoses as intermediates.

Renal Threshold. That concentration of a substance in blood above which it appears in the urine.

Somatostatin. A hormone of the hypothalamus that inhibits or slows the release of glucagon and insulin from the pancreas.

SELF-TESTING QUESTIONS

COMPLETION

1. The synthesis of glucose from starting materials that do not include any mono-, di-, or polysaccharides is called _____.

2. The synthesis of glycogen may be called _____.

3. The release of glucose from glycogen is called _____.

4. A hormone that can help release glucose from glycogen in a hurry is _____.

5. Events involved in the distribution, storage, mobilization, and usage of glucose may be summarized by the _____ Cycle.

6. If the _____ rises above normal, a condition of hyperglycemia is said to exist.

7. The appearance of glucose in urine is called _____.

8. Hyperinsulinism or starvation may produce a condition of _____ —emia.

9. Clinically, an insufficiency of effective insulin is associated with _____.

10. If the kidneys are releasing glucose into the urine, the _____ for glucose has probably been exceeded.

11. In a situation of sudden stress, the adrenal medulla secretes a trace of the hormone _____. Its target cells are in _____ and to some extent in _____. At these cells it launches a series of enzyme activations that lead finally to the enzyme _____ that catalyzes the _____.

12. Because liver cells have an enzyme that catalyzes the hydrolysis of _____ to glucose, the glucose stored as _____ in liver can be made available for release into the _____ and thereby help to raise the _____.

13. A hormone that is an even better activator of glycogenolysis at the liver than epinephrine is called _____.

14. The hormone secreted by the pancreas in response to a rise in blood sugar level is _____, and it stimulates removal of _____ from the bloodstream.

15. The "signal" for the release of insulin is a rise in _____.

16. The ability of the pancreas to respond to glucose is measured by the _____ test.

17. The effect of reduced insulin activity while glucagon activity remains normal is to reduce the supply of _____ inside cells. In this situation, the cells increase their rate of _____.

18. A sequence in glucose catabolism that can be run anaerobially is also called _____. Its end product is _____ which (when there is enough oxygen available) can lose the pieces of the element _____ and be thereby oxidized to _____. This, in turn, can be broken down oxidatively to acetyl _____.

19. The anaerobic sequence enables a tissue to make a fresh supply of _____ for energy-demanding processes without oxygen being immediately available.

20. When a tissue is operating anaerobically, we say it is running an oxygen _____, and an accumulation of _____, an end-product of _____, occurs.

21. When a tissue is operating aerobically, the change of an NADH-enzyme in glycolysis to its NAD^+-enzyme form is accomplished by passing the $H:^-$ from the NADH to _____; but when the tissue is operating anaerobically, the $H:^-$ (plus H^+) is transferred to _____ to make _____.

22. The $H:^-$ unit in NADH is used eventually to reduce _____, whereas the $H:^-$ in _____ is used to reduce organic substances that later become fatty acids.

23. The body's chief source of NADPH is the _____ _____.

24. A large fraction of the lactate produced in the anaerobic sequence is converted to _____. The energy for accomplishing this is provided by sending the smaller fraction of the lactate (or some other metabolite) through the

_____ sequence.

MULTIPLE-CHOICE

1. The end product of glycolysis under anaerobic conditions is
   - (a) phosphoenolpyruvate
   - (b) pyruvate
   - (c) lactate
   - (d) acetyl CoA

2. The end product of glycolysis under aerobic conditions is
   - (a) phosphoenolpyruvate
   - (b) pyruvate
   - (c) lactate
   - (d) acetyl CoA

3. All dietary monosaccharides undergo anaerobic catabolism in which the steps are for the most part the same as the steps in the catabolism of
   - (a) glucose
   - (b) acetyl CoA
   - (c) insulin
   - (d) pyruvate

4. The principal reducing agent needed to make fatty acids from carbohydrates is
   - (a) NADH   (b) NADPH   (c) NAD^+ (d) $NADP^+$

5. If lactate is generated by hard work,
   - (a) glycolysis has occurred
   - (b) an oxygen debt exists
   - (c) a fraction is eventually converted back to glycogen
   - (d) the blood will become hyperglycemic

6. One important purpose of the pentose phosphate pathway of glucose catabolism is to make
   - (a) ATP   (b) NADPH   (c) NADH   (d) glucose 1-phosphate

7. A hormone made in the α-cells of the pancreas that is a powerful activator of adenylate cyclase at liver cells is
   - (a) insulin   (b) epinephrine   (c) somatostatin   (d) glucagon

8. A hormone whose target tissue is the pancreas and that acts to slow down the release of insulin from the pancreas is
   - (a) glucagon
   - (b) somatostatin
   - (c) epinephrine
   - (d) human growth hormone

9. The carbon atoms in glucose most directly made by gluconeogenesis come from

(a) HCO_3^- (b) lactate   (c) acetyl coenzyme A   (d) CO_2

10. The breakdown of glucose to lactate is called
    (a) gluconeogenesis
    (b) lactolysis
    (c) glycolysis
    (d) glycogenolysis

11. The principal site of gluconeogenesis is the
    (a) muscle   (b) liver   (c) pancreas   (d) adrenal medulla

12. The series of changes that move from glucose to lactate and back to glucose is called the
    (a) Cori Cycle   (c) Krebs' Cycle
    (b) citric acid cycle   (d) tricarboxylic acid cycle

13. The concentration of monosaccharides in blood is called
    (a) diabetes mellitus   (c) the blood sugar level
    (b) the renal threshold   (d) the normal fasting level

14. Hormones that tend to raise the blood sugar level include
    (a) insulin   (c) glucagon
    (b) epinephrine   (d) somatostatin

15. If too much insulin somehow appears in the bloodstream,
    (a) glucosuria will result
    (b) the blood will become hypoglycemic
    (c) the blood will become hyperglycemic
    (d) the renal threshold for glucose will be exceeded

ANSWERS

ANSWERS TO SELF-TESTING QUESTIONS

Completion
  1. gluconeogenesis
  2. glycogenesis
  3. glycogenolysis
  4. epinephrine
  5. Cori
  6. blood sugar level
  7. glucosuria

8. hypoglycemia
9. diabetes mellitus
10. renal threshold
11. epinephrine; muscles; liver; phosphorylase; conversion of glyocgen into glucose 1-phosphate
12. glucose 6-phosphate; glycogen; bloodstream; blood sugar level
13. glucagon
14. insulin; glucose
15. the blood sugar level
16. glucose tolerance
17. glucose; gluconeogenesis
18. glycolysis; lactate, hydrogen, pyruvate; coenzyme A
19. ATP
20. debt, lactate, glycolysis
21. the respiratory chain (or, to FMN in the respiratory chain); pyruvate, lactate
22. oxygen, NADP
23. pentose phosphate pathway of glucose catabolism
24. glucose; aerobic

Multiple-Choice

1. c
2. b
3. a
4. b
5. a, b, and c
6. b
7. d
8. b

9. b
10. c
11. b
12. a
13. c
14. b and c
15. b

28 Metabolism of Lipids

OBJECTIVES

Our study of the molecular basis of the energy required for living is essentially completed in this chapter, as is our study of the molecular basis of diabetes. When you have finished studying the chapter and have answered the Review Exercises, you should be able to do the following.

1. Describe how lipids are transported in the bloodstream.
2. State the function of chylomicrons.
3. Describe what occurs to chylomicrons in adipose tissue.
4. List the ways in which the liver can deal with cholesterol.
5. Name the ways in which cholesterol is used outside of the liver.
6. Describe the function of VLDL complexes.
7. Explain how VLDL changes to IDL.
8. State how IDL changes to LDL.
9. Explain the uses of LDL and how it becomes HDL.
10. Describe the function of HDL.
11. Explain the function of the liver receptors for IDL and LDL and how they help the body regulate the blood cholesterol level.
12. Contrast lipid reserves and carbohydrate reserves in man in terms of roughly how long they would last during a period of starvation.
13. Name the principal sources of lipids that are eventually stored in depot fat.
14. Explain the advantage of storing chemical energy in the form of lipids rather than in the form of carbohydrates.

377

15. Outline the steps in the mobilization of the energy reserves in adipose tissue.
16. Using palmitic acid as an example, describe the overall result of the degradation of such an acid by the fatty acid cycle.
17. Describe by equations how acetyl CoA is converted to butyryl ACP.
18. Describe by equations how acetyl CoA is used to lengthen the chain of butyryl ACP.
19. Name the ketone bodies and explain how they are produced in greater than normal amounts when effective insulin is missing.
20. Explain the relations between (a) ketonemia and ketonuria, (b) ketonemia and ketone breath, and (c) ketonemia and ketoacidosis.
21. Discuss the step-by-step progression of events from a condition in which effective insulin is lacking to the coma that results if the condition remains untreated.
22. Define the terms in the Glossary.

GLOSSARY

Beta Oxidation. The fatty acid cycle of catabolism.

Chylomicron. A microdroplet of lipid material with a trace amount of protein that carries lipids picked up from the digestive tract to adipose tissue and to the liver.

Fatty Acid Cycle. The catabolism of a fatty acid by a series of repeating steps that produce acetyl units (in acetyl CoA).

Ketoacidosis. The acidosis caused by untreated ketonemia.

Ketone Bodies. Acetoacetate, β-hydroxybutyrate, or their parent acids, and acetone.

Ketonemia. An elevated concentration of ketone bodies in the blood.

Ketonuria. An elevated concentration of ketone bodies in the urine.

Ketosis. The combination of ketonemia, ketonuria, and acetone breath.

Lipigenesis. The synthesis of fatty acids from two-carbon acetyl units.

Lipoprotein Complex. A combination of a lipid molecule with a protein molecule that serves as the vehicle for carrying the lipid in the bloodstream.

SELF-TESTING QUESTIONS

COMPLETION

1. After triacylglycerols have been completely digested, the products are a mixture of _____ and _____. As these migrate out of the digestive tract they are changed into _____ that then become incorporated into particles called _____, which are carried in the blood.

2. The principal storage site for triacylglycerols is _____ _____, but when they are in the bloodstream they are carried by complexes called _____.

3. In adipose tissue, chylomicrons unload some of their molecules of _____, and this leaves the chylomicron remnants richer in molecules of the more dense, nonsaponifiable lipid, _____.

4. The organ that has receptors that can recognize chylomicron remnants is the _____.

5. When these receptors are defective or absent, the individual is likely to have _____.

6. The following things can happen to cholesterol in the liver:

   _____

   _____

   _____

7. The lipoprotein complex that the liver makes to export cholesterol into circulation is symbolized as _____, which stands for _____.

8. Tissues that can remove triacylglycerols from VLDL are _____ and _____.

9. This removal transforms the VLDL into _____, which stands

for _____.

10. The IDL can be reabsorbed by the _____, but some experiences further loss of _____ so that the IDL changes into _____, which stands for _____ _____.

11. One function of IDL is to carry _____ to the liver.

12. Another function of IDL is to carry cholesterol to _____ tissue and _____ glands.

13. Left-over cholesterol is carried back to the liver as _____, which stands for _____.

14. Storing chemical energy as lipid rather than as wet glycogen or dissolved glucose is advantageous because lipids have a particularly low _____, which means a low quantity of _____ per each calorie stored. In a 70-kg adult male, there is roughly _____ of triacylglycerol, enough to last about _____ if it has to serve as the sole source of caloric needs.

15. When insulin is in circulation, the release of fatty acids from adipose tissue is _____. Otherwise the hor-
   (suppressed or activated)
   mones _____ and _____ can act to activate an enzyme called _____ that catalyzes the hydrolysis of triacylglycerols in adipose tissue so that fatty acids can be released.

16. When the acyl group of a fatty acid is catabolized inside a mito-chondrion, it is attached to a coenzyme called _____.
   The first step in the oxidation is the loss of _____ from the α- and β-carbon atoms to give a functional group adjacent to the carbonyl group having the name _____.
   In the next step a molecule of _____ adds to this functional group to give a β-_____acyl system. This is then oxidized in the third step to give the _____-group.

Finally a molecule of coenzyme A interacts with this compound to split out _____ and leave an acyl-CoA system having _____ fewer carbon atoms than it originally had. Then this shorter version is subjected to the same series of steps, and the process is called the _____ cycle or _____ oxidation.

17. The hydrogen removed from the acyl CoA in the fatty acid cycle to give the α,β-unsaturated acyl CoA derivative is accepted by an enzyme having as its coenzyme _____, whose reduced form is _____. This enzyme is part of the _____ _____.

18. The hydrogen removed in the oxidation of the 2° alcohol group in the third step of the fatty acid cycle is accepted by an enzyme having the coenzyme _____, whose reduced form is symbolized as _____.

19. The acetyl CoA units produced by the fatty acid cycle enter the metabolic pathway having the name _____.

20. If glycolysis produces acetyl CoA that the cell doesn't need for energy, this excess acetyl CoA can be made into _____ by a series of reactions called _____ that take place in the _____ of the cell.

21. In lipigenesis, one acetyl CoA molecule first combines with HCO_3^- to give _____, which has the structure:

_____

Another molecule of acetyl CoA is hooked to a unit called _____ of the synthase complex. The malonyl unit of malonyl CoA is joined to another part of the synthase and becomes malonyl ACP, where ACP stands for _____. Malonyl ACP now reacts with acetyl-E to give _____, which has the structure:

_____

22. The keto group of this compound is next _____ by NADPH + H$^+$, which changes the keto group into _____ _____.

23. In the next step this group is removed to leave the ACP derivative of an unsaturated acid with the structure:

    _____

24. This unsaturated acyl derivative of coenzyme A is now hydrogenated by the action of an enzyme with _____ as its coenzyme. We now have the ACP derivative of a simple fatty acid with the name _____.

25. The raw material for making the steroid nucleus is _____ _____. One principal end product in steroid synthesis is an alcohol called _____ from which such other steroids as the _____ salts and the _____ hormones are made.

26. One way that the body controls how much cholesterol it makes is by regulating the activity of an enzyme called _____. One inhibitor of this enzyme is _____, so if the diet is rich in this substance, the body itself makes little if any more.

27. An enhanced rate of fatty acid oxidation in the liver may produce excessive amounts of _____ in the blood, a condition known as _____. This will slowly lower the pH of the blood, a condition known as _____ _____.

28. Both starvation and untreated _____ can lead to acidosis.

29. Efforts by the body to eliminate ketone bodies and associated positive ions, nitrogen wastes, and excess glucose from the bloodstream mean that more _____ than usual is also removed.

30. If insufficient water is drunk per day, the blood may _____ _____, its circulation in the brain in

sufficient quantities may become more difficult, and some blood flow may be diverted from the _____ in order to supply the brain. This diversion only makes it more difficult for the body to eliminate wastes.

31. As the anions from ketonemia become part of the urine being made, the chief cation to leave with them has the formula _____. The loss of this cation is sometimes called " the loss of _____ _____," but in reality this ion is not a base. However, each of these cations that is lost represents the neutralization of one _____ ion, so the net effect is the loss of base within the system.

32. A simplified statement of one sequence of events in undetected (and therefore untreated) diabetes mellitus would be as follows. The excessive release of _____ from adipose tis-
$$\text{(a)}$$
sue and the increased rate of catabolism of fatty acids in the _____ lead to the production of _____
$$\text{(b)}\text{(c)}$$
at a rate faster than can normally be handled. As a result there is a slow rise in the concentration of _____
$$\text{(d)}$$
in the bloodstream (a condition called _____).
$$\text{(e)}$$
Because two of them are acids, the _____
$$\text{(f)}$$
of the blood slowly drops.

MULTIPLE–CHOICE

1. The fatty acid cycle
   (a) produces lactic acid
   (b) requires the β-form of oxygen
   (c) involves extensive glycogenolysis
   (d) produces units of acetyl CoA

2. If the amount of energy taken into a healthy body in the form of carbohydrates is greater than the body needs for energy, the body will experience
   (a) glucosuria
   (b) a greater rate of running the lipigenesis cycle
   (c) a greater rate of running the fatty acid cycle

(d) an enhanced rate of glycogenolysis

3. The citric acid cycle is fed two carbon units from
   (a) the catabolism of glucose   (c) the lipigenesis cycle
   (b) the fatty acid cycle   (d) the Cori Cycle

4. The average, adequately nourished adult male has enough chemical
   energy in storage as lipids to sustain life for how long:
   (a) 1 day   (b) 1 week   (c) 1 month   (d) 2 months

5. Fatty acids are transported in the bloodstream bound as complexes
   to molecules of
   (a) protein   (b) triacylglycerol   (c) FFA   (d) cholesterol

6. The densities of the various lipoprotein complexes increase in the
   order:
   (a) chylomicron < LDL < VLDL < IDL < HDL
   (b) chylomicron < VLDL < IDL < LDL < HDL
   (c) VLDL < chylomicron
   (d) LDL < chylomicron < VLDL < IDL < HDL

7. The chief carriers of triacylglycerols to adipose tissue are
   (a) chylomicrons   (b) HDL   (c) LDL   (d) IDL

8. The chief carrier of cholesterol from peripheral tissue to the
   liver is
   (a) LDL   (b) IDL   (c) chylomicron   (d) HDL

9. During each "turn," the fatty acid cycle produces

   (a) $FADH_2$, NADH, acetyl CoA

   (b) $FADH_2$, NADH, acetoacetyl CoA

   (c) $FADH_2$, NADP, ADP

   (d) $FADH_2$, NADH, β-hydroxybutyrate

$$\overset{\displaystyle O}{\underset{\displaystyle \|}{}}$$

10. Once the substance $R - C - S - CoA$ has been formed, the next
    step in the fatty acid cycle is
    (a) the addition of water   (c) dehydration
    (b) an attack by CoASH   (d) dehydrogenation

11. Once the substance
$$R - CH=CH - \overset{\overset{\text{O}}{\|}}{C} - S - CoA$$
has formed in the fatty acid cycle, the next step is
(a) the addition of water   (c) dehydration
(b) an attack by CoASH   (d) dehydrogenation

12. The fatty acid cycle degrades fatty acids by how many carbons per "turn"?
(a) 1   (b) 2   (c) 3   (d) 4

13. To make the butyryl-ACP needed for the lipigenesis cycle, acetoacetyl-ACP is first made from
(a) acetoacetic acid
(b) two molecules of acetyl-ACP
(c) butyric acid
(d) malonyl-ACP and acetyl-E

14. A fat-free, high carbohydrate diet promotes
(a) the lipigenesis cycle
(b) beta oxidation
(c) the fatty acid cycle
(d) lipolysis

15. Which one of these structures is not a ketone body?

(a) $CH_3\overset{\overset{\text{O}}{\|}}{C}CH_2\overset{\overset{\text{O}}{\|}}{C}O^-$ (c) $CH_3\overset{\overset{\text{O}}{\|}}{C}CH_3$

(b) $CH_3\overset{\overset{\text{O}}{\|}}{C}O^-$ (d) $CH_3\overset{\overset{\text{HO}}{|}}{C}HCH_2\overset{\overset{\text{O}}{\|}}{C}O^-$

16. Which one of the conditions given is most closely linked to acidosis?
(a) acetone breath   (c) ketonuria
(b) ketonemia   (d) glycosuria

17. The uses of acetyl CoA in the body include
(a) the synthesis of fatty acids
(b) the synthesis of cholesterol
(c) the synthesis of certain amino acids
(d) fuel for the citric acid cycle

ANSWERS

ANSWERS TO SELF-TESTING QUESTIONS

Completion

1. fatty acids; glycerol; triacylglycerols; chylomicrons
2. adipose tissue; lipoprotein complexes
3. triacylglycerols; cholesterol
4. liver
5. hypercholesterolemia
6. excretion via the bile
   conversion to bile salts
   export via the blood to other tissues
7. VLDL; very low density lipoprotein complex
8. adipose tissue; muscle tissue
9. IDL; intermediate density lipoprotein complex
10. liver; triacylglycerol; LDL; low density lipoprotein complex
11. cholesterol
12. peripheral; adrenal
13. HDL; high density lipoprotein complexes
14. energy density; grams; 12 kg; 43 days
15. suppressed; epinephrine and glucagon; lipase
16. coenzyme A; hydrogen; alkene group; water; hydroxy; keto; acetyl CoA; two; fatty acid; beta
17. FAD; $FADH_2$; respiratory chain
18. NAD^+, NADH
19. citric acid cycle
20. fatty acids; lipigenesis; cytosol
21. malonyl CoA; $^-OCCH_2C$ — S — CoA (with two C=O groups); E; acyl carrier protein;

    acetoacetyl ACP; CH_3CCH_2C — S — ACP (with two C=O groups)
22. reduced (or hydrogenated); a $2°$ alcohol group
23. $CH_3CH=CHC$ — S — ACP (with C=O group)
24. NADPH; butyryl ACP
25. acetyl CoA; cholesterol; bile; sex
26. HMG–CoA reductase; cholesterol
27. ketone bodies; ketonemia; acidosis (or ketoacidosis)

28. diabetes
29. water
30. thicken (or become more viscous); kidneys

31. Na^+; base; H^+

32. (a) fatty acids
    (b) liver
    (c) ketone bodies
    (d) hydrogen ions (or, hydronium ions)
    (e) acidosis (or, in this case, ketoacidosis)
    (f) pH

Multiple-Choice

1. d	9. a
2. b	10. d
3. a and b	11. a
4. c	12. b
5. a	13. d
6. b	14. a
7. a	15. b
8. d	16. b
	17. a, b, c, and d

29 Metabolism of Nitrogen Compounds

OBJECTIVES

Although the catabolism of amino acids inevitably helps generate ATP and therefore contributes to the molecular basis of energy for living, this chapter brings us back to the molecular basis of materials for living.

After you have studied the material in this chapter and have answered the Review Exercises, you should be able to do the following.

1. List the four main fates of amino acids in the body.
2. Describe the nitrogen pool.
3. By means of illustrative equations, show how reductive amination and transamination contribute to the synthesis of some amino acids.
4. Name the principal end products of the catabolism of amino acids.
5. Write an equation illustrating a specific example for each of the catabolic reactions of oxidative deamination, direct deamination, and decarboxylation.
6. Describe the overall result of the urea cycle.
7. Describe an origin of hyperammonemia and state the principal problem associated with it.
8. Describe the overall result of the catabolism of the purine bases (A or G).
9. In general terms, explain how gout and kidney stones are related to purine metabolism.

10. Give the relations between: (a) heme and biliverdin, (b) biliverdin and bilirubin, (c) bilirubin and bilinogen, and (d) bilirubin and jaundice.
11. Define each of the terms in the Glossary.

GLOSSARY

Bile Pigment. Colored products of the partial catabolism of heme that are transferred from the liver to the gall bladder for secretion via the bile.

Bilin. The brownish pigment that is the end product of the catabolism of heme and that contributes to the characteristic colors of feces and urine.

Bilinogen. A product of the catabolism of heme that contributes to the characteristic colors of feces and urine and some of which is oxidized to bilin.

Bilirubin. A reddish-orange substance that forms from biliverdin during the catabolism of heme and which enters the intestinal tract via the bile and is eventually changed into bilinogen and bilin.

Biliverdin. A greenish pigment that forms when partly catabolized hemoglobin (as verdohemoglobin) is further broken down, and which is changed in the liver to bilirubin.

Deamination. The removal of an amino group from an amino acid.

Decarboxylation. The removal of a carboxyl group.

Gout. A disease of the joints in which deposits of salts of uric acid accumulate and cause inflammation, swelling, and pain.

Nitrogen Pool. The sum total of all nitrogen compounds in the body.

Oxidative Deamination. The change of an amino group to a keto group with loss of nitrogen.

Reductive Amination. The conversion of a keto group to an amino group by the action of ammonia and a reducing agent.

Transamination. The transfer of an amino group from an amino acid to a receptor with a keto group such that the keto group changes to an amino group.

Urea Cycle. The reactions by which urea is made from amino acids.

SELF-TESTING QUESTIONS

COMPLETION

1. Amino acids and other nitrogenous substances in the body, wherever they are, make up the ___nitrogen pool___.

2. Nitrogen enters the system largely as nitrogen compounds that are products of the digestion of ___Protein___.

3. Nitrogen leaves the system largely in the form of compounds called ___urea___, ___uric acid___, and a trace of ___Ammonia___.

4. The reducing agent in reductive amination has the short formula of ___NADPH___, and the source of nitrogen is the ___Ammonia___ ion.

   If the compound: $^-O_2CCH_2CCO_2^-$ (with O double-bonded to middle C) underwent reductive amination, the following amino acid would form: _____

5. The same amino acid could be made by _____
   using the same keto acid but using glutamate instead of NH_4^+ as the source of nitrogen. It leaves glutamate in the form of the following:

   _____

6. The reverse of reductive amination is called ___oxidation___
   _____. The nitrogen of the amino acids undergoing this process emerges in the form of the _____,
   which can be changed to the chief nitrogen waste, _____.

7. When serine undergoes the following change:

$$NH_3^+CHCO_2^- \quad \longrightarrow \quad \longrightarrow \quad CH_3CCO_2^- + NH_3$$
$$| \qquad\qquad\qquad\qquad\qquad\qquad\qquad\quad O$$
$$CH_2OH$$

serine   pyruvate

the overall reaction is called ___direct deamination___.

8. When tyrosine undergoes the following change:

$$NH_3^+CHCO_2^- \quad \longrightarrow \quad NH_2CH_2CH_2-\bigcirc-OH \;+\; CO_2$$
$$|$$
$$CH_2$$

tyrosine   tyramine

the change is called _____.

9. When transamination occurs to alanine, the product is called _____, and it can be used to make acetyl CoA, the "fuel" for the _____ or as a building block for making _____ in lipigenesis.

10. When the amino group of an amino acid is destined to become part of urea, it is first removed from the amino acid by the process of _____. The oxaloacetate ion is the amino group acceptor, and it changes to _____. Then, a second transamination puts the amino group into the _____ cycle. If enzymes needed for this cycle have reduced activity, a condition of _____ results.

11. The production of urea is accomplished by a series of reactions known as the _____Krebs_____ cycle.

12. The nitrogen in the purine bases (_____ and _____) is excreted in the form of _____.

13. The formation of sodium urate at a rate faster than it is excreted may lead to a condition known as _____.

14. When heme is catabolized, the products are colored compounds generally known as the _____.
    The one that is reddish-orange is called _____.

15. Bilinogen that is excreted in feces is called _____.

16. Bilin that is excreted in urine is called _____.

MULTIPLE-CHOICE

1. The following reaction is an example of

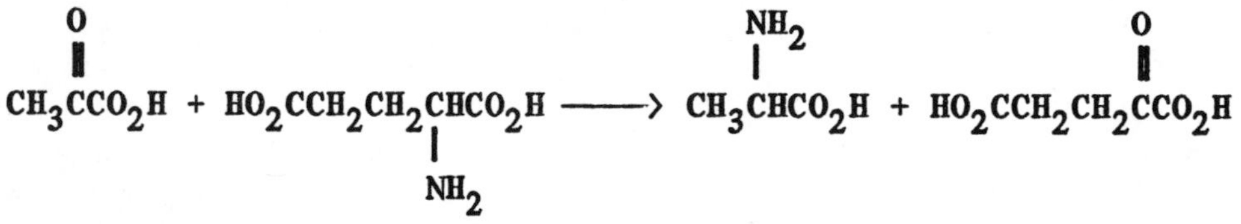

$$CH_3CCO_2H + HO_2CCH_2CH_2CHCO_2H \longrightarrow CH_3CHCO_2H + HO_2CCH_2CH_2CCO_2H$$

   (a) oxidative deamination   (c) decarboxylation
   (b) transamination   (d) gluconeogenesis

2. The following reaction is an example of

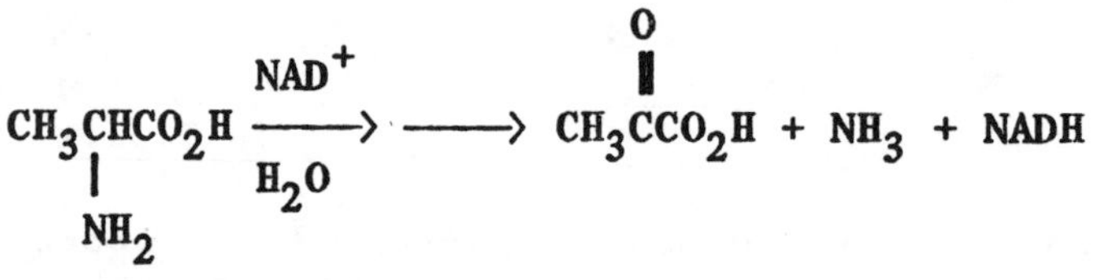

$$CH_3CHCO_2H \xrightarrow[H_2O]{NAD^+} \longrightarrow CH_3CCO_2H + NH_3 + NADH$$

   (a) oxidative deamination   (c) decarboxylation
   (b) transamination   (d) gluconeogenesis

3. The following reaction illustrates

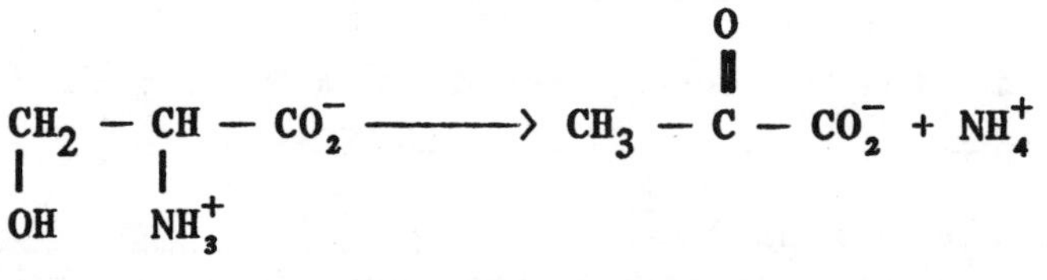

$$CH_2-CH-CO_2^- \longrightarrow CH_3-C-CO_2^- + NH_4^+$$

   (a) oxidative deamination   (c) decarboxylation
   (b) direct deamination   (d) transamination

4. If transamination occurred to $CH_3CH_2\underset{\underset{NH_3^+}{|}}{\overset{\overset{CH_3}{|}}{C}}HCHCO_2^-$, it would become

(a) $CH_3CH_2\overset{\overset{CH_3}{|}}{C}HCH_2NH_3^+$

(c) $CH_3CH_2\underset{\underset{H_3C\ NH_3^+}{|\ \ |}}{C}HCHCO_2^-$

(b) $CH_3CH_2\underset{\underset{O}{\|}}{\overset{\overset{CH_3}{|}}{C}}HCCO_2^-$

(d) $CH_3CH_2\underset{\underset{CH_3}{|}}{C}HCH_2CO_2^-$

5. The nitrogen waste made from the purine bases of nucleic acids is
   (a) ammonia   (b) urea   (c) uric acid   (d) bilinogen

6. The end products in the catabolism of proteins in the body are
   (a) amino acids
   (b) nitrogen, water, and carbon dioxide
   (c) ammonia, water, and carbon dioxide
   (d) urea, water, and carbon dioxide

7. In a dietary sense, alanine is classified as a nonessential amino
   acid. This is because
   (a) the body can make its own alanine
   (b) the body has no need for alanine
   (c) the body excretes alanine as fast as it can be ingested
   (d) alanine cannot be catabolized

8. A non-protein nitrogen compound made from molecules of amino acids
   is
   (a) triacylglycerol   (c) nucleic acid
   (b) creatine   (d) heme

9. Intermediates in the catabolism of amino acids may be used to make
   (a) glucose   (c) ketone bodies
   (b) fatty acids   (d) other amino acids

10. The major site of the catabolism of amino acids is the
    (a) liver   (b) kidneys   (c) adipose tissue   (d) gall bladder

11. In a condition of hyperammonemia, the level of concentration of what substance rises in the blood?
    (a) ammonia   (b) amino acids   (c) uric acid   (d) DOPA

12. The greenish pigment produced from heme is called
    (a) urobilinogen   (c) chlorophyll
    (b) urobilin   (d) biliverdin

ANSWERS

ANSWERS TO SELF-TESTING QUESTIONS
Completion
1. nitrogen pool
2. protein
3. urea, uric acid, ammonia
4. NADPH; ammonium;

$$^-O_2CCH_2\underset{\underset{NH_3^+}{|}}{C}HCO_2^-$$

5. transamination: $^-O_2CCH_2CH_2\underset{\underset{O}{\|}}{C}CO_2^-$

6. oxidative deamination; ammonium ion; urea
7. direct deamination
8. decarboxylation
9. pyruvate; citric acid cycle; fatty acids
10. transamination; aspartate; urea; hyperammonemia
11. urea (or Krebs' ornithine)
12. adenine, guanine, uric acid (or urate ions)
13. gout (also, kidney stones or the aggravation of arthritis)
14. tetrapyrrole pigments (or bile pigments); bilirubin
15. stercobilinogen
16. urobilin

Multiple-Choice
1. b   7. a
2. a   8. b, c, and d
3. b   9. a, b, c, and d
4. b   10. a
5. c   11. a
6. d   12. d

SELECTED ANSWER MANUAL

to accompany

FUNDAMENTALS OF GENERAL, ORGANIC,

AND BIOLOGICAL CHEMISTRY

3rd Edition

INTRODUCTION

This Answer Manual is a supplement to <u>Fundamentals of General, Organic, and Biological Chemistry,</u> 3rd edition by John R. Holum. It provides answers for all of the Review Exercises at the ends of the chapters for which answers are not provided in the text itself.

The Review Exercises provide a detailed review of terms and concepts in each chapter. They should always be answered <u>in writing</u> rather than by mentally noting something such as "Oh, I know that." Then the answers can be checked against this Answer Manual. If your answer is correct in substance although certainly rarely in identical words, then there is one important remaining step — learning the answers. This does not mean rote memorization, but it does mean reaching a point where the answer to a question is well enough known that you could give it independently.

All answers have been checked several times, but long experience has taught me that an error can escape the notice of as many as 18 people — and probably many more. If you would be so kind as to bring such errors to my attention, I'd be very grateful.

John R. Holum
Augsburg College
Minneapolis, MN 55454

CHAPTER 1

GOALS, METHODS, AND MEASUREMENTS

Review Exercises

1.1   Both kittens and humans thrive on milk.

1.2   Anatomical, physiological, psychological, sociological, philosophical, religious.

1.3   Hypothesis. Theories are much broader.

1.4   Experimental data can be measured with accuracy and precision and they can be rechecked by independent observers.

1.5   Facts must control reasoning.

1.6   First, form a hypothesis about the cause — perhaps the bulb is burned out. Second, test the hypothesis.

1.7   It's a characteristic of something by means of which we can identify it and recognize it when we see it again.

1.8   A physical property can be studied and measured without changing the substance into some other substance. A chemical property can be observed only as the substance changes chemically into another substance.

1.9   It is a number times a unit.

1.10   An object's inertia is its inherent resistance to any change in its motion or direction. The mass is the quantitative measure of this inertia.

1.11   Volume, a derived unit, can be expressed as a particular product of a base unit; volume = length x length x length.

1.12   Derived quantity

1.13   Meter, for length. Kilogram, for mass. Second, for time.
        Kelvin, for temperature. Mole, for quantity of substance.

1.14   (a) second   (b) kilogram
        (c) kelvin   (d) meter
        (e) mole

1.15   A <u>reference</u> <u>standard</u> of measurement is the physical description
        or the physical embodiment of a <u>base</u> <u>unit</u>. For example, the
        physical embodiment – the reference standard – for the base
        unit called the kilogram is a block of platinum–iridium alloy.

1.16   A reference standard should be free of risks of corrosion,
        fire, war, theft, and vandalism, and it should be accessible at
        any time to scientists in any country.

1.17   The kilogram mass

1.18   (a) centimeter, cm
            millimeter, mm
        (b) milliliter, mL
            microliter, μL
        (c) gram, g
            milligram, mg

1.19   100 cm = 1 m

1.20   10 mm = 1 cm

1.21   1000 g = 1 kg

1.22   1000 mg = 1 g

1.23   Slightly shorter

1.24   Slightly smaller

1.25   It is the lowest degree of coldness possible when described in
        terms of Celsius degrees.

1.26   0 $^{\circ}$C.   32 $^{\circ}$F.   273 K (before rounding, 273.15 K)

1.27   100 °C.   212 °F.   373 K (before rounding, 373.15 K)

1.28   (a) 180 degrees   (b) 100 degrees   (c) 100 degrees

1.29   They are identical.

1.30   The °C is 9.5 times the °F.

1.32   21 °C

1.34   −40 °C

1.36   No.   101 °F

1.38   34.6 °C

1.39   (a) 26 μg of vitamin E
        (b) 28 mm wide
        (c) 5.0 dL of solution
        (d) 55 km in distance
        (e) 46 μL of solution
        (f) 64 cm long

1.40   (a) 125 milligrams of water   (b) 25.5 milliliters of blood
        (c) 15 kilograms of salt   (d) 12 deciliters of solution
        (e) 2.5 micrograms of insulin   (f) 16 microliters of fluid

1.41   (a) 1.3×10^{-2} L   (b) 6×10^{-6} g

        (c) 4.5×10^{-3} m   (d) 1.455×10^{3} s

1.42   (a) 2.4605×10^{4} m   (b) 6.54115×10^{5} g

        (c) 9.5×10^{-9} L   (d) 5.68×10^{-6} s

1.43   (a) 1.3 cL   (b) 6 μg
        (c) 4.5 mm   (d) 1.455 ks

1.44   (a) 24.605 km   (b) 654.115 kg
        (c) 9.5 mL   (d) 5.68 μs

1.45 3.94×10^{6} people

1.46   (a) Yes, the measurements correspond closely to the true value and the average of all measurements; 172.7 (correctly rounded), equals the true value.

(b) Yes, the data are precise to four significant figures, and successive measurements agree closely with each other.

1.47   (a) F, H, and I
        (b) B, C, D, and G
        (c) A and E

1.48   Three

1.49   Two

1.50   (a) 144,549.1   (b) 1.4455×10^5 (c) 1.445×10^5

        (d) 1.45×10^5 (e) 1×10^5 (f) 1.4×10^5

1.51   An infinite number

1.52   (a) $\dfrac{5280 \text{ ft}}{1 \text{ mile}}$ or $\dfrac{1 \text{ mile}}{5280 \text{ ft}}$

        (b) $\dfrac{60 \text{ grains}}{1 \text{ dram}}$ or $\dfrac{1 \text{ dram}}{60 \text{ grains}}$

        (c) $\dfrac{453.6 \text{ g}}{1 \text{ lb}}$ or $\dfrac{1 \text{ lb}}{453.6 \text{ g}}$

        (d) $\dfrac{480 \text{ grains}}{1 \text{ ounce}}$ or $\dfrac{1 \text{ ounce}}{480 \text{ grains}}$

        (e) $\dfrac{39.37 \text{ in.}}{1 \text{ m}}$ or $\dfrac{1 \text{ m}}{39.37 \text{ in.}}$

        (f) $\dfrac{480 \text{ minims}}{1 \text{ liquid ounce}}$ or $\dfrac{1 \text{ liquid ounce}}{480 \text{ minims}}$

1.53   (a) 0.520 liquid ounce

        (b) 1.68×10^3 grain

        (c) 159 g

1.54 $\dfrac{\text{kg} \times \text{m}^2}{\text{sec}^2}$

1.56   (a) 154 lb   (b) 163 cm

1.58   60.5 L

1.60 2.02×10^3 lb

1.62   150 mg

1.64 2.032×10^4 ft; 3.848 mi

1.65   Because the density of a substance is independent of the size of the sample.

1.66   The density would not be affected because the ratio of new mass to new volume, 0.5 g/0.5 mL or 1.0 g/mL, is the same as that of water.

1.67   The density increases.

1.68   Density is always expressed as a ratio of mass to volume, typically in units of g/mL, whereas specific gravity has no units.

1.69   The density of water (to two significant figures) is 1.0 g/mL over the range of ordinary temperatures, so when the density of water is divided into the density of anything else, also expressed in g/mL, the units cancel and the numerical result is the same as dividing anything by 1 – there's no change.

1.70   The urinometer sinks to a lower level in a fluid of <u>low density</u>.

1.72 2.56×10^3 g, 2.56 kg, 5.63 lb of aluminum

1.74   223 g of methyl alcohol

CHAPTER 2

MATTER AND ENERGY

Review Exercises

2.1   Energy is not considered to have mass. Matter both has mass and occupies space.

2.2   Solid, liquid, and gas

2.3   Solids have definite shapes and volumes, and neither depends on the specific container. Gases have indefinite shapes and volumes; both depend on the specific container.

2.4   The formula units in water are relatively quite mobile, but those in a solid are not.

2.5   The formula units in air are relatively much more mobile than those in a liquid.

2.6   Metal

2.7   Nonmetal

2.8   In a chemical change a substance changes into one or more different substances. Physical changes do not involve this feature.

2.9   (a) Physical   (e) Physical
        (b) Chemical   (f) Physical
        (c) Chemical   (g) Physical
        (d) Physical

2.10   No element can be further broken down into simpler (and stable) substances.

2.11   100

2.12   13

2.13   Compounds are made from two or more elements that occur together in <u>definite</u> <u>proportions</u>.

2.14   Pure substances have constant compositions and obey the law of definite proportions. Elements and compounds are <u>pure</u> <u>substances</u> in this sense.

2.15   Chemical reaction

2.16   Hydrogen and oxygen are the <u>reactants</u>. Water is the <u>product</u>.

2.17   An alloy is an intimate mixture of two or more metals prepared by mixing the metals in their liquid forms and then cooling the system.

2.18   The law of conservation of mass

2.19   1:1

2.20   Atoms of different elements have different masses.

2.21   2:1, a ratio of small whole numbers

2.22   (a) 0.6013 g of Cl/1.000 g of Sn
        (b) 1.195 g of Cl/1.000 g of Sn

        (c) $\dfrac{\text{Cl in compound B}}{\text{Cl in compound A}} = \dfrac{1.195 \text{ g Cl}}{0.6013 \text{ g Cl}} = \dfrac{1.987}{1}$

        (d) Law of multiple proportions, because 1.987/1 is acceptably close to a ratio of small whole numbers, 2/1.

2.23   The reactants must undergo changes into new substances that obey the law of definite proportions.

2.24   The change is very likely a chemical change if it is accompanied by a change in color, odor, or physical state, or if heat is released as the change occurs.

2.25   The second letter in BN is not lower case, as it would have to be if it stood for an element.

2.26   (a) I   (b) Li   (c) Zn   (d) Pb   (e) N   (f) Ba

2.27   (a) C   (b) Cl   (c) Cu   (d) Ca   (e) F   (f) Fe

2.28   (a) H   (b) Al   (c) Mn   (d) Mg   (e) Hg   (f) Na

2.29   (a) O   (b) Br   (c) K   (d) Ag   (e) P   (f) Pt

2.30   One atom of sulfur

2.31   (a) Phosphorus   (b) Platinum   (c) Lead
       (d) Potassium   (e) Calcium   (f) Carbon
       (g) Mercury   (h) Hydrogen   (i) Bromine
       (j) Barium   (k) Fluorine   (1) Iron

2.32   (a) Sulfur   (b) Sodium   (c) Nitrogen
       (d) Zinc   (e) Iodine   (f) Copper
       (g) Oxygen   (h) Lithium   (i) Manganese
       (j) Magnesium   (k) Silver   (1) Chlorine

2.33   (b) H_2O

2.34   No natural law is violated. Only the convention of using the
       simplest whole numbers to show the ratio of 1:1 is violated.

2.35   The definite atom-to-atom proportion of Fe to S, 1:1.

2.36   Fe + 2S ———→ FeS_2

2.37   Energy is an <u>ability</u>, an ability to do work.

2.38   Energy of motion.   KE = (1/2)$\underline{mv}^2$

2.39   Energy can be neither created nor destroyed; it can only be
       transformed.

2.40   The kinetic energy of the moving materials is changed into heat
       energy (from friction) and sound energy as the slide eventually
       stops.

2.41   The dynamite has chemical energy.

2.42   Heat and light energy

2.43   Potential energy

2.45   35.9 m/s

2.46   When the two objects are in contact and have different
        temperatures, but one temperature rises as the other falls; and
        when the contact results in the change of physical state of one
        as it remains at a constant temperature.

2.47   The melting point

2.48   The units of specific heat are $\frac{cal}{g \, {}^oC}$.

2.49   Its heat capacity. The specific heat is the heat capacity per
        unit of mass – usually the gram.

2.50   Its mass

2.51   For 10.00 g of water, heat capacity = 10.0 cal/^oC.

2.52   The temperature of the gold will increase most.
        Final temperature of the gold = 101 ^oC
        Final temperature of the olive oil = 25.3 ^oC

2.54   The units are cal/g.

2.55   Gold has the lower heat of fusion, so a given amount of
        calories will melt a larger mass of gold than of ice.

2.56   The energy of the heat of fusion is now part of the energy of
        the water in its liquid state and will be returned to the
        environment when the water freezes.

2.57   The change of each gram of ice from 0 ^oC to 1 ^oC removes two
        quantities of heat from the environment – the heat of fusion
        for this mass of ice to melt (80 cal) and the heat needed to
        raise the temperature of the resulting 1 g of liquid water by
        one degree (1 cal) for a total of 81 cal of heat. To change
        the temperature of 1 g of liquid water from essentially 0 ^oC to
        1 ^oC takes only 1 cal from the environment.

2.58   The heat in the <u>steam</u> because it holds the heat of vaporization and can deliver this to the ice.

2.59   Evaporation and vaporization

2.60   The heat is released back to the surroundings.

2.61   When the temperature is quite high, such as near or above the boiling point, water vapor is called steam. Around ordinary temperature, it's water vapor.

2.62   Exothermic

2.63   Endothermic

2.64   Combustion is one example.

2.66 $\underline{t} = 474\ ^{\circ}C$

2.68 $0\ ^{\circ}C$ or $32\ ^{\circ}F$

2.69   Maintain muscle tone
       Control body temperature
       Circulate the blood
       Breathe
       Make chemical substances
       Operate tissues and glands

2.70   The sum total of all of the chemical reactions that supply the energy for the basal activities.

2.71   The rate at which chemical energy is used for basal activities.

2.72   1 kcal/min

2.73   As drink, 1.2 L; in food, 1.0 L; and as made by chemical reactions of metabolism, 0.30 L.

2.74   40%

2.75   Insensible perspiration

2.76   Radiation, conduction, and convection

2.77   Radiation is the transfer of heat in the same manner as heat is lost from a warm iron. Conduction is the direct transfer of heat to a colder object.

2.78   Radiation

2.79   Hypothermia is a decrease in body temperature, and the reactions that sustain metabolism slow down dangerously.

2.80   Hyperthermia is an increase in body temperature, and the reactions that sustain metabolism increase, which increases the demand for oxygen, and which increases the work load of the heart.

2.82 8.7×10^3 kcal

2.84   (a) 6.3×10^2 kcal/serving
       (b) 2.1 hr
       (c) 7.4 miles

CHAPTER 3

ATOMIC THEORY AND THE PERIODIC SYSTEM
OF THE ELEMENTS

Review Exercises

3.1   Proton, 1+, 1 amu
        Neutron, 0 charge, 1 amu
        Electron, 1–

3.2   The electron and proton

3.3   Protons repel protons and electrons repel electrons. (Atomic nuclei, if considered as subatomic particles, also repel atomic nuclei and protons.)

3.4   3+

3.5   (a) This change is the removal of an electron from a neutral particle, whereas change (b) is the removal of a negatively charged particle (the electron) from a particle that already is oppositely charged (1+).

3.6   They attract each other because $\underline{M}$ has a charge of 3+ and $\underline{Y}$ is oppositely charged with a charge of 2–.

3.7 10^{-23} g

3.8   1.0000 g, which is almost identical with the atomic weight of hydrogen.

3.10   Most elements consist of a mixture of a few isotopes that occur together in nature in definite proportions. The isotopes of any given element all have identical numbers of protons in their nuclei but the numbers of neutrons vary from isotope to isotope.

3.11 Protium (hydrogen): 0 neutrons, 1 proton
      Deuterium: 1 neutron, 1 proton
      Tritium: 2 neutrons, 1 proton
          They have the same number of protons, but different
          numbers of neutrons. The atoms of these three isotopes
          have identical numbers of electrons.

3.12 $\underline{M}$ and $\underline{X}$ are isotopes. $\underline{Q}$ and $\underline{Z}$ are isotopes.

3.13 $^{18}_{8}O$

3.14 $^{60}_{27}Co$

3.15 (a) They have identical numbers of protons in their nuclei and
           identical numbers of electrons outside their nuclei.
      (b) They have different numbers of neutrons in their nuclei.

3.16 These two isotopes have identical chemical properties, so no
      special symbols that distinguish isotopes have to be used.

3.17 The electrons of an atom move about the hard, dense atomic
      nucleus in just a few allowed energy states. Their movements
      resemble those of the planets about the sun. As long as the
      electrons do not change their energy states, the atom does not
      emit or absorb energy.

3.18 An atom's electrons are confined to allowed energy states and
      the atom does not emit or absorb energy as long as the
      electrons do not change states.

3.19 It is a location in which one or two electrons of an atom can
      reside.

3.20 It is the electron configuration of lowest energy for an atom.

3.21 It emits energy that corresponds to the difference in energy
      between the higher state and the ground state.

3.22 No, the energy used to obtain data on location modifies the
      energy of the electron.

3.23 The location and the energy of an electron in the atom.

3.24   The probability of finding the electron at a certain location when it has a particular value of energy.

3.25   One sublevel, the 1s, which consists of one orbital.

3.26   Two sublevels, with one orbital in one – the 2s – and three orbitals in the other – the $2p_x$, $2p_y$, and the $2p_z$.

3.27   Three sublevels, s, p, and d, with one orbital in one – the 3s – three orbitals in another – the $3p_x$, the $3p_y$, nd the $3p_z$ – and five orbitals of the d type.

3.28 $4s^1$

3.29   (a) 8   (b) 6   (c) 2   (d) 18

3.30 $1s^2 2s^2 2p_x^2 2p_y^2 2p_z^2 3s^2 3p_x^1 3p_y^1$

3.31 $1s^2 2s^2 2p^6 3s^2 3p^6 3d^{10} 4s^2 4p^6$

3.32   (a) 26
        (b) No, sublevel 3d is partially filled.
        (c) Yes, according to Hund's rule, the configuration in sublevel 3d is: ↑↓ ↑ ↑ ↑ ↑

3.33   (a) Electrons reside in the available orbitals of lowest energy and, by the Pauli Exclusion Principle, there can be up to two electrons in an orbital provided that they have opposite spins.
        (b) Hund's Rule says that electrons at the same sublevel spread out among the sublevel's orbitals as much as possible.

3.34   (a) Of the same element, because they have identical numbers of electrons and so would have identical numbers of protons and identical atomic numbers.
        (b) Electron configuration 1 is more stable since its lone level-3 electron is in the 3s orbital, not the 3p orbital, which corresponds to higher energy.
        (c) State 1 can be changed to state 2 by the absorption of the exact amount of energy that is the difference in the energies of the 3p and the 3s orbitals.

3.35 (a) $1s^2 2s^2 2p_x^2 2p_y^2 2p_z^2 3s^1$ (b) $1s^2 2s^2 2p_x^2 2p_y^1 2p_z^1$

(c) $1s^2 2s^2 2p_x^2 2p_y^2 2p_z^2 3s^2 3p_x^2 3p_y^2 3p_z^1$

(d) $1s^2 2s^2 2p_x^2 2p_y^2 2p_z^2 3s^2 3p_x^2 3p_y^2 3p_z^2 4s^2$

3.36 (a) $1s^2 2s^2 2p_x^2 2p_y^2 2p_z^2 3s^2$

(b) $1s^2 2s^2 2p_x^2 2p_y^2 2p_z^2 3s^2 3p_x^2 3p_y^2 3p_z^2 4s^1$

(c) $1s^2 2s^2 2p_x^2 2p_y^2 2p_z^1$ (d) $1s^2 2s^1$

3.37 Mass numbers are the sums of whole numbers – numbers of protons and neutrons per atom of an isotope. The atomic weight of an element (in amu) is the average off the relative masses of the isotopes of the element taking into account the percentages that the various isotopes have in naturally occurring samples of the element.

3.38 127   3.39 80

3.40 (a) Twice as heavy
(b) The mass of the pile of sulfur atoms would be twice as much as that of the pile of magnesium atoms.
(c) A mass that is twice as much as that of the 2.0 g of magnesium – 4.0 g of sulfur.

3.43 The periodic law

3.44 Periods are horizontal rows of elements and groups are vertical columns of elements.

3.45 A representative element (one of the A-elements)

3.46 A representative element (because period 2 has no transition elements)

3.47 Mendeleev arranged the elements in order of increasing atomic weight, but in the modern form of the table they are arranged in order of increasing atomic number.

3.48 The positions of elements 52 (Te) and 53 (I) would be reversed.

3.49   (a) Transition element. The 3$\underline{d}$ sublevel is not filled.
        (b) Level 4; 1 electron
        (c) Metal, because it has just 1 electron in its outside level.

3.50   (a) Group VA – V, because its outside level has 5 electrons,
             and A because its inner levels are filled.
        (b) Nonmetal, because its outside level has too many electrons
             to let the element be a metal.

3.51   (a) Group IA, alkali metal family
        (b) Group VIIA, halogen family
        (c) Group VIA, oxygen family
        (d) Group IIA, alkaline earth metal family

3.52   (a) Group VIIA, halogen family
        (b) Group VA, nitrogen family
        (c) Group IIA, alkaline earth metal family
        (d) Group IA, alkali metal family

3.53   (a) KH   (b) CaH_2 (c) GaH_3 (d) GeH_3 (e) AsH_3

        (f) SeH_2 (usually written as H_2Se)   (g) BrH (usually written

        as HBr)

3.54   (a) 9, 17, and 35
        (b) 15, 16, 17, and 18
        (c) In order, from left to right: VA, VIA, VIIA, 0
        (d) Element 15:   5
             Element 16:   6
             Element 36:   8
        (e) $\underline{HY}$
        (f) Element 36
        (g) Nonmetals

CHAPTER 4

CHEMICAL COMPOUNDS AND BONDS

Review Exercises

4.1 A force of electrical attraction between two atoms.

4.2 (3 p$^+$ / 4 n) $1s^2 2s^1$ + (9 p$^+$ / 10 n) $1s^2 2s^2 2p_x^2 2p_y^2 2p_z^1$ $\longrightarrow$

lithium atom, Li   fluorine atom, F

[(3 p$^+$ / 4 n) $1s^2$]$^+$ + [(9 p$^+$ / 10 n) $1s^2 2s^2 2p_x^2 2p_y^2 2p_z^2$]$^-$

lithium ion, Li$^+$ fluoride ion, F$^-$

4.3 (12 p$^+$ / 12 n) $1s^2 2s^2 2p_x^2 2p_y^2 2p_z^2 3s^2$ + 2[(9 p$^+$ / 10 n) $1s^2 2s^2 2p_x^2 2p_y^2 2p_z^1$] $\longrightarrow$

magnesium atom, Mg   2 fluorine atoms, F

[(12 p$^+$ / 12 n) $1s^2 2s^2 2p_x^2 2p_y^2 2p_z^2$]$^{2+}$ + 2[(9 p$^+$ / 10 n) $1s^2 2s^2 2p_x^2 2p_y^2 2p_z^2$]$^-$

magnesium ion, Mg^{2+} 2 fluoride ions, 2F$^-$

4.4 2[(11 p$^+$ / 12 n) $1s^2 2s^2 2p_x^2 2p_y^2 2p_z^2 3s^1$] + (8 p$^+$ / 8 n) $1s^2 2s^2 2p_x^2 2p_y^1 2p_z^1$ $\longrightarrow$

2 sodium atoms, 2Na   oxygen atom, O

$$2 \left[\left(\begin{array}{c} 11 \ p^+ \\ 12 \ n \end{array} \right) 1\underline{s}^2 2\underline{s}^2 2\underline{p}_x^2 2\underline{p}_y^2 2\underline{p}_z^2 \right]^+ \quad + \quad \left[\left(\begin{array}{c} 8 \ p^+ \\ 8 \ n \end{array} \right) 1\underline{s}^2 2\underline{s}^2 2\underline{p}_x^2 2\underline{p}_y^2 2\underline{p}_z^2 \right]^{2-}$$

$$\text{2 sodium ions, } 2Na^+ \qquad\qquad\qquad \text{oxide ion, } O^{2-}$$

4.5   Group IIIA

4.6   Group VIA

4.7 2^+

4.8   Group VA

4.9   1. A filled level-1 as the <u>outside</u> level.
2. Eight electrons — an octet — in any other outside level.

4.10   Charge = 1−

4.11   Charge = 2+

4.12   No charge, because the outer level (5) has an octet.

4.13   Positively charged, because all transition elements are metals, which generally form positively charged ions.

4.14   (a) $1\underline{s}^2 2\underline{s}^2 2\underline{p}_x^2 2\underline{p}_y^2 2\underline{p}_z^2 3\underline{s}^2 3\underline{p}_x^2 3\underline{p}_y^2 3\underline{p}_z^2$

(b) $1\underline{s}^2 2\underline{s}^2 2\underline{p}_x^2 2\underline{p}_y^2 2\underline{p}_z^2$

4.15 $1\underline{s}^2$ H^-

4.16   Ionic compounds

4.17   The sodium <u>ion</u>, Na^+

4.18   1+   4.19   4+

4.20   (a) 3+, Bi^{3+} (b) 2+, Cd^{2+} (c) 3+, Cr^{3+}

(d) 3+, Gd^{3+} (e) 2+, Sn^{2+} (f) 3+, Ti^{3+}

4.21   (a) 7+   (b) 8+   (c) 5+   (d) 1+

4.22   (a) S   (b) S   (c) Al   (d) Al

4.23   (a) Mg   (b) Br_2 (c) Mg   (d) Br_2

4.24   (a) 2+   (b) 3+   (c) Yes   (d) O_2 (e) O_2

4.25

(a) K^+	(f) S^{2-}	(k) Ag^+	(p) Fe^{2+}
(b) Al^{3+}	(g) Na^+	(l) Mg^{2+}	(q) F^-
(c) I^-	(h) Fe^{3+}	(m) Cu^{2+}	(r) Cl^-
(d) Cu^+	(i) O^{2-}	(n) Br^-	(s) Zn^{2+}
(e) Ba^{2+}	(j) Li^+	(o) Ca^{2+}	(t) Ba^{2+}

4.26
(a) Sodium ion   (k) Zinc ion
(b) Iron(III) ion, ferric ion   (l) Fluoride ion
(c) Lithium ion   (m) Iron(II) ion, ferrous ion
(d) Oxide ion   (n) Chloride ion
(e) Sulfide ion   (o) Calcium ion
(f) Barium ion   (p) Copper(II) ion, cupric ion
(g) Copper(I) ion, cuprous ion
(h) Iodide ion   (q) Bromide ion
(i) Aluminum ion   (r) Magnesium ion
(j) Potassium ion   (s) Silver ion

4.27   Mercurous ion

4.28 Sn^{2+}

4.29   Lead(II) ion

4.30   Gold(III) ion

4.31
(a) LiCl   (c) Al_2S_3 (e) CuO
(b) BaO   (d) NaBr   (f) $FeCl_3$

4.32
(a) Cu_2S (c) Na_2S (e) $MgCl_2$
(b) KF   (d) CaI_2 (f) $FeBr_2$

4.33
(a) Iron(III) bromide, ferric bromide
(b) Magnesium chloride
(c) Sodium fluoride
(d) Zinc oxide
(e) Copper(II) bromide, cupric bromide
(f) Lithium oxide

4.34   (a) Potassium iodide
       (b) Calcium sulfide
       (c) Barium chloride
       (d) Aluminum oxide
       (e) Iron(II) chloride, ferrous chloride
       (f) Silver iodide

4.35   A molecule is a small particle that consists of two or more
       atomic nuclei and enough electrons to ensure overall electrical
       neutrality. An atom is a small, electrically neutral particle
       that has just one nucleus. An ion is a small particle that is
       not electrically neutral. It can have one nucleus or it can
       have two or more, but it always has too many or too few
       electrons for overall electrical neutrality.

4.36   Molecular orbital

4.37   The attraction of the atomic nuclei toward the shared pair of
       electrons between them.

4.38   The two atomic orbitals of the separate atoms partially merge
       (overlap) to create a new space that encompasses both of the
       nuclei and within which the two electrons reside.

4.39   Two filled atomic orbitals do not overlap.

4.40   [The figures to be drawn should be identical to those in Figure
       4.2, except that the two p_z orbitals that overlap are at level
       3 in chlorine, not in level 2.] In the newly formed molecular
       orbital, the electron density of the shared pair of electrons
       concentrates between the two nuclei. These nuclei are
       attracted toward this electron-dense region, and this
       attraction constitutes the covalent bond between the two
       chlorine atoms.

4.41   [The figures that should be drawn should be identical with
       those in Figure 4.4, and the discussion should be like that of
       the legend to this Figure.]

4.42   (a) Yes, it can have one covalent bond.
       (b) H — X

4.43   (a) 4

418

(b)
$$H - \underset{\underset{H}{|}}{\overset{\overset{H}{|}}{Z}} - H$$

4.44   B

4.45   A

4.46   (a) 4   (b) IVA

4.47
$$H - \underset{\overset{|}{H}}{P} - H$$

4.48   S=C=S

4.49
$$H - \underset{\overset{|}{H}}{Sb} - H$$

4.50   H — O — Cl

4.51   (a) Bicarbonate ion   (j) Permanganate ion
       (b) Sulfate ion   (k) Hydrogen sulfate ion
       (c) Nitrate ion   (1) Hydrogen sulfite ion
       (d) Hydroxide ion   (m) Nitrite ion
       (e) Ammonium ion   (n) Phosphate ion
       (f) Cyanide ion   (o) Dichromate ion
       (g) Monohydrogen phosphate ion   (p) Dihydrogen phosphate ion
       (h) Chromate ion   (q) Hydronium ion
       (i) Carbonate ion   (r) Acetate ion

4.52   (a) CO_3^{2-} (g) H_3O^+ (m) $Cr_2O_7^{2-}$

       (b) NO_3^- (h) HPO_4^{2-} (n) HSO_3^-

       (c) OH^- (i) NO_2^- (o) CrO_4^{2-}

       (d) NH_4^+ (j) HCO_3^- (p) SO_4^{2-}

       (e) PO_4^{3-} (k) HSO_4^- (q) SO_3^{2-}

       (f) CN^- (1) $H_2PO_4^-$ (r) $C_2H_3O_2^-$

4.53   (a) $(NH_4)_3PO_4$ (c) $MgSO_4$ (e) $LiHCO_3$ (g) NH_4Br
        (b) K_2HPO_4 (d) $CaCO_3$ (f) $K_2Cr_2O_7$ (h) $Fe(NO_3)_3$

4.54   (a) NaH_2PO_4 (e) $KHSO_4$
        (b) $CuCO_3$ (f) $(NH_4)_2CrO_4$
        (c) $AgNO_3$ (g) $Ca(C_2H_3O_2)_2$
        (d) $Zn(HCO_3)_2$ (h) $Fe_2(SO_4)_3$

4.55   (a) Sodium nitrate   (e) Ammonium cyanide
        (b) Calcium sulfate   (f) Sodium phosphate
        (c) Potassium hydroxide   (g) Potassium permanganate
        (d) Lithium carbonate   (h) Magnesium dihydrogen phosphate

4.56   (a) Potassium monohydrogen phosphate
        (b) Sodium bicarbonate   (f) Calcium acetate
        (c) Ammonium nitrite   (g) Potassium dichromate
        (d) Zinc chromate   (h) Sodium hydrogen sulfate
        (e) Lithium hydrogen sulfate

4.57   (a) 15   (b) 20   (c) 11

4.58   (a) 20   (b) 17   (c) 15

4.59   53

4.60 H^+

4.61   Bases

4.62   An acid and a base

4.63   Yes $\overset{\delta-\ \ \delta+}{X\!-\!Y}$

4.64   (a) Not polar   (c) Not polar   (e) Not polar
        (b) $\overset{\delta+\ \ \delta-}{H\!-\!F}$ (d) Not polar   (f) $\overset{\delta+\ \ \delta-}{H\!-\!I}$

4.65   X, because electronegativities increase as one moves upward
        within the same group or family in the Periodic Table.

4.66   Z, because electronegativities increase as one moves from left
        to right within the same period of the Periodic Table.

CHAPTER 5

QUANTITATIVE RELATIONSHIPS IN CHEMICAL REACTIONS

Review Exercises

5.1 Carbon and oxygen react to give carbon monoxide in the proportion of 2 atoms of carbon used to 1 molecule of oxygen used and 2 molecules of carbon monoxide produced.

5.2 Nitrogen and hydrogen react to give ammonia in the proportion of 1 molecule of nitrogen used to 3 molecules of hydrogen used and 2 molecules of ammonia produced.

5.3 H_3PO_4 + 2NaOH $\longrightarrow$ Na_2HPO_4 + $2H_2O$

5.4 (a) $2SO_2$ + O_2 $\longrightarrow$ $2SO_3$

(b) CaO + $2HNO_3$ $\longrightarrow$ $Ca(NO_3)_2$ + H_2O

(c) $2AgNO_3$ + $MgCl_2$ $\longrightarrow$ $2AgCl$ + $Mg(NO_3)_2$

(d) 2HCl + $Ca(OH)_2$ $\longrightarrow$ $CaCl_2$ + $2H_2O$

(e) $2C_2H_6$ + $7O_2$ $\longrightarrow$ $4CO_2$ + $6H_2O$

5.5 (a) $2NaHCO_3$ + H_2SO_4 $\longrightarrow$ Na_2SO_4 + $2H_2O$ + $2CO_2$

(b) Fe_2O_3 + $3H_2$ $\longrightarrow$ 2Fe + $3H_2O$

(c) $Ca(OH)_2$ + $2HNO_3$ $\longrightarrow$ $Ca(NO_3)_2$ + $2H_2O$

(d) 2NO + O_2 $\longrightarrow$ $2NO_2$

(e) Al_2O_3 + $3H_2SO_4$ $\longrightarrow$ $Al_2(SO_4)_3$ + $3H_2O$

5.6 6.02×10^{23}

5.7 Because this number equals the number of formula units of a substance whose total mass in grams is numerically equal to the formula weight of the substance.

420

5.9 6.02×10^{20} molecules of aspirin

5.11   1.20 mg of impurity

5.12   The law of conservation of mass in chemical reactions.

5.14   (a) 146   (b) 142   (c) 63.0
        (d) 98.1   (e) 132   (f) 310

5.15   Calculate the formula weight and write the unit <u>grams</u> after it.

5.16   Avogadro's number equals the number of formula units in 1 mol
        of the substance.

5.17   A different balance would be needed for every value of formula
        weight.

5.18   g H_2O/mol H_2O

5.19 $\dfrac{159.8 \text{ g } Br_2}{1 \text{ mol } Br_2}$ $\dfrac{1 \text{ mol } Br_2}{159.8 \text{ g } Br_2}$

5.21   (a) 84.0 g   (b) 81.7 g   (c) 36.2 g
        (d) 56.4 g   (e) 75.9 g   (f) 178 g

5.23   (a) 0.196 mol   (b) 0.201 mol   (c) 0.454 mol
        (d) 0.292 mol   (e) 0.217 mol   (f) 0.0923 mol

5.25   (a) $\dfrac{2 \text{ mol } C_2H_6}{7 \text{ mol } O_2}$ and $\dfrac{7 \text{ mol } O_2}{2 \text{ mol } C_2H_6}$

        (b) $\dfrac{4 \text{ mol } CO_2}{2 \text{ mol } C_2H_6}$ and $\dfrac{2 \text{ mol } C_2H_6}{4 \text{ mol } CO_2}$

        (c) $\dfrac{6 \text{ mol } H_2O}{7 \text{ mol } O_2}$ and $\dfrac{7 \text{ mol } O_2}{6 \text{ mol } H_2O}$

        (d) $\dfrac{2 \text{ mol } C_2H_6}{6 \text{ mol } H_2O}$ and $\dfrac{6 \text{ mol } H_2O}{2 \text{ mol } C_2H_6}$

5.27   (a) 525 mol of CO   (b) 70 mol of Fe   (c) 375 mol of CO

5.29   (a) 989 g of oxygen (when calculated in steps, rounding after each step); 991 g of oxygen (when found by a chain calculation)

      (b) 171 g of CO_2 (step-wise calculation)
           170 g of CO_2 (chain calculation)

      (c) 52.0 g of O_2 (step-wise calculation)
           52.1 g of O_2 (chain calculation)

5.31   (a) 7.24×10^2 kg of P_4O_{10}

      (b) 2.75×10^2 kg of H_2O (step-wise calculation with rounding after each step)

           2.76×10^2 kg of H_2O (when found by a chain calculation)

5.33   The solution is unsaturated, but very nearly saturated since at 50 $^\circ$C 37.0 g of NaCl can dissolve in 100 g of water.

5.34   Saturated, but very dilute

5.35   Start with some desired volume of water, warm it and then add solid KCl until no more dissolves (and some remains undissolved). Then cool this system to room temperature, and the solution above the undissolved solute is saturated.

5.36   To make it unsaturated, warm it.

      To try to make it supersaturated, carefully cool the solution and hope that excess solute does not precipitate out.

5.37   A <u>molecule</u> is a tiny particle, the smallest representative sample of a compound. A <u>mole</u> is Avogadro's number of molecules. <u>Molarity</u> is the ratio of the moles of a solute per liter of its solution.

5.38   Molar concentration

5.39   No, because when the solution is used, the <u>solution</u> is measured out, not the pure solvent that was used to prepare the solution.

5.41   (a) 3.18 g of Na_2CO_3 (c) 7.50 g of $KHCO_3$

      (b) 2.00 g of NaOH   (d) 8.55 g of $C_{12}H_{22}O_{11}$

5.43   2.5 mL of 1.0 $\underline{M}$ H_2SO_4 solution

5.45   7.5 x 10^2 mL of 0.10 $\underline{M}$ H_2SO_4 solution

5.47   1.0 x 10^2 mL of 0.50 $\underline{M}$ H_2SO_4 solution

5.49   80.2 mL of 0.100 $\underline{M}$ HCl solution (when calculated step-wise with rounding after each step). By a chain-calculation, 80.3 mL.

5.51   21.1 g of Na_2CO_3 is needed. Since 40.0 g are taken, more than enough is   supplied.

5.53   Dissolve 31 mL of 16 $\underline{M}$ HNO_3 in water and make the final volume equal to 500 mL.

CHAPTER 6

STATES OF MATTER, KINETIC THEORY
AND EQUILIBRIA

Review Exercises

6.1   Pressure, volume, temperature, and moles

6.2   Diffusion

6.3   Pressure is the ratio of force to area. Pressure = $\dfrac{\text{force}}{\text{area}}$.

6.4   The gravitational attraction that the earth exerts on the gases in the atmosphere gives the atmosphere, at any given point, a weight per unit area, or a pressure.

6.5   The weight at sea level and 0 $^{\circ}$C of a column of air uniformly 1 in.^{2} in cross section and that extends to the end of the earth's atmosphere.   This is 14.7 lb/in.^{2}.   The standard atmosphere is also defined as the pressure that supports a column of mercury 760 mm high in a vacuum at 0 $^{\circ}$C.

6.6   It means that the glass tube doesn't have to be as long.

6.7   1 atm = 760 mm Hg

6.8   1 torr = 1 mm Hg

6.9   The pressure of the atmosphere supports the column of mercury.

6.11   The higher pressure is exerted by 50 lb on 5 in.^{2} (or 10 lb/in.^{2}), not by 150 lb on 25 in.^{2} (or 6 lb/in.^{2}).

6.13   0.197 atm

6.15   757 mm Hg

6.16   That the pressure is inversely proportional to the volume, and that this is true for all gases (all that Boyle studied, that is).

6.17   It means 1 divided by. The pressure is proportional to 1 divided by the volume.

6.18   That the same mass of gas sample is used and that the temperature is the same.

6.19 $P_{total} = P_a + P_b + P_c + \ldots$

6.20   Nitrogen, oxygen, carbon dioxide, and water vapor

6.21   (a) n, and T (b) n, and P (c) T, and V
        (d) V, P, and T (e) n, and V

6.22   We first subtract the partial pressure of the water vapor (obtained from a table) from the total pressure to get the net pressure or the partial pressure of the gas being measured. Then we do a Boyle's law calculation with this pressure, the original pressure, and the original volume to find the net volume of the gas.

6.23   We pick whichever ratio of pressures, will, when multiplied by the initial volume, give a new value of volume that corresponds either to an increase or to a decrease as predicted by Boyle's law.

6.24   Whenever any given mass of any gas is cooled at constant pressure, there is obtained a series of data for gas volumes at a number of values of temperature. When these data are plotted, the points fall on a straight line (Charles' law, in effect), and when such lines for various gases are extrapolated to zero volume, they intercept the temperature axis at -273.15 $^{\circ}C$. Since it's impossible to have a negative volume, -273.15 $^{\circ}C$ must be the lowest attainable temperature, and it's renamed as 0 K.

6.25   The volume of a fixed mass of any gas is directly proportional to its Kelvin temperature, if the gas pressure is kept constant.

6.26 Pick a ratio of Kelvin temperatures which, when multiplied by the initial volume, will produce a new value for the volume that is either greater or less than the original volume according to the dictates of Charles' law.

6.27 Equal volumes of gases have equal numbers of moles when compared at the same pressure and temperature.

6.29 532 mm Hg

6.31 31 mm Hg

6.32 707 mm Hg

6.33 727 mm Hg

6.35 314 mL of methane

6.37 −124 $^{\circ}$C

6.38 1 atm (760 mm Hg) and 273 K

6.39 The volume occupied by one mole of a gas

6.40 At STP

6.41 $\underline{PV} = \underline{nRT}$

6.42 $\underline{P}$ in mm Hg; $\underline{V}$ in mL; $\underline{T}$ in kelvins; and $\underline{n}$ in moles

6.43 6.02×10^{23} molecules of H_2

6.45 (a) 7.08×10^{-2} mol of H_2 (b) 1.75×10^3 mL

6.47 (a) 0.310 mol of CO_2 (b) 7.29×10^3 mL of CO_2

6.48 The postulates that describe an ideal gas; the postulates of the kinetic theory of gases.

6.49 It obeys all of the gas laws exactly.

6.50 The particles of a gas neither attract nor repel each other.

6.51   Gas pressure is the net result of the innumerable forces exerted on the fixed area of the container by the collisions made at the walls by the gas particles.

6.52   When the volume of the container is reduced, the area of its walls also decreases, so if <u>area</u> becomes smaller in the equation

$$\text{pressure} = \frac{\text{force}}{\text{area}}$$

the pressure must increase.

6.53   (a)   The Kelvin temperature
        (b)   The Kelvin temperature is proportional to the average kinetic energy.

6.54   They cease.

6.55   A higher temperature means a higher average kinetic energy, and this means a higher average velocity. In a fixed–volume container, this would mean a higher pressure, so if the pressure is to be kept constant (a condition of Charles' law), the volume must be allowed to expand with increasing temperature.

6.56   Since a higher temperature means a higher average velocity of the gas particles, they beat on the fixed walls of the container with greater average force. This means a higher pressure, since pressure = force/area.

6.57   The distances between particles in liquids and solids are virtually zero, so physical properties of liquids and solids are much more sensitive to the chemical natures of the particles.

6.58   (a)   Rapidly moving molecules of the liquid escape into the space above the liquid and thus create a partial pressure called the vapor pressure.
        (b)   Vaporization is endothermic, so the addition of heat (by raising the temperature) shifts the following equilibrium to the right (Le Chatelier's principle):

$$\text{liquid} + \text{heat} \rightleftharpoons \text{vapor}$$

The extra vapor in a closed space exerts a higher pressure.

6.59 Less volatile. With a higher boiling point than water, DMSO will at any given temperature have a lower vapor pressure than that of water. Lower vapor pressure, in this context, means a lower volatility.

6.60 (a) B   (b) B   (c) A   (d) C   (e) A, B   (f) 100 $^{\circ}$C
(g) Their rates are equal.   (h) Equilibrium

6.61 The rate of evaporation equals the rate of condensation.

6.62 Using CO_2 as the example: $CO_2(\underline{s}) \rightleftharpoons CO_2(\underline{g})$

6.63 At equilibrium, the rates of escape and return change identically as you change the volume of the liquid, so the equilibrium vapor pressure is unchanged.

6.64 Boiling occurs whenever the liquid's vapor pressure equals the atmospheric pressure. If the latter is lowered, the temperature needed to create a vapor pressure equal to it is also lowered.

6.65 The temperature of the boiling water is lower in Denver, and the rate of a reaction (such as cooking) is therefore lower in Denver. Hence, the reaction takes longer in Denver.

6.66 Collisions and near misses of nitrogen molecules create temporary polarities in the otherwise nonpolar molecules and these temporary polarities, as a net effect, give rise to weak, net forces of attraction called London forces.

6.67 The London forces that can develop are generally greater the greater the sizes of the molecules involved.

6.68 If a system is in equilibrium and something happens to change the conditions, the system will shift in whatever way will help to restore equilibrium.

6.69 (a) Forward
(b) Forward

6.70   The melting point

6.71   (a) Water molecules in the vapor state can't return to the
             liquid state when they are blown away by the breeze.
        (b) Crushing the ice greatly increases the surface area from
             which water molecules can go into the liquid state, so the
             water molecules in ice can escape into their liquid state
             at many more places at the same time.
        (c) Water molecules can pass directly from the solid to the
             vapor state.

6.72   Electrical forces between the particles are weaker between
        nitrogen molecules and stronger between Na^+ and Cl^- ions.

6.73   The electron clouds about each of the reacting particles must
        interpenetrate before the chemical change can happen, and this
        takes energy.

6.74   (a) $CO(g)$ and $O_2(g)$
        (b) $CO_2(g)$
        (c) C
        (d) B
        (e) Exothermic, because the product is at a lower value of
             energy than the reactants, so this difference in energy
             must have been released to the environment.
        (f) E

6.75

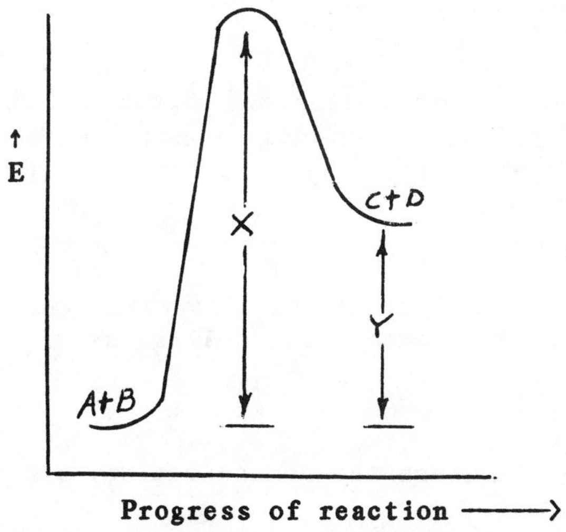

X = Energy of activation

Y = Heat of reaction

6.76

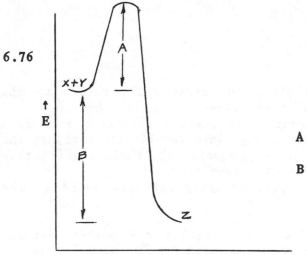

A = 3 kcal, energy of activation

B = 10 kcal, heat of reaction

6.77   A rise in temperature increases the energies of the collisions of the reacting particles and raises the frequency with which their electron clouds successfully interpenetrate and give rise to a chemical reaction.

6.78   10 °C

6.79   As rates of metabolic reactions increase with increasing temperature, the rate of oxygen consumption rises. This demand for oxygen makes the heart pump harder. It has to try to circulate the blood through the lungs and then deliver the oxygen to the tissues that need it at a higher rate.

6.80   Increase their concentrations

6.81   By increasing the frequency of collisions between reactant particles, which we do by increasing their concentrations, we increase the frequency at which <u>successful</u> collisions occur.

6.82   C, D

6.83   The catalyst lowers (b), the energy of activation, and so increases (d), the frequency of successful collisions.

6.84   Enzymes

6.85   Increase the temperature, increase the concentration, and use a catalyst

CHAPTER 7

WATER, SOLUTIONS AND COLLOIDS

Review Exercises

7.1 The outer level electron pairs of the central atom (O) repel each other with the result that the two pairs involved in the bonds to the two hydrogen atoms have axes at the bond angle for this molecule.

7.2 The individual bond polarities would cancel each other's effects.

7.3 A. Only in this planar form with perfectly symmetrical bond angles can the bond polarities exactly cancel and give a net zero polarity.

7.4 The three electron pairs get as far away from each other as they can (while still being associated with the central atom) when their axes are in the same plane and make angles of 120°.

7.5 In both, the bond polarities (which are small) cancel each other.

7.6 According to the VSEPR theory, the electron pairs in the bonds are farther away from each other in the tetrahedral array than in the square planar geometry.

7.7 In BF_3 there are only three pairs in the outer level of the central atom, but in NH_3 there are four pairs (with three involved in bonds and one pair unshared), so the geometry is more like that of methane.

7.8 (a) Covalent bond
(b) It is a nonpolar molecule and there is no $\delta+$ or $\delta-$ on H.

7.9 The relative electronegativities of C and H are so similar that the C — H bonds are nearly nonpolar. Hence there is no significant δ+ or δ− anywhere.

7.10 The ionic bond between the ions, Na^+ and OH^-, and the covalent bond that holds H to O within the hydroxide ion.

7.11

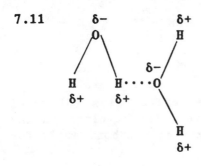

7.12

$$H - N - H \cdots N - H$$

with H atoms and δ+ δ− labels as shown

7.13 (a) The boiling point of ammonia is much less than that of water. Since the sizes of their molecules are about the same, we can't explain this difference in boiling point by differences in London forces between the molecules. Instead, there must be differences in the sizes of partial charges that exist more permanently than those involved in London forces.
(b) Both δ+ and δ− are larger in the water molecule than in the ammonia molecule.
(c) Oxygen is more electronegative than nitrogen.

7.14 5 kcal/mol

7.15 (a) A relatively large quantity of energy is needed to overcome the relatively strong forces of attraction between the molecules in these states.
(b) The relatively large forces of attraction between water molecules in the liquid state make for a strong jamming together of these molecules at the surface.

7.16   On a waxy surface, wax molecules don't attract water molecules so the latter jam in on each other to form a bead. On a glass surface, the polar particles in the glass attract water molecules more strongly than the water molecules can attract each other, so this attraction by the glass for the water molecules makes the water spread out and form a film rather than beads.

7.17   Reduces it.

7.18   Detergents and soaps

7.19   Bile salts. Bile. The action of the surfactant helps to wash oils and fats from undigested food particles and it helps to break up the oil and fat globules into very fine globules that are more rapidly attacked by water in the process of digestion.

7.20   It is harder to digest fats and oils in the absence of the bile salts.

7.21   A small sample taken in one place has the same composition and properties as a small sample from any other place.

7.22   The relative sizes of the dispersed or dissolved particles.

7.23   (a) Suspensions   (d) Solutions
        (b) Colloidal dispersions (e) Suspensions
        (c) Colloidal dispersions

7.24   Those that have the same kind of electrical charge, because they repel each other and can't coalesce into larger particles that would separate out.

7.25   Dissolved ions and small molecules make it a solution; dispersed protein molecules make it a colloidal dispersion; and various kinds of blood cells make blood a suspension.

7.26   The solute particles are too small to scatter light.

7.27   The uneven buffeting that randomly moving solvent molecules give to the particles in colloidal dispersion.

7.28   Test for the Tyndall effect.

7.29   It is a colloidal dispersion of one liquid in another. Examples are cream, mayonnaise, and milk.

7.30   A colloidal dispersion of a solid in a liquid. Examples are starch in water, jellies, and paints.

7.31   A sol that has adopted semisolid, semirigid form. Examples are gelatin desserts and fruit jellies.

7.32   The $\delta-$ ends of polar water molecules

7.33   It is surrounded by polar water molecules with their $\delta+$ ends pointing at it. (The drawing should be like that in the lower right corner of Figure 7.10.)

7.34   Water molecules are strongly attracted to each other and they find nothing in molecules of CCl_4 with which to establish forces of attraction of comparable strength, so the water molecules stay together.

7.35   (a)   This increases the surface area against which the solvent molecules can simultaneously attack and dissolve the solid.
        (b)   This moves freshly dissolved solute molecules away from the surfaces of the solid, which continuously exposes the surfaces to dissolving action.
        (c)   This increases the violence with which solvent molecules hit the solid's surface and cause the solid particles to go into solution.

7.36   The rate at which salt dissolves and returns to the crystals

$$NaCl(\underline{s}) \rightleftharpoons Na^+(\underline{aq}) + Cl^-(\underline{aq})$$

7.37   Make sure that the solution has undissolved $NaNO_3$ present.

7.38   The dissolving of a solid and a liquid is usually an endothermic change, so heat shifts the following equilibrium to the right, in favor of more solute being in solution.

        undissolved substance + heat $\rightleftharpoons$ dissolved substance

7.39 $NH_4Cl(\underline{s}) + heat \rightleftharpoons NH_4^+(\underline{aq}) + Cl^-(\underline{aq})$

The removal of heat (by lowering the temperature) shifts this equilibrium to the left and more solid NH_4Cl forms.

7.40 $CaSO_4 \cdot 2H_2O(\underline{s})$ +   heat $\longrightarrow$ $CaSO_4(\underline{s})$ + $2H_2O(g)$

7.41   They obey the law of definite proportions

7.42 $CuSO_4(\underline{s})$ + $5H_2O$ $\longrightarrow$ $CuSO_4 \cdot 5H_2O(\underline{s})$

7.43   It can remove moisture from the air by forming a hydrate. Yes.

7.44   The pellets can draw enough moisture from the air to form a solution of sodium hyroxide that coats every exposed pellet.

7.45 $\underline{x} = 2$. The formula is $\underline{M} \cdot 2H_2O$.

7.47   0.0375 g/L

7.49   A percentage of the carbon dioxide molecules in solution in water have reacted with the water to give carbonic acid. This reaction increases the solubility of CO_2 in water over that of oxygen.

7.50   Molecules of ammonia can enter into hydrogen bonding with water molecules. Nitrogen moecules can't do this.

7.51   Using oxygen as an example, the equilibrium between its gaseous (undissolved) state and its solution is:

$$O_2(g) \quad + \quad pressure \rightleftharpoons O_2(\underline{aq})$$

When the pressure is increased, this equilibrium shifts to the right in favor of more oxygen in solution.

7.52   The air in contact with this blood under a total pressure of 1 atm has a partial pressure of CO_2 equal to 30 mm Hg.

7.53   The first region (where the gas tension is 80 mm Hg).

7.54 $\dfrac{1.2 \text{ g NaOH}}{100 \text{ g NaOH solution}}$ and $\dfrac{100 \text{ g NaOH solution}}{1.2 \text{ g NaOH}}$

7.55 $\dfrac{1.5 \text{ g NaCl}}{100 \text{ mL NaCl solution}}$ and $\dfrac{100 \text{ mL NaCl solution}}{1.5 \text{ g NaCl}}$

7.56   1.5 mg/L

7.57 $\dfrac{12 \text{ volumes alcohol}}{100 \text{ volumes solution}}$ and $\dfrac{100 \text{ volumes solution}}{12 \text{ volumes alcohol}}$

7.58   (a) 4.50 g of NaCl   (c) 5.00 g of NH_4Cl
       (b) 3.13 g of $NaC_2H_3O_2$ (d) 17.5 g of Na_2CO_3

7.60   (a) 12.5 g of NaCl   (c) 7.50 g of $CaCl_2$
       (b) 5.00 g of KBr   (d) 6.75 g of NaCl

7.62   75.0 mL of methyl alcohol

7.64   (a) 70.0 mL of NaOH solution   (d) 80.0 mL of NaOH solution
       (b) 25.0 mL of Na_2CO_3 solution (e) 720 mL of glucose solution
       (c) 20.0 mL of glucose solution

7.66   8.71 g of $NaC_2H_3O_2 \cdot 3H_2O$

7.68   125 g of stock solution

7.70   (a) 35.9%(w/w) HCl   (b) 118 mL of concentrated solution

7.72 $-1.86\ ^{\circ}C$

7.73   The solute in the second solution breaks up (ionizes) into two
       ions per formula unit when it dissolves, so the effective
       concentration is 2.00 mol/1000 g H_2O.

7.74   The rate at which water can move <u>into</u> the solution is greater
       than the rate at which it can return, because solute particles
       tend to block the return. The figure should resemble part b of
       Figure 7.15.

7.75   The osmotic membrane doesn't let anything in the dissolved
       state go through, but the dialyzing membrane lets solutes
       through and blocks colloidally dispersed particles.

7.76   The blocking action of the solute particles is a function of
       their <u>presence</u> and <u>relative numbers</u>, not their chemical
       properties.

7.77   A higher temperature gives greater average velocities to the solvent molecules. They therefore can move more rapidly through the membrane, but this increase in rate is greater from the side of the dilute solution into the concentrated solution than the increase in the rate of return.

7.78   This solution is 2 $\underline{M}$ in all solute particles, and osmolarity is a function of the molar concentrations of all osmotically active solute particles.

7.79   0.080 $\underline{M}$ Na_2SO_4 solution, because it is   3 x 0.080 $\underline{M}$ = 0.240 $\underline{M}$ in its osmolarity.

7.81   Solution A. While the osmolarities of A and B are identical with respect to their dissolved salts and sugars, A has the higher concentration of starch.

7.82   186 mm Hg

7.84   They swell and split open.

7.85   (a) Hypertonic
        (b)   (1) Hemolysis   (2) crenation

7.86   The loss of large molecules lowers the colloidal osmotic pressure of the blood to a value less than the osmotic pressure of the fluids just outside of the blood vessels. As a result, there is a net flow of water from the blood to the outside of the blood vessels, and the blood volume decreases.

CHAPTER 8

ACIDS, BASES, AND SALTS

Review Exercises

8.1   Acids, bases, and salts

8.2   (a) No
      (b) Yes, any polyatomic ion such as SO_4^{2-}
      (c) No
      (d) No
      (e) The concentration of potassium <u>ion</u>

8.3 $2H_2O \underset{\longleftarrow}{\longrightarrow} H_3O^+ \quad + \quad OH^-$

      Hydronium   Hydroxide
         ion   ion

8.4   They furnish hydronium ions in water.

8.5   Hydrogen ion and hydronium ion

8.6   Acids are compounds that supply hydrogen ions in water. Bases
      are compounds that supply hydroxide ions in water.

8.7   Salts do not supply a <u>common</u> ion in solution, whereas aqueous
      acids furnish hydrogen ions and aqueous bases supply hydroxide
      ions.

8.8   (a) Acidic   (b) Neutral   (c) Basic

8.9   They consist of oppositely charged ions, and between these are
      strong forces of attraction that set up rigid solids.

8.10 (1) An aqueous solution that conducts electricity, such as a
           salt solution
      (2) An ionic compound in its pure form which, if either melted

or put into solution in water, conducts electricity. Examples are NaCl, KBr, NaOH, and $Mg(NO_3)_2$.

8.11 Cathode

8.12 Positive

8.13 Cations

8.14 Cations take electrons from the cathode simultaneously as anions deliver electrons to the anode, so the net effect is just as if electrons transferred directly from the cathode to the anode — an electron flow.

8.15 Methyl alcohol does not furnish any ions of any kind in water.

8.16 Strong electrolyte

8.17 It is a molecular compound. It does not consist of ions in the liquid state.

8.18 Hydrochloric acid is the aqueous solution of the gas, hydrogen chloride.

8.19 In H_3O^+, because the weaker bond in H — Cl breaks to form the stronger bond in the hydronium ion by the following reaction.

$$HCl \ (g) \ + \ H_2O \ \longrightarrow \ H_3O^+(aq) \ + \ Cl^-(aq)$$

8.20 Weak acid

8.21 Monoprotic. Acetic acid

8.22 Hydrofluoric acid, HF(aq)
Hydrochloric acid, HCl(aq)
Hydrobromic acid, HBr(aq)
Hydriodic acid, HI(aq)

8.23 $H_2A(aq) \ + \ H_2O \ \rightleftharpoons \ HA^-(aq) \ + \ H_3O^+(aq)$

Or, $H_2A(aq) \rightleftharpoons HA^-(aq) \ + \ H^+(aq)$

$HA^-(aq) \ + \ H_2O \rightleftharpoons A^{2-}(aq) \ + \ H_3O^+(aq)$

Or, $HA^-(aq) \rightleftharpoons A^{2-}(aq) \ + \ H^+(aq)$

8.24 The second H^+ ion has to pull away from a particle that has more negative charge than the first H^+ ion. Unlike charges attract, so the more unlike they are the more energy is needed.

8.25 The extra oxygen atom provides more electron–withdrawing action, and this weakens the H — O bond more.

8.26 Sulfuric acid is the stronger acid. Its molecules have one more oxygen to exert electron–withdrawing action on the electron pairs of the H — O bonds.

8.27 $CO_2(aq) + H_2O \rightleftharpoons H_2CO_3(aq)$

8.28 Letting H^+ represent the hydronium ion:

$$H_2CO_3(aq) \rightleftharpoons H^+(aq) + HCO_3^-(aq)$$
$$HCO_3^-(aq) \rightleftharpoons H^+(aq) + CO_3^{2-}(aq)$$

8.29 What does dissolve in water ionizes 100 percent.

8.30 Only a very low percentage of all ammonia molecules in solution react with water to generate hydroxide ions in the equilibrium:

$$NH_3(aq) + H_2O \rightleftharpoons NH_4^+(aq) + OH^-(aq)$$

8.31 Sodium hydroxide, NaOH, and potassium hydroxide, KOH

8.32 $CO_2(aq) + NaOH(aq) \longrightarrow NaHCO_3(aq)$

8.33 A solution of ammonia in water. Since it contains only a very low percentage of ammonium and hydroxide ions, it's inappropriate to call it ammonium hydroxide (although some chemical suppliers do).

8.34 Hydrochloric acid, $HCL(aq)$
Hydrobromic acid, $HBr(aq)$
Hydriodic acid, $HI(aq)$
Nitric acid, $HNO_3(aq)$
Sulfuric acid, $H_2SO_4(aq)$

8.35 Sodium hydroxide, NaOH (very soluble)
Potassium hydroxide, KOH (very soluble)
Magnesium hydroxide, $Mg(OH)_2$ (slightly soluble)
Calcium hydroxide, $Ca(OH)_2$ (slightly soluble)

8.36   (a) Yes   (b) No

8.37

(a) $HNO_3(aq)$ + $NaOH(aq)$ $\longrightarrow$ $NaNO_3(aq)$ + H_2O

$H^+(aq)$ + $OH^-(aq)$ $\longrightarrow$ H_2O

(b) $2HCl(aq)$ + $K_2CO_3(aq)$ $\longrightarrow$ $2KCl(aq)$ + H_2O + $CO_2(g)$

$2H^+(aq)$ + $CO_3^{2-}(aq)$ $\longrightarrow$ H_2O + $CO_2(g)$

(c) $2HBr(aq)$ + $CaCO_3(s)$ $\longrightarrow$ $CaBr_2(aq)$ + H_2O + $CO_2(g)$

$2H^+(aq)$ + $CaCO_3(s)$ $\longrightarrow$ $Ca^{2+}(aq)$ + H_2O + $CO_2(g)$

(d) $HNO_3(aq)$ + $NaHCO_3(aq)$ $\longrightarrow$ $NaNO_3(aq)$ + H_2O + $CO_2(g)$

$H^+(aq)$ + $HCO_3^-(aq)$ $\longrightarrow$ H_2O + $CO_2(g)$

(e) $HI(aq)$ + $NH_3(aq)$ $\longrightarrow$ $NH_4I(aq)$

$H^+(aq)$ + $NH_3(aq)$ $\longrightarrow$ $NH_4^+(aq)$

(f) $2HNO_3(aq)$ + $Mg(OH)_2(s)$ $\longrightarrow$ $Mg(NO_3)_2(aq)$ + $2H_2O$

$2H^+(aq)$ + $Mg(OH)_2(s)$ $\longrightarrow$ $Mg^{2+}(aq)$ + $2H_2O$

(g) $2HBr(aq)$ + $Zn(s)$ $\longrightarrow$ $ZnBr_2(aq)$ + $H_2(g)$

$2H^+(aq)$ + $Zn(s)$ $\longrightarrow$ $Zn^{2+}(aq)$ + $H_2(g)$

8.38

(a) $2KOH(aq)$ + $H_2SO_4(aq)$ $\longrightarrow$ $K_2SO_4(aq)$ + $2H_2O$

$OH^-(aq)$ + $H^+(aq)$ $\longrightarrow$ H_2O

(b) $Na_2CO_3(aq)$ + $2HNO_3(aq)$ $\longrightarrow$ $2NaNO_3(aq)$ + $CO_2(g)$ + H_2O

$CO_3^{2-}(aq)$ + $2H^+(aq)$ $\longrightarrow$ $CO_2(g)$ + H_2O

(c) $KHCO_3(aq)$ + $HCl(aq)$ $\longrightarrow$ $KCl(aq)$ + $CO_2(g)$ + H_2O

$HCO_3^-(aq)$ + $H^+(aq)$ $\longrightarrow$ $CO_2(g)$ + H_2O

(d) $MgCO_3(s)$ + $2HI(aq)$ $\longrightarrow$ $MgI_2(aq)$ + $CO_2(g)$ + H_2O

$MgCO_3(s)$ + $2H^+(aq)$ $\longrightarrow$ $Mg^{2+}(aq)$ + $CO_2(g)$ + H_2O

(e) $NH_3(aq)$ + $HBr(aq)$ $\longrightarrow$ $NH_4Br(aq)$

$$NH_3(\underline{aq}) + H^+(\underline{aq}) \longrightarrow NH_4^+(\underline{aq})$$

(f) $Ca(OH)_2(\underline{s}) + 2HCl(\underline{aq}) \longrightarrow CaCl_2(\underline{aq}) + H_2O$

$Ca(OH)_2(\underline{s}) + 2H^+(\underline{aq}) \longrightarrow Ca^{2+}(\underline{aq}) + H_2O$

(g) $2Al(\underline{s}) + 6HCl(\underline{aq}) \longrightarrow 2AlCl_3(\underline{aq}) + 3H_2(\underline{g})$

$2Al(\underline{s}) + 6H^+(\underline{aq}) \longrightarrow 2Al^{3+}(\underline{aq}) + 3H_2(\underline{g})$

8.39

(a) $OH^-(\underline{aq}) + H^+(\underline{aq}) \longrightarrow H_2O$

(b) $HCO_3^-(\underline{aq}) + H^+(\underline{aq}) \longrightarrow CO_2(\underline{g}) + H_2O$

(c) $CO_3^{2-}(\underline{aq}) + 2H^+(\underline{aq}) \longrightarrow CO_2(\underline{g}) + H_2O$

(d) $NH_3(\underline{aq}) + H^+(\underline{aq}) \longrightarrow NH_4^+(\underline{aq})$

8.40 $\underline{M}CO_3(\underline{s}) + 2H^+(\underline{aq}) \longrightarrow \underline{M}^{2+}(\underline{aq}) + CO_2(\underline{g}) + H_2O$

8.41 $\underline{M}(OH)_2(\underline{s}) + 2H^+(\underline{aq}) \longrightarrow \underline{M}^{2+}(\underline{aq}) + 2H_2O$

8.42 $\underline{M}(\underline{s}) + 2H^+(\underline{aq}) \longrightarrow \underline{M}^{2+}(\underline{aq}) + H_2(\underline{g})$

8.43 (a) To be higher in the activity series means, in a qualitative sense, that the atoms of the element have a greater tendency to become ions than those of elements lower in the series.

(b) The lower ionization energies of sodium and potassium also indicate this greater tendency to become ions.

8.44 Zinc reacts more rapidly with 1 $\underline{M}$ HNO_3. The reaction is with the hydronium ion, and the molar concentration of this ion is much higher in 1 $\underline{M}$ HNO_3 (which is 100% ionized) than in 1 $\underline{M}$ $HC_2H_3O_2$ which is a weak acid and weakly ionized in water.

8.46 0.800 mol of KOH

8.48 5.46 g of $CaCO_3$

8.50 1.40 g of K_2CO_3

8.52 35.2 mL of KOH solution

8.54 $Na_2CO_3(\underline{s}) + 2HCl(\underline{aq}) \longrightarrow CO_2(\underline{g}) + H_2O + 2NaCl(\underline{aq})$

$$Na_2CO_3(\underline{s}) + 2H^+(\underline{aq}) \longrightarrow CO_2(g) + H_2O + 2Na^+(\underline{aq})$$

18.3 L of CO_2

79.5 g of Na_2CO_3

8.55   Arrhenius would pick the hydrogen ion, because it is the only species in his view that is responsible for the acidic properties. Brønsted would pick the most abundant proton-donor, intact molecules of acetic acid.

8.56   Arrhenius viewed the OH^- ion as the only basic species. Brønsted would call the most abundant proton-acceptor, NH_3, as the most abundant base.

8.57   (a) H_2SO_4 (b) H_2CO_3 (c) HI   (d) HNO_2

8.58   (a) H_2SO_3 (b) HBr   (c) H_3O^+ (d) $HC_2H_3O_2$

8.59   (a) HCO_3^- (b) HPO_4^{2-} (c) NH_3 (d) O^{2-}

8.60   (a) NH_2^- (b) NO_2^- (c) SO_3^{2-} (d) HSO_3^-

8.61   (a) HCO_3^- (b) $H_2PO_4^-$ (c) NO_2^-

8.62   (a) NH_2^- (b) OH^- (c) S^{2-}

8.63   (a) $H_2PO_4^-$ (b) H_2SO_3 (c) NH_4^+

8.64   (a) HCl   (b) H_2O (c) HSO_4^-

8.65 $HPO_4^{2-}(\underline{aq}) + SO_4^{2-}(\underline{aq}) \longleftrightarrow PO_4^{3-}(\underline{aq}) + HSO_4^-(\underline{aq})$

8.66 $(\underline{Asp})^-(\underline{aq}) + H_3O^+(\underline{aq}) \longrightarrow\!\!\!\leftarrow H(\underline{Asp})(\underline{aq}) + H_2O$

8.67   The test reagent is aqueous sodium hydroxide. When it is added to a solution of an ammonium salt, the odor of ammonia would soon be noted because the following reaction produces aqueous ammonia:

$$OH^-(\underline{aq}) + NH_4^+(\underline{aq}) \longrightarrow H_2O + NH_3(\underline{aq})$$

8.68   Potassium hydroxide, KOH

$$KOH(\underline{aq}) + HCl(\underline{aq}) \longrightarrow KCl(\underline{aq}) + H_2O$$

Potassium bicarbonate, $KHCO_3$
$$KHCO_3(\underline{aq}) + HCl(\underline{aq}) \longrightarrow KCl(\underline{aq}) + CO_2(\underline{g}) + H_2O$$

Potassium carbonate, K_2CO_3
$$K_2CO_3(\underline{aq}) + 2HCl(\underline{aq}) \longrightarrow 2KCl(\underline{aq}) + CO_2(\underline{g}) + H_2O$$

8.69   Lithium hydroxide, LiOH
$$LiOH(\underline{aq}) + HBr(\underline{aq}) \longrightarrow LiBr(\underline{aq}) + H_2O$$

Lithium bicarbonate, $LiHCO_3$
$$LiHCO_3(\underline{aq}) + HBr(\underline{aq}) \longrightarrow LiBr(\underline{aq}) + CO_2(\underline{g}) + H_2O$$

Lithium carbonate, Li_2CO_3
$$Li_2CO_3(\underline{aq}) + 2HBr(\underline{aq}) \longrightarrow 2LiBr(\underline{aq}) + CO_2(\underline{g}) + H_2O$$

8.70   (c), (d)

8.71   (d), (e)

8.72   (b), (c)

8.73   (d), (f)

8.74   (a) $Ag^+(\underline{aq}) + Cl^-(\underline{aq}) \longrightarrow AgCl(\underline{s})$

      (b) No reaction

      (c) $H^+(\underline{aq}) + OH^-(\underline{aq}) \longrightarrow H_2O$

      (d) $Pb^{2+}(\underline{aq}) + 2Cl^-(\underline{aq}) \longrightarrow PbCl_2(\underline{s})$

      (e) No reaction

      (f) $Cu^{2+}(\underline{aq}) + S^{2-}(\underline{aq}) \longrightarrow CuS(\underline{s})$

      (g) $Ba^{2+}(\underline{aq}) + SO_4{}^{2-}(\underline{aq}) \longrightarrow BaSO_4(\underline{s})$

      (h) $H^+(\underline{aq}) + OH^-(\underline{aq}) \longrightarrow H_2O$

      (i) $Ni^{2+}(\underline{aq}) + S^{2-}(\underline{aq}) \longrightarrow NiS(\underline{s})$

      (j) $Ag^+(\underline{aq}) + Cl^-(\underline{aq}) \longrightarrow AgCl(\underline{s})$

(k) $H^+(\underline{aq}) + HCO_3^-(\underline{aq}) \longrightarrow CO_2(\underline{g}) + H_2O$

(l) $Ca^{2+}(\underline{aq}) + 2OH^-(\underline{aq}) \longrightarrow Ca(OH)_2(\underline{s})$

8.75 (a) $Cd^{2+}(\underline{aq}) + S^{2-}(\underline{aq}) \longrightarrow CdS(\underline{s})$

(b) $H^+(\underline{aq}) + OH^-(\underline{aq}) \longrightarrow H_2O$

(c) $Ba^{2+}(\underline{aq}) + SO_4^-(\underline{aq}) \longrightarrow BaSO_4(\underline{s})$

(d) $Pb^{2+}(\underline{aq}) + SO_4^{2-}(\underline{aq}) \longrightarrow PbSO_4(\underline{s})$

(e) No reaction

(f) $H^+(\underline{aq}) + HCO_3^-(\underline{aq}) \longrightarrow CO_2(\underline{g}) + H_2O$

(g) $Ni^{2+}(\underline{aq}) + S^{2-}(\underline{aq}) \longrightarrow NiS(\underline{s})$

(h) $H^+(\underline{aq}) + OH^-(\underline{aq}) \longrightarrow H_2O$

(i) No reaction

(j) $H^+(\underline{aq}) + HCO_3^-(\underline{aq}) \longrightarrow CO_2(\underline{g}) + H_2O$

(k) No reaction

(l) $Pb^{2+}(\underline{aq}) + CrO_4^{2-}(\underline{aq}) \longrightarrow PbCrO_4(\underline{s})$

8.76 (a) $Na(\underline{Ste})(\underline{s}) \rightleftharpoons Na^+(\underline{aq}) + (\underline{Ste})^-(\underline{aq})$

(b) It would shift to the left, and more solid sodium stearate would separate out of the solution.

(c) Common ion effect, as explained in parts (a) and (b).

8.77 $\underline{K}_{eq} = \dfrac{[H_2O]^2}{[H_2]^2[O_2]}$

8.78 $\underline{K}_{eq} = \dfrac{[HI]^2}{[H_2][I_2]}$

(b) To the right, in favor of more product

(c) No change

8.79 (a) $HNO_2(\underline{aq}) \rightleftharpoons H^+(\underline{aq}) + NO_2^-(\underline{aq})$

(b) $\underline{K}_a = \dfrac{[H^+][NO_2^-]}{[HNO_2]}$

(c) HNO_2

(d) It would not change.

8.80 (a) $HPO_4^{2-}(\underline{aq}) + H_2O \rightleftharpoons H_2PO_4^-(\underline{aq}) + OH^-(\underline{aq})$

(b) $\underline{K}_{eq} = \dfrac{[H_2PO_4^-][OH^-]}{[HPO_4^{2-}][H_2O]}$

(c) $\underline{K}_b = \dfrac{[H_2PO_4^-][OH^-]}{[HPO_4^{2-}]}$

(d) HPO_4^{2-}

8.81 (a) The new term, $[AgCl(\underline{s})] \times \underline{K}_{eq}$, itself will be a constant if the value of $[AgCl(\underline{s})]$ is a constant. And this must be so since AgCl is a <u>solid</u> and therefore its "concentration" in moles per liter actually is akin to the density of the solid, <u>which is constant</u>. Thus, we can translate moles/liter into grams/liter by converting the moles of AgCl that would be present in one liter of this solid into grams of AgCl in one liter of this solid. Remember that density is defined as mass/volume, so grams/liter represents mass/volume, or density (although a density not in g/mL, the usual but not the required units). Thus, the term $[AgCl(\underline{s})] \times \underline{K}_{eq}$ is a constant because it is the product of two constants.

(b) $\underline{K}_{sp} = [Ba^{2+}][SO_4^{2-}]$

(c) $BaSO_4$, since its $\underline{K}_{sp}$ is larger than that of AgCl.

ACIDITY: DETECTION, CONTROL, MEASUREMENT

Review Exercises

9.1 $\underline{K_w}$ = $[H^+][OH^-]$

9.2 1.00×10^{-14}

9.3   The forward reaction in the equilibrium: $H_2O \rightleftharpoons H^+ + OH^-$ is endothermic, so the addition of heat shifts the equilibrium to the right in accordance with Le Châtelier's principle.

9.5 9.0×10^{-16}

9.6 $[H^+]$ = 1×10^{-pH}

9.7   Acidic

9.8   Acidic

9.9   2.0

9.10   pOH = 2   pH = 12

9.11   Because in water at 25 $^\circ$C, $[H^+]$ = 1.00×10^{-7}

9.12   More alkaline. Alkalosis

9.13   More acidic. Acidosis

9.14 1×10^{-5} mol/L. Slightly acidic

9.15 1×10^{-4} to 1×10^{-5} mol/L

9.16   Strong. At a pH of 1, the value of $[H^+]$ is 1×10^{-1} mol/L, which is numerically identical to the initial concentration of the acid, so all of the acid is ionized.

9.17 Weak. At a pH of 4.56, the value of $[H^+]$ is 1×10^{-4} to 1×10^{-5} mol/L, but the initial concentration of the acid is much higher (1×10^{-2} mol/L), over 100 times higher than $[H^+]$. Therefore, only a small percent of the acid molecules can be ionized.

9.18 Basic (pH = 7.9)

9.19 pH = 7.20. This means acidosis since 7.20 is less than 7.35, the normal pH of blood.

9.20 The acetate ions react slightly with water as follows, which produces a slight excess of hydroxide ions in the solution.

$$C_2H_3O_2^-(aq) \; + \; H_2O \; \rightleftharpoons \; HC_2H_3O_2(aq) \; + \; OH^-(aq)$$

9.21 (a) Neutral   (b) Acidic   (c) Basic   (d) Acidic   (e) Basic

9.22 (a) Neutral   (b) Basic   (c) Basic   (d) Acidic   (e) Basic

9.23 It maintains a steady pH even as slight amounts of acids or bases are added.

9.24 HCO_3^- and H_2CO_3

9.25 Acid is neutralized by: $H^+(aq) \; + \; HCO_3^-(aq) \; \longrightarrow \; H_2CO_3(aq)$
Base is neutralized by:
$\quad OH^-(aq) \; + \; H_2CO_3(aq) \; \longrightarrow \; H_2O \; + \; HCO_3^-(aq)$

9.26 The permanent removal of CO_2 at the lungs means that the following equilibria all shift to the right so that one H^+ disappears for each CO_2 that is lost.

$$H^+(aq) \; + \; HCO_3^-(aq) \; \rightleftharpoons \; H_2CO_3(aq) \; \rightleftharpoons \; H_2O \; + \; CO_2(g)$$

9.27 (a) Alkalosis
(b) For each CO_2 lost at the lungs, one H^+ ion is permanently neutralized (as described in the answer to Review Exercise 9.26). The loss of H^+ ion, of course, means a rise in the

value of the pH of the blood.

9.28   (a)   Acidosis
       (b)   Hypoventilation helps the body to retain CO_2, which helps
             it retain H_2CO_3 that is made from CO_2 and H_2O. Retaining
             H_2CO_3 means retaining the H^+ ion that results from the
             slight ionization of H_2CO_3 as an acid.

9.29 $HPO_4{}^{2-}$ and $H_2PO_4{}^-$

9.30   Slightly acidic

9.31   3.5 – 5.0 meq K^+/L

9.32   4.2 – 5.2 meq Ca^{2+}/L

9.34   3.11 g or 3.11 x 10^3 mg of Na^+

9.36   2.00 meq Mg^{2+}

9.38   Anion gap = 19 meq/L. This is above the normal range of 5 – 14
       meq/L, so this anion gap suggests a disturbance in metabolism.

9.39   It means that the concentration of the solution is accurately
       known.

9.40   When the color change of the indicator occurs

9.41   The indicator should undergo its color change at a pH that
       matches the pH of the solution that would result if it were
       prepared from the salt that is made by the titration.

9.42   Titration of any strong acid (such as HCl) with any strong base
       (such as NaOH)

9.43   Titration of acetic acid (or any weak acid) with sodium
       hydroxide (or any strong base)

9.44   Titration of ammonia solution with a strong acid such as HCl.

9.45   Each mole of a diprotic acid corresponds to two equivalents of
       acid.

9.46 (a) $\dfrac{1 \text{ eq HCl}}{36.46 \text{ g HCl}}$ or $\dfrac{36.46 \text{ g HCl}}{1 \text{ eq HCl}}$

(b) $\dfrac{0.115 \text{ eq } H_2SO_4}{1000 \text{ mL } H_2SO_4 \text{ solution}}$ or $\dfrac{1000 \text{ mL } H_2SO_4 \text{ solution}}{0.115 \text{ eq } H_2SO_4}$

9.47 $\dfrac{eq}{L} = \dfrac{1000 \text{ meq}}{1000 \text{ mL}} = \dfrac{meq}{mL}$

9.48 $\dfrac{mol}{L} = \dfrac{1000 \text{ mmol}}{1000 \text{ mL}} = \dfrac{mmol}{mL}$

9.50 (a) 63.04 g/eq   (b) 42.16 g/eq
     (c) 90.09 g/eq   (d) 36.00 g/eq

9.52 (a) 0.04997 eq   (b) 0.6724 eq
     (c) 1.225 eq   (d) 1.575 eq

9.54 (a) 0.50 meq   (b) 109 meq
     (c) 21 meq   (d) 35 meq

9.56 (a) 0.4000 $\underline{M}$ HI   (b) 0.2000 $\underline{M}$ H_2SO_4
     0.4000 $\underline{N}$ HI   0.4000 $\underline{N}$ H_2SO_4

9.58 64.00

9.60 (a) 0.8091 g of HBr   (b) 1.226 g of H_2SO_4
     (c) 33.13 g of Na_2CO_3

9.62 (a) 0.1267 $\underline{N}$ NaOH   (b) 5.068 g of NaOH/L

CHAPTER 10

RADIOACTIVITY AND NUCLEAR CHEMISTRY

Review Exercises

10.1   It emits radiations such as alpha radiation, beta radiation, or gamma radiation (and sometimes two of these).

10.2   Transmutation

10.3   Gamma radiation

10.4   (a) $^{4}_{2}He$ (b) $^{0}_{-1}e$ (c) $^{0}_{0}\gamma$

10.5   It is the most massive of the particles and it carries the largest charge. Therefore, it collides very quickly with a molecule in the air or other matter that it enters.

10.6   The mass number is reduced by 4 units and the atomic number is reduced by 2 units.

10.7   There is no change in mass number, because the mass number of the beta particle is 0. The atomic number increases by one because the loss of $^{0}_{-1}e$ from a neutron creates an additional proton.

10.8   No change in either occurs.

10.9   No. No transmutation occurs if only gamma radiation is emitted.

10.10 A neutron changes into a proton as an electron is ejected.

10.12 (a) $^{216}_{84}Po$ (b) $^{140}_{57}La$

10.14 (a) $^{211}_{83}Bi \longrightarrow ^{0}_{-1}e + ^{211}_{84}Po$

451

(b) $^{242}_{94}Pu \longrightarrow\ ^{4}_{2}He\ +\ ^{238}_{92}U\ +\ ^{0}_{0}\gamma$

(c) $^{30}_{13}Al \longrightarrow\ ^{0}_{-1}e\ +\ ^{30}_{14}Si$

(d) $^{243}_{96}Cm \longrightarrow\ ^{4}_{2}He\ +\ ^{239}_{94}Pu$

10.15 Lead–214 forms by successive decays of U–238. The quantity of lead–214 in the sample diminishes by half each 19.7 minutes.

10.16 The beta and gamma emitter, because its radiations are more penetrating.

10.18 2.20×10^{-3} ng

10.19 The ions that radiations produce are strange, unstable, highly reactive ions that initiate undesirable reactions in the body.

10.20 Any particle with an unpaired electron. The particle lacks an outer octet and so is reactive.

10.21 They can cause birth defects.

10.22 Its radiations can cause cancer.

10.23 Their intensities diminish with the square of the distance; and they can be blocked by dense absorbing materials such as lead.

10.24 There is no level of exposure below which no damage is possible.

10.25 In low doses over a long period, radiations can initiate cancer. In well–focused, massive doses over a short period, radiations can kill cancer cells.

10.26 4

10.27 They move in straight lines and spread out in the way that light from a light bulb spreads out.

10.28 Radionuclides in natural materials such as the soil, fallout from nuclear testing, cosmic rays, medical X rays, radioactive wastes released from nuclear power plants, television tubes, and medical radionuclides.

10.29   Cosmic rays are more intense at higher altitudes.

10.30   Sr-90 is a bone seeker. I-131 is taken up by the thyroid gland. Cs-137 gets as widely distributed as the sodium ion.

10.31   In decay, only a tiny part of the atom breaks away. In fission, the whole atom splits roughly in half.

10.32   Each fission event produces more neutron initiators than were needed to cause the fission event.

10.33   The rods absorb neutrons and prevent them from causing fissions.

10.35   4000 millirad

10.36   The becquerel

10.37   The sample is undergoing $1.5 \times (3.7 \times 10^{10}) \times 10^{-3} = 5.6 \times 10^7$ disintegrations per second.

10.38   The roentgen

10.39   The gray. The older unit is the rad.

10.40   650 rad

10.41   They are basically equivalent.

10.42   Different kinds of tissue have different responses to the same quantity of rads.

10.43   The rad doses are adjusted for the kinds of tissues and radiations and expressed as rems.

10.44   100 mrem

10.45   The electron-volt

10.46   100 keV or less

10.47   Radiations can cause cancer, so for diagnosis the lowest usable energies are in order.

10.48 Radiations fog the film, and the degree of fogging is proportional to the exposure.

10.49 Radiations enter the tube and generate ions in the low-pressure gas in the tube. This makes the gas a conductor, so a circuit is thereby closed, which leads to a count or a click. Alpha particles can't penetrate the tube's window.

10.51 $^{109}_{47}Ag + ^{4}_{2}He \longrightarrow ^{113}_{49}In$

10.53 $^{66}_{30}Zn + ^{1}_{1}H \longrightarrow ^{67}_{31}Ga$

10.55 $^{19}_{9}F + ^{4}_{2}He \longrightarrow ^{23}_{11}Na \longrightarrow ^{22}_{11}Na + ^{1}_{0}\underline{n}$

10.57 $^{14}_{7}N + ^{2}_{1}H \longrightarrow ^{16}_{8}O \longrightarrow ^{15}_{8}O + ^{1}_{0}\underline{n}$

10.59 $^{32}_{16}S + ^{1}_{0}\underline{n} \longrightarrow ^{32}_{15}P + ^{1}_{1}H$

10.60 The shorter the half-life, the more active is the radionuclide and hence the smaller is the dose that is needed to get results.

10.61 We can get by with a smaller quantity.

10.62 Gamma radiation can penetrate the entire body and reach a detector and thereby serve a diagnostic purpose. Alpha and beta radiation would be absorbed in the body, causing harm, without giving diagnostic value.

10.63 It emits only gamma radiation, and it has a shorter half-life.

10.64 It has a more intense radiation and it involves only gamma radiation.

10.65 As phosphate ion, because the hydroxyapatite in bone contains this ion.

CHAPTER 11

INTRODUCTION TO ORGANIC CHEMISTRY

Review Exercises

11.1   They occur widely in living organisms.

11.2   Organic compounds cannot be made without the presence of a vital force that is available only in living organisms.

11.3   All efforts to make organic compounds in the laboratory had failed.

11.4   He prepared urea, a waste product from animals, by heating a solution of ions that had been obtained from mineral sources.

11.5   Covalent bond.

11.6   (b), (d), and (e)

11.7   Ionic. Inorganic

11.8   Very few organic compounds consist of ions or furnish ions in water.

11.9   (a) Molecular (All ionic compounds are <u>solids</u> at room temperature.)
        (b) Ionic (The compound is a carbonate or a bicarbonate.)
        (c) Molecular (Most ionic compounds melt above this temperature, and few burn in air.)
        (d) Ionic (It has a high melting point and does not burn.)
        (e) Molecular (All ionic compounds are solids at room temperature.)

11.10   Straight-chain. No carbon has <u>more</u> than two other carbon atoms attached to it.

11.11 None is possible.

11.12 (a) 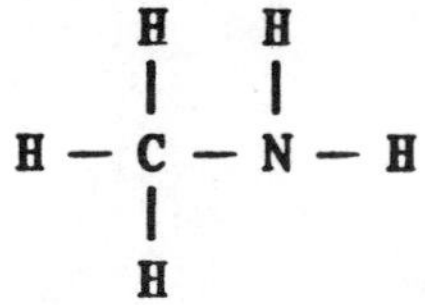
```
       H   H
       |   |
   H — C — N — H
       |
       H
```

(g) 
```
       H
       |
   H — N — O — H
```

(b) 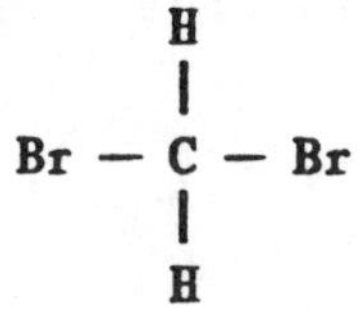
```
       H
       |
  Br — C — Br
       |
       H
```

(h)   H — C ≡ C — H

(c) 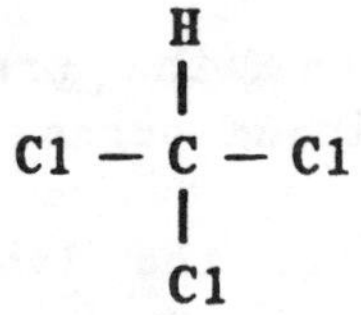
```
       H
       |
  Cl — C — Cl
       |
       Cl
```

(i) 
```
       H   H
       |   |
   H — N — N — H
```

(d) 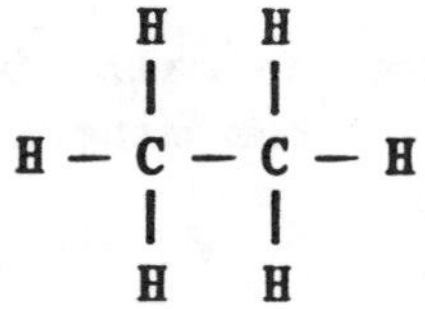
```
       H   H
       |   |
   H — C — C — H
       |   |
       H   H
```

(j)   H — C ≡ N

(e) 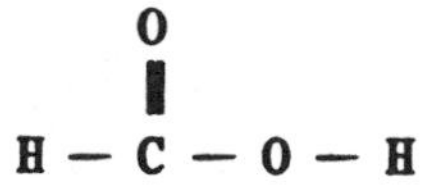
```
       O
       ‖
   H — C — O — H
```

(k) 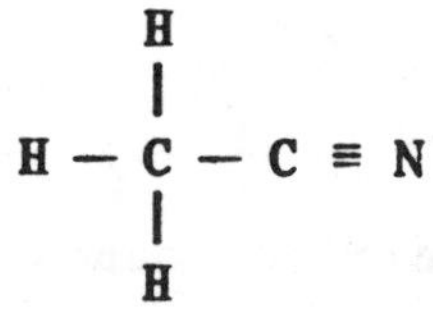
```
       H
       |
   H — C — C ≡ N   or   H — C=C=N — H
       |
       H
```
with

```
                                    H
                                    |
                              H — C=C=N — H
```

(f) 
```
       O
       ‖
   H — C — H
```

(1)
```
       H
       |
   H — C — O — H
       |
       H
```

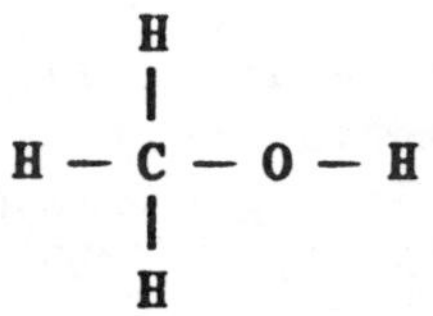

11.13 (a) $CH_3CH_2CH_2CH_3$

(b) 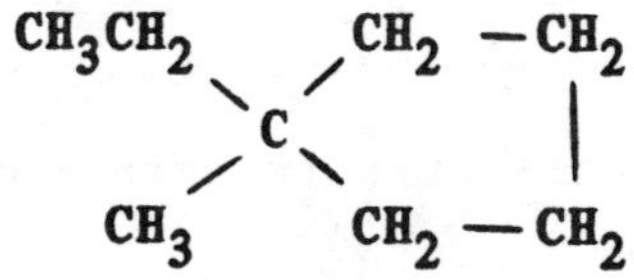
```
  CH3CH2        CH2 — CH2
         \           |
          C          |
         /    \       |
  CH3     CH2 — CH2
```

(c)

CH$_2$—CH$_2$ C—CH$_2$=CH   CH$_2$—CH$_2$
CH$_2$ CH=C—CH$_2$=CH$_2$ CH$_2$

11.14 (a) Identical   (h) Identical
       (b) Identical   (i) Isomers
       (c) Unrelated   (j) Isomers
       (d) Isomers   (k) Identical
       (e) Identical   (l) Identical
       (f) Identical   (m) Isomers
       (g) Isomers   (n) Unrelated

11.15 (a) Alkene   (f) Aldehyde
       (b) Alcohol   (g) Carboxylic acid
       (c) Thioalcohol   (h) Ketone
       (d) Alkyne   (i) Amine
       (e) Ester   (j) Ether

11.16 (a) Alkane   (h) Carboxylic acid
       (b) Alkane   (i) Ester and carboxylic acid
       (c) Alcohols   (j) Ester and alcohol – carboxylic
                                acid and alcohol
       (d) Alkene and cycloalkane   (k) Ketone
       (e) Aldehyde   (l) Alkane
       (f) Alkane   (m) Amides
       (g) Amines   (n) These are inorganic compounds.

11.17 Enzyme-catalyzed reactions depend on a fitting of a molecule of
       the compound that is to react to the surface of the enzyme
       molecule, and such fitting depends on the shapes being right.

11.18 It is an atomic orbital that has two lobes, one small and one
       large. It, and three others like it, are made by mixing an s-
       orbital and three p-orbitals. It resembles a p-orbital in that
       it has two lobes. It resembles an s-orbital in that one lobe
       is large and is the only lobe used in overlapping to form a
       bond.

11.19 The sp^3 hybrid orbitals of carbon each overlap with a 1s
       orbital of hydrogen.

11.20 An $\underline{sp}^3$ hybrid orbital of one carbon atom overlaps with an $\underline{sp}^3$ hybrid orbital of another carbon atom.

11.21 A single covalent bond about which there can be free rotation of groups (assuming that the molecule is open-chain).

11.22 An $\underline{sp}^3$ hybrid orbital in ammonia and two $\underline{sp}^3$ hybrid orbitals in water.

11.23 Probably the plain $3p_x$, $3p_y$, and $3p_z$ orbitals and not hybrid orbitals, because the observed bond angles of nearly 90° fit this model better.

CHAPTER 12

SATURATED HYDROCARBONS
ALKANES AND CYCLOALKANES

Review Exercises

12.1 (a) and (c)

12.2 None

12.3

```
              H           O   H
              |           ||  |
   H          C           C — N — H
    \        / \\        /
     C ===== C   C ===== C
     |           |
     C           C
    / \\        / \
   H   N ====== C   H
   |              \
   H               H
```

12.4

```
     H                    H              H   H
     |                    |              |   |
 H — C           N ====== C — N — H  H   C   S   C — C — O — H
     |          //         \         \  / \\ / \  |   |
     H           C          N — H      C    C   C H   H
                /|          |         ||        ||
               N |          C         N+        C   H
                \\         /| \      /  \        |
                 C ====== C |  H    /    \       C — H
                 |           H  H  H      \      |
                 H                         C     H
                                               
```

12.5 B. It has the higher formula weight.

12.6 B. It has more polar, — OH — groups, and gasoline is a nonpolar solvent.

459

460

12.7 When water is added to pentane, <u>two</u> liquid layers appear because pentane won't dissolve in water. When water is added to hydrochloric acid, only one layer forms because the two mix in all proportions.

12.8 Methyl alcohol dissolves in water but hexane doesn't.

12.9 $CH_3CH_2CH_2CH_2CH_2CH_3$ Hexane

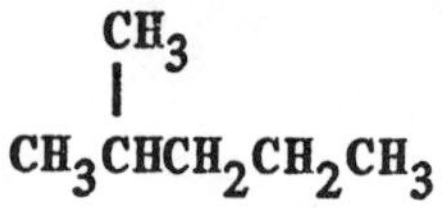

$$CH_3CHCH_2CH_2CH_3 \quad \overset{CH_3}{\underset{|}{}}$$

      CH₃
      |
$CH_3CHCH_2CH_2CH_3$ 2-Methylpentane

        CH₃
        |
$CH_3CH_2CHCH_2CH_3$ 3-Methylpentane

      CH₃
      |
$CH_3CCH_2CH_3$ 2,2-Dimethylbutane 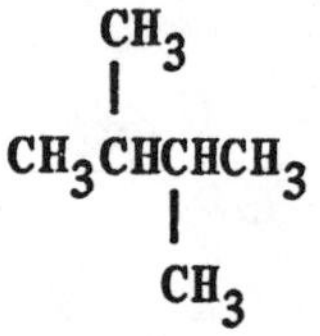
      |
      CH₃

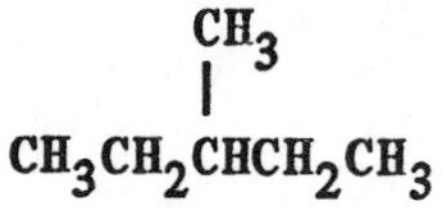

      CH₃
      |
$CH_3CHCHCH_3$ 2,3-Dimethylbutane
        |
        CH₃

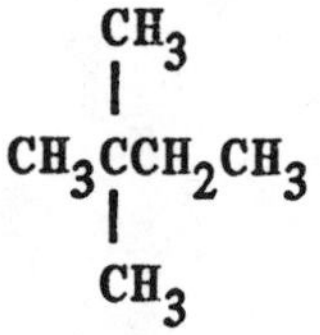

12.10 $CH_3CH_2CH_2CH_2CH_2CH_3$

12.11
      CH₃
      |
$CH_3CHCH_2CH_2CH_3$

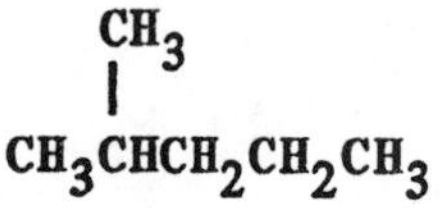

12.12 $CH_3CH_2CH_2CH_2CH_2CH_2CH_3$ Heptane

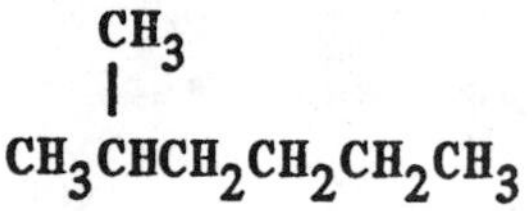

      CH₃
      |
$CH_3CHCH_2CH_2CH_2CH_3$ 2-Methylhexane

$$\underset{\displaystyle CH_3CH_2CHCH_2CH_2CH_3}{\overset{\displaystyle \overset{CH_3}{|}}{}}$$ 3-Methylhexane

$$CH_3\underset{\underset{\displaystyle CH_3}{|}}{\overset{\overset{\displaystyle CH_3}{|}}{C}}CH_2CH_2CH_3$$ 2,2-Dimethylpentane

$$CH_3\overset{\overset{\displaystyle CH_3}{|}}{C}H\underset{\underset{\displaystyle CH_3}{|}}{C}HCH_2CH_3$$ 2,3-Dimethylpentane

$$CH_3\overset{\overset{\displaystyle CH_3}{|}}{C}HCH_2\overset{\overset{\displaystyle CH_3}{|}}{C}HCH_3$$ 2,4-Dimethylpentane

$$CH_3CH_2\underset{\underset{\displaystyle CH_3}{|}}{\overset{\overset{\displaystyle CH_3}{|}}{C}}CH_2CH_3$$ 3,3-Dimethylpentane

$$CH_3CH_2\overset{\overset{\displaystyle CH_2CH_3}{|}}{C}HCH_2CH_3$$ 3-Ethylpentane

$$CH_3\underset{\underset{\displaystyle CH_3}{|}}{\overset{\overset{\displaystyle CH_3}{|}}{C}} —\!\!— \overset{\overset{\displaystyle CH_3}{|}}{C}HCH_3$$ 2,2,3-Trimethylbutane

12.13   (a) $CH_3CH_2CH_2CH_2CH_2CH_2CH_3$ (b) $CH_3\overset{\overset{\displaystyle CH_3}{|}}{C}HCH_2CH_2CH_2CH_3$

12.14   (a) 2,3,3,9-Tetramethyl-5-ethyl-5-<u>sec</u>-butyldecane

(b) 2-Methyl-6-propyl-5-isobutyl-7-<u>t</u>-butyldecane

$$\overset{\displaystyle \text{CH}_3}{\underset{\displaystyle |}{}}$$

12.15 (a) 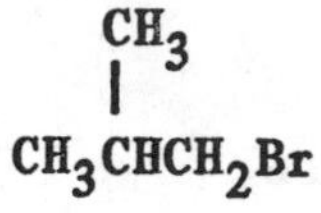 CH_3CHCH_2Br 　　　　　(b) CH_3CH_2I

(c) $CH_3CH_2CH_2Cl$ 　　　　　(d) 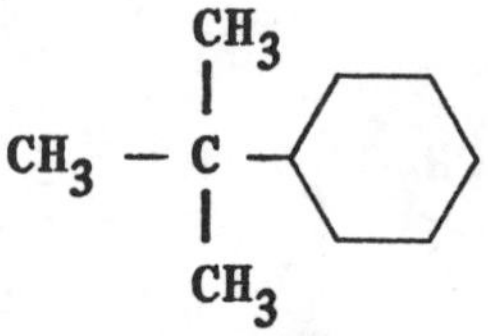 $CH_3 - \overset{\displaystyle \overset{\text{CH}_3}{|}}{\underset{\displaystyle \underset{\text{CH}_3}{|}}{C}} - C_6H_{11}$

12.16 (a) 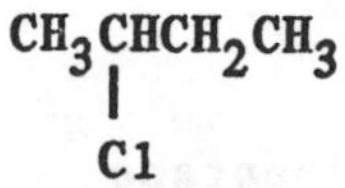 $CH_3\overset{\displaystyle \underset{\text{Cl}}{|}}{CH}CH_2CH_3$ 　　　　　(b) $CH_3CH_2CH_2CH_2I$

(c) 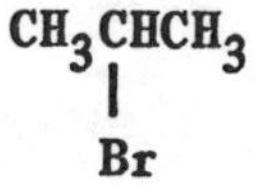 $CH_3\overset{\displaystyle \underset{\text{Br}}{|}}{CH}CH_3$ 　　　　　(d) 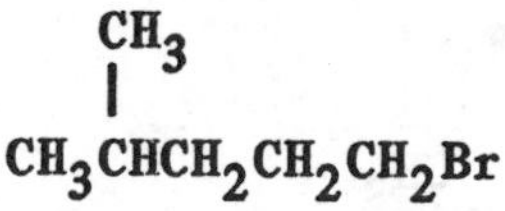 $CH_3\overset{\displaystyle \overset{\text{CH}_3}{|}}{CH}CH_2CH_2CH_2Br$

12.17 (a)

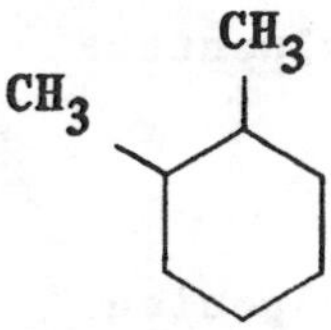

1,2-Dimethylcyclohexane

(b) 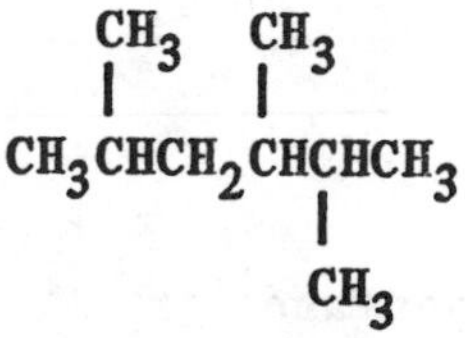 $CH_3\overset{\displaystyle \overset{\text{CH}_3}{|}}{CH}CH_2\overset{\displaystyle \overset{\text{CH}_3}{|}}{CH}\overset{\displaystyle \underset{\text{CH}_3}{|}}{CH}CH_3$

2,3,5-Trimethylhexane

(c) $CH_3CH_2CH_2CH_2Cl$ 　　　　　1-Chlorobutane

(d) $CH_3CH_2CH_3$ 　　　　　Propane

12.18 (a) 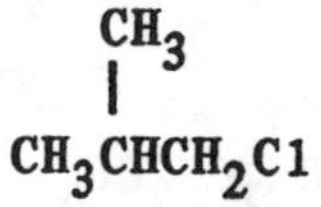 $CH_3\overset{\displaystyle \overset{\text{CH}_3}{|}}{CH}CH_2Cl$ 　　　　　1-Chloro-2-methylpropane

(b)

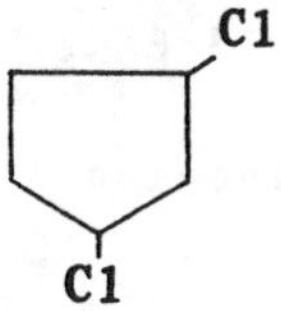

1,3-Dichlorocyclopentane

(c) $CH_3CH_2CHCH_2CH_3$ 3-Methylpentane
$\qquad\qquad\quad|$
$\qquad\qquad\;\;CH_3$

(d)

$\qquad\qquad\qquad$ 1,2,4-Trimethylcyclohexane

12.19 $2C_8H_{18}$ + $25O_2$ $\longrightarrow$ $16CO_2$ + $18H_2O$

12.20 C_2H_6O + $3O_2$ $\longrightarrow$ $2CO_2$ + $3H_2O$

12.21 CH_3Cl Methyl chloride

$\qquad$ CH_2Cl_2 Methylene chloride

$\qquad$ $CHCl_3$ Chloroform

$\qquad$ CCl_4 Carbon tetrachloride

12.22 $ClCH_2CH_2Cl$ 1,2-Dichloroethane

$\qquad\quad Cl$
$\qquad\quad |$
$\qquad$ CH_3CHCl 1,1-Dichloroethane

12.23 $CH_3CH_2CH_2Cl_2$ 1,1-Dichloropropane

$\qquad\quad Cl$
$\qquad\quad |$
$\qquad$ CH_3CHCH_2Cl 1,2-Dichloropropane

$\qquad$ $ClCH_2CH_2CH_2Cl$ 1,3-Dichloropropane

$\qquad\quad Cl$
$\qquad\quad |$
$\qquad$ CH_3CCH_3 2,2-Dichloropropane
$\qquad\quad |$
$\qquad\quad Cl$

CHAPTER 13

UNSATURATED HYDROCARBONS

Review Exercises

13.1   (a) Identical   (b) Identical
       (c) Identical   (d) Identical
       (e) Isomers

13.2   (a) No cis-trans isomers

      Trans-isomer   Cis-isomer

  (c) No cis-trans isomers
  (d)

       Cis-isomer   Trans-isomer

13.3   (a) No cis-trans isomers

    (b)

               Trans-isomer   Cis-isomer

(c) Cl

![cyclohexene with Cl substituents - Trans-isomer]

Cl   Cl

![cyclohexene with Cl substituents - Cis-isomer]

Cl

Trans-isomer   Cis-isomer

13.4 (a) $CH_3CH=CH_2$

(b) $CH_3\overset{\overset{\displaystyle CH_3}{|}}{C}=CH_2$

(c) $\underset{\underset{\displaystyle H}{|}}{\overset{\overset{\displaystyle CH_3}{|}}{C}}=\underset{\underset{\displaystyle H}{|}}{\overset{\overset{\displaystyle CH_2CH_2CH_3}{|}}{C}}$

(d) $CH_3CH=\overset{\overset{\displaystyle Cl}{|}}{C}CH_2CH_3$

(e)

CH₃CH₂

CH₃CH₂

(f)

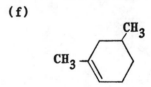

13.5 (a) 1-Decene
      (b) 1-Chloro-3-methyl-1-butene

(c) 4-Methyl-2-propyl-1-hexene
(d) 4,4-Dimethyl-2-pentene

13.6 $CH_2=CHCH_2CH_2CH_3$ 1-Pentene

$\underset{\underset{\displaystyle H}{}}{\overset{\overset{\displaystyle CH_3}{}}{C}}=\underset{\underset{\displaystyle H}{}}{\overset{\overset{\displaystyle CH_2CH_3}{}}{C}}$ cis-2-Pentene

$\underset{\underset{\displaystyle H}{}}{\overset{\overset{\displaystyle CH_3}{}}{C}}=\underset{\underset{\displaystyle CH_2CH_3}{}}{\overset{\overset{\displaystyle H}{}}{C}}$ trans-2-Pentene

$$CH_2=CCH_2CH_3$$
with CH_3 substituent

2-Methyl-1-butene

$$CH_2=CHCHCH_3$$
with CH_3 substituent

3-Methyl-1-butene

$$CH_3C=CHCH_3$$
with CH_3 substituent

2-Methyl-2-butene

13.7

1-Methylcyclopentene

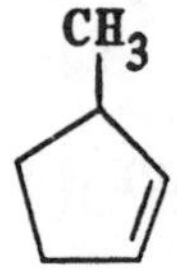

3-Methylcyclopentene

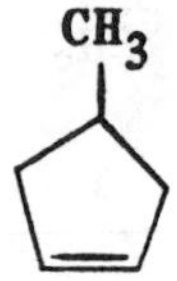

4-Methylcyclopentene

13.8 $HC\equiv CCH_2CH_2CH_3$

1-Pentyne

$CH_3C\equiv CCH_2CH_3$

2-Pentyne

$$HC\equiv CCHCH_3$$
with CH_3 substituent

3-Methyl-1-butyne

13.9

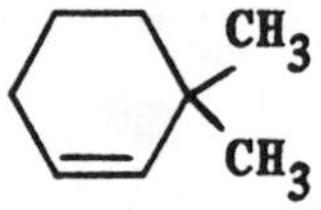

3,3-Dimethylcyclohexene

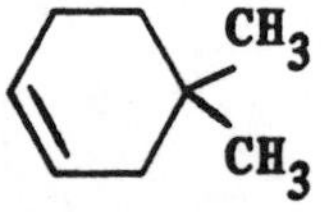

4,4-Dimethylcyclohexene

1,2-Dimethylcyclohexene

2,3-Dimethylcyclohexene

trans-3,4-Dimethylcyclohexene

cis-3,4-Dimethylcyclohexene

trans-4,5-Dimethylcyclohexene

cis-4,5-Dimethylcyclohexene

1,3-Dimethylcyclohexene

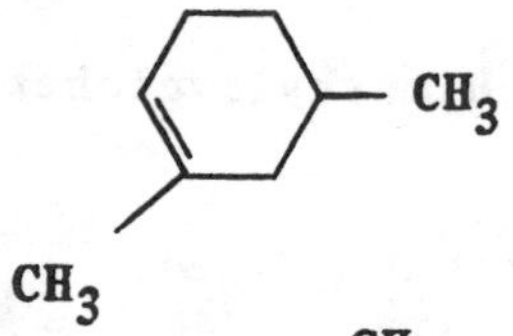

2,4-Dimethylcyclohexene

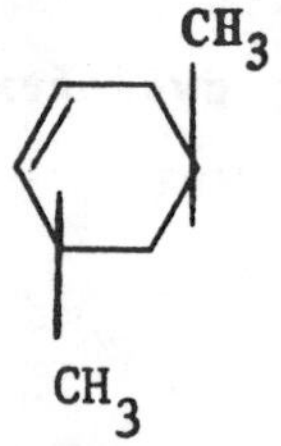

<u>trans</u>-3,5-Dimethylcyclohexene

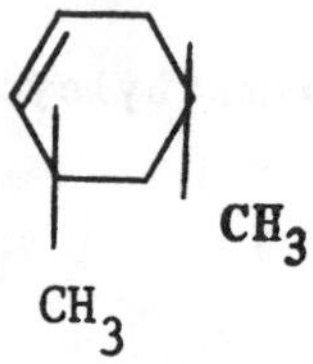

<u>cis</u>-3,5-Dimethylcyclohexene

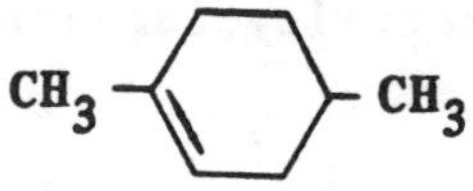

1,4-Dimethylcyclohexene

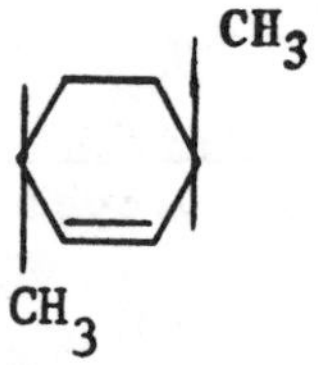

<u>trans</u>-3,6-Dimethylcyclohexene

<u>cis</u>-3,6-Dimethylcyclohexene

13.10 CH_2=C=$CHCH_2CH_3$ 1,2-Pentadiene

CH_2=CHCH=$CHCH_3$ 1,3-Pentadiene

CH_2=$CHCH_2$CH=CH_2 1,4-Pentadiene

CH_3CH=C=$CHCH_3$ 2,3-Pentadiene

$\underset{\displaystyle CH_2=CCH=CH_2}{\overset{\displaystyle CH_3}{|}}$ 2-Methyl-1,3-butadiene

$$\begin{array}{c} CH_3 \\ | \\ CH_3C{=}C{=}CH_2 \end{array}$$ 3-Methyl-1,2-butadiene

13.11   The products of the reactions are as follows

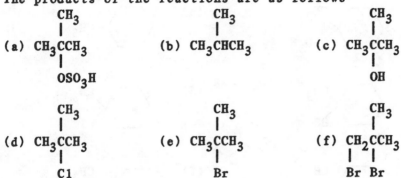

(a) $\begin{array}{c} CH_3 \\ | \\ CH_3CCH_3 \\ | \\ OSO_3H \end{array}$

(b) $\begin{array}{c} CH_3 \\ | \\ CH_3CHCH_3 \end{array}$

(c) $\begin{array}{c} CH_3 \\ | \\ CH_3CCH_3 \\ | \\ OH \end{array}$

(d) $\begin{array}{c} CH_3 \\ | \\ CH_3CCH_3 \\ | \\ Cl \end{array}$

(e) $\begin{array}{c} CH_3 \\ | \\ CH_3CCH_3 \\ | \\ Br \end{array}$

(f) $\begin{array}{c} CH_3 \\ | \\ CH_2CCH_3 \\ |\ \ | \\ Br\ Br \end{array}$

13.12   The products of the reactions are as follows.

(a) CH_3
           OSO_3H

(b) CH_3

(c) CH_3
           OH

(d) CH_3
           Cl

(e) CH_3
           Br

(f) Br
      CH_3
      Br

13.13   The products of the reactions are as follows.

(a) $(CH_3)_2CCH_2CH_3$
         $|$
      OSO_3H

(b) $(CH_3)_2CHCH_2CH_3$

(c) $(CH_3)_2CCH_2CH_3$
         $|$
       OH

(d) $(CH_3)_2CCH_2CH_3$
         $|$
       Cl

(e) 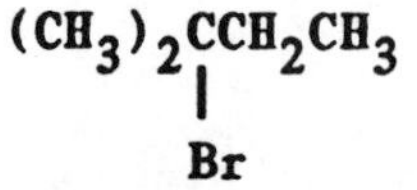 $(CH_3)_2CCH_2CH_3$
  |
  Br

(f) 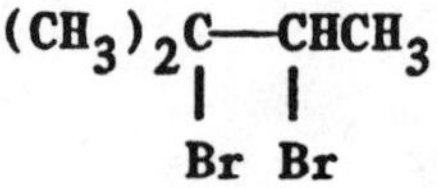 $(CH_3)_2C-CHCH_3$
  | |
  Br Br

13.14 The products of the reactions are as follows.

(a)

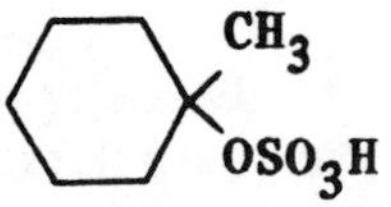

(b)

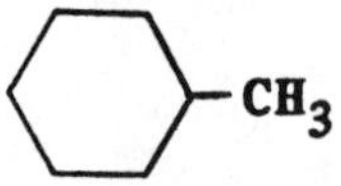

(c)

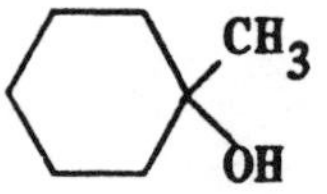

(d)

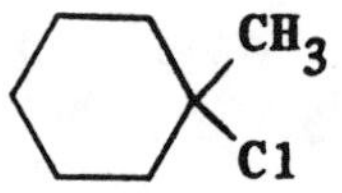

(e)

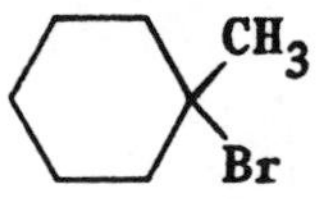

(f)

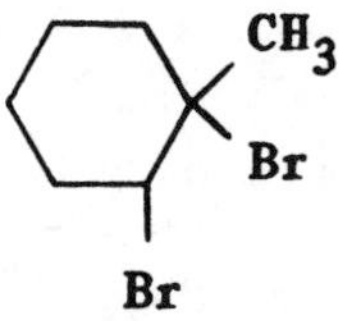

13.15 $CH_2=CH_2 + H_2SO_4 \longrightarrow CH_3CH_2-OSO_3H$

Ethyl hydrogen sulfate (This is very polar, so it is very soluble in the very polar, concentrated H_2SO_4.)

13.16

Cyclohexane
(C_6H_{12})

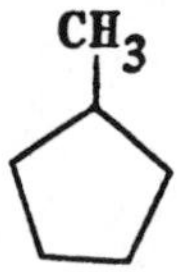

Methylcyclopentane
(C_6H_{12})

13.17 51.4 g of $KMnO_4$

13.18 12.7 g of $K_2C_6H_8O_4$

13.19 The more stable $CH_3\overset{+}{C}HCH_2CH_3$, not the less stable $^+CH_2CH_2CH_2CH_3$, is the intermediate carbocation, so the chlorine goes to C-2.

13.20 Since the two cations that can form as intermediates are of roughly the same energy, we have to expect that both form.

Chloride ions can combine with either about equally easily, so some combine with one cation and some with the other. In this way, both products form.

13.21 (a) etc. $- \underset{\underset{CH_3}{|}}{\overset{\overset{CH_3}{|}}{C}} - CH_2 - \underset{\underset{CH_3}{|}}{\overset{\overset{CH_3}{|}}{C}} - CH_2 - \underset{\underset{CH_3}{|}}{\overset{\overset{CH_3}{|}}{C}} - CH_2 - \underset{\underset{CH_3}{|}}{\overset{\overset{CH_3}{|}}{C}} - CH_2 -$ etc.

(b) $\left(\underset{\underset{CH_3}{|}}{\overset{\overset{CH_3}{|}}{C}} - CH_2 \right)_{\underline{n}}$

13.22 etc. $- CH_2 - \underset{\underset{OCCH_3}{|}}{\overset{\overset{O}{\|}}{CH}} - CH_2 - \underset{\underset{OCCH_3}{|}}{\overset{\overset{O}{\|}}{CH}} - CH_2 - \underset{\underset{OCCH_3}{|}}{\overset{\overset{O}{\|}}{CH}} - CH_2 - \underset{\underset{OCCH_3}{|}}{\overset{\overset{O}{\|}}{CH}} -$ etc.

$\left(CH_2 - \underset{\underset{OCCH_3}{|}}{\overset{\overset{O}{\|}}{CH}} \right)_{\underline{n}}$

13.23 Probably a trace of alkenes

13.24 Polymerization of the trace of alkenes

13.25 One $\underline{s}$ orbital and two $\underline{p}$ orbitals mix to give three new, equivalent orbitals - $\underline{sp^2}$ hybrid orbitals. Each consists of two lobes, one larger than the other. The axes of these three orbitals all lie in the same plane and make angles of 120° with each other. The lobes of the remaining unhybridized $\underline{p}$ orbital lie on opposite sides of the plane. The axis of this p orbital is perpendicular to the plane and it intersects this place where the axes of the sp^2 intersect. (Drawings very similar to those in Figures 13.3 and 13.4 should accompany the discussion and be referred to.

13.26 The drawing should resemble part (c) of Figure 13.4, except that the lobes would be omitted and the lines would be extended to their common point of intersection.

13.27 120°. They are the same.

13.28 Free rotation about a sigma bond doesn't reduce the overlap of the orbitals that were used to make the bond, so such rotation costs little to no energy and therefore can happen.

13.29 Free rotation about a pi bond would break the side-to-side overlap of the two p orbitals. (The drawings of either part b or c of Figure 13.6 should serve as models for your drawings.) Breaking this overlap costs considerable energy, so it doesn't normally happen.

13.30 Sigma electrons

13.31 Its ring is not a benzene ring.

13.32 It has a benzene ring.

13.33 (a) $C_6H_6 + SO_3 \xrightarrow{H_2SO_4} C_6H_5SO_3H$

(b) $C_6H_6 + HNO_3 \xrightarrow{H_2SO_4} C_6H_5NO_2 + H_2O$

(c) No reaction
(d) No reaction
(e) No reaction without an iron catalyst
(f) No reaction

(g) $C_6H_6 + Br_2 \xrightarrow{FeBr_3} C_6H_5Br + HBr$

13.34 C_6H_5R, where R is any alkyl group such as the CH_3 group in toluene.

13.35   CH$_3$ —⬡— CH$_3$

13.36 (a) CH$_3$—⬡   **(b)** NH$_2$ —⬡

(c) HO —⬡   **(d)** ⬡— CO$_2$H

(e) ⬡— CH=O   **(f)** ⬡— NO$_2$

13.37 (a) ⬡— SO$_3$H   **(b)** Cl —⬡— CH$_3$
            |
           NO$_2$

(c) ⬡— NO$_2$ **(d)** HO —⬡— NO$_2$
       | |
      Br   NO$_2$

13.38 The unhybridized p orbitals at the six carbons of the ring, one
at each carbon, undergo side–by–side overlap to give a large,
double doughnut shaped molecular shell. This constitutes the
delocalized pi bond network. Underneath this (between the two
"doughnuts") the carbons hold each other by molecular orbitals
formed by the overlapping of sp^2 hybrid orbitals. The hydrogen
atoms are held to the carbon atoms by molecular orbitals formed
by the overlapping of the sp^2 hybrid orbital at each carbon to
the s orbital of the hydrogen. (The drawing should be like
Figure 13.8.)

13.39 With more space in which to exist, the electrons are in a more stable arrangement.

13.40 Addition reactions would permanently destroy the cyclic pi electron system that gives the ring so much stability.

13.41 (a) $CH_3CH_2CH_2CH_3$

(b) $CH_3CHCH_2CH_3$
$\quad\quad\ |$
$\quad\quad OH$

(c) 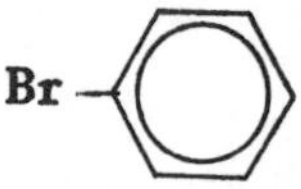 Br ─◯

(d) No reaction

(e)

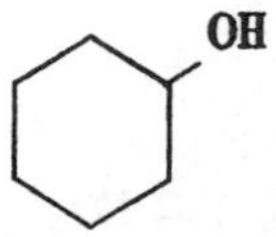

(f) $CH_3CH_2CH_2CH_2CH_3$

(g) No reaction

(h) 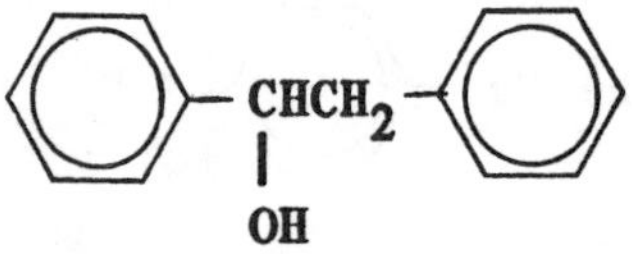 ◯─ CHCH$_2$ ─◯
$\quad\quad\ |$
$\quad\quad OH$

(i) No reaction

(j) $2C_4H_{10} + 13O_2 \longrightarrow 8CO_2 + 10H_2O$

(k) No reaction

13.42 (a) CH_3CHCH_3
$\quad\quad\ |$
$\quad\quad OH$

(b) $CH_3CH_2CH_2CH_2CH_3$

(c) No reaction

(d) $\quad\quad\quad\ CH_3 \quad\ CH_3$
$\quad\quad\quad\ | \quad\quad\ |$
$CH_3CH \!-\!\!-\! CCH_3$
$\quad\quad\quad\quad\quad\ |$
$\quad\quad\quad\quad\quad Cl$

(e)

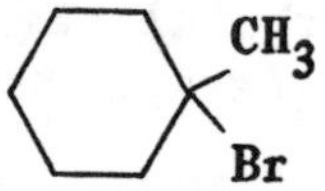

(f) No reaction

(g) 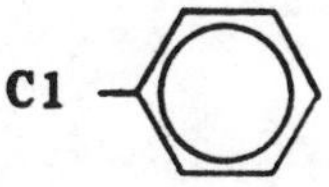   Cl —⬡   (h) No reaction

(i) C_7H_8 + $9O_2$ $\longrightarrow$ $7CO_2$ + $4H_2O$

(j) No reaction   (k) Cl — CH_2CH — $CHCH_2$ — Cl
                                              | |
                                              Cl   Cl

(l) 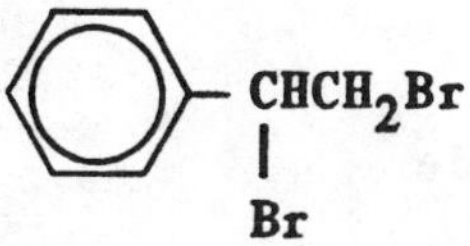   ⬡— $CHCH_2Br$
             |
             Br

CHAPTER 14

ALCOHOLS, THIOALCOHOLS, PHENOLS, AND ETHERS

Review Exercises

14.1   1, phenol.   2, ether.   3, alkene

14.2   1, alkene.   2, alkene.   3, aldehyde

14.3   1, 1° alcohol.   2, ketone   3, 2° alcohol.   4, ketone.
       5, alkene

14.4   1, ketone.   2, carboxylic acid.   3, 2° alcohol.   4, 2° alcohol

14.5   (a) $CH_3\overset{\overset{\displaystyle CH_3}{|}}{C}HCH_2OH$ (b) $CH_3\overset{\overset{\displaystyle OH}{|}}{C}HCH_3$

       (c) $CH_3CH_2CH_2OH$ (d) $HOCH_2\underset{\underset{\displaystyle OH}{|}}{C}HCH_2OH$

14.6   (a) CH_3OH (b) $CH_3\overset{\overset{\displaystyle CH_3}{|}}{\underset{\underset{\displaystyle CH_3}{|}}{C}}OH$

       (c) CH_3CH_2OH (d) $CH_3CH_2CH_2CH_2OH$

14.7   (a) Isopropyl alcohol   (b) Ethylene glycol
       (c) t-Butyl alcohol   (d) Glycerol

14.8   (a) Ethyl alcohol   (b) Propyl alcohol
       (c) Isobutyl alcohol   (d) Butyl alcohol

14.9 $HOCH_2CH_2OH$, 1, 2-ethanediol

14.10 $HOCH_2CHCH_2OH$, 1,2,3-propanetriol
         |
         OH

14.11 (a) 2-Methyl-1-propanol   (b) 2-Propanol
         (c) 1-Propanol   (d) 1,2,3-Propanetriol

14.12 (a) Methanol   (b) 2-Methyl-2-propanol
         (c) Ethanol   (d) 1-Butanol

14.13 (a) 2-Propanol   (b) 1,2-Ethanediol
         (c) 2-Methyl-2-propanol   (d) 1,2,3-Propanetriol

14.14 (a) Ethanol   (b) 1-Propanol
         (c) 2-Methyl-1-propanol   (d) 1-Butanol

14.15 3,4-Dimethyl-2-ethyl-1-pentanol

14.16 3-Methyl-4-ethyl-2-isopropyl-3-t-butyl-1-heptanol

14.17

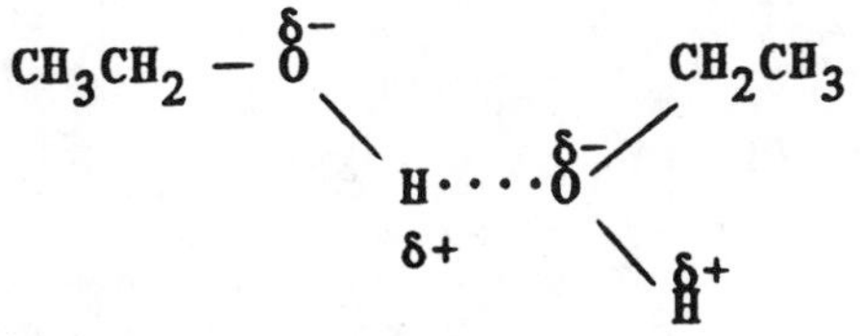

14.18

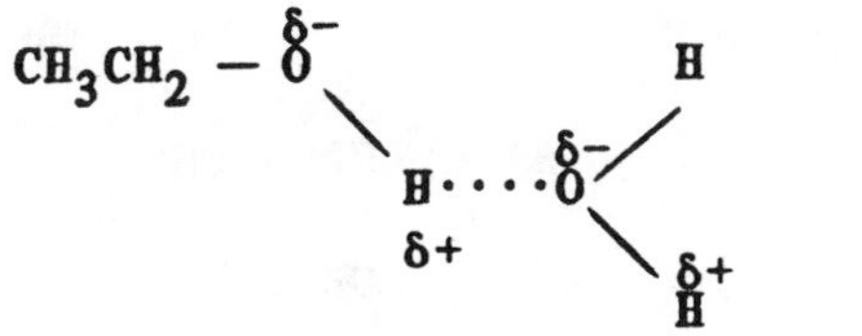

14.19 D < B < A < C

14.20 C < B < A

14.21 (a) $CH_3CH=CHCH_3$ (plus some $CH_2=CHCH_2CH_3$)

(b) $CH_3C=CHCH_2CH_3$ with CH_3 substituent on the carbon (plus some $CH_2=CCH_2CH_2CH_3$ with CH_3 substituent)

(c)  1-methylcyclohexene ring — CH_3 (Plus some methylenecyclohexane =CH_2)

(d) 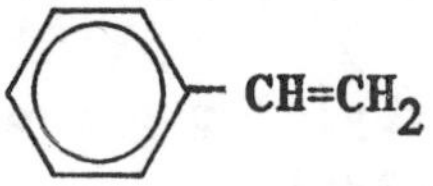 phenyl— $CH=CH_2$

(e) 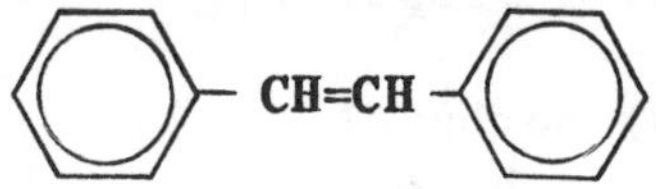 phenyl— $CH=CH$ —phenyl

(f) 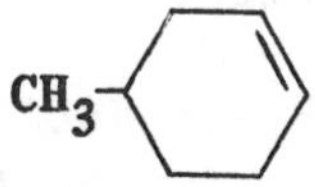 4-methylcyclohexene, CH_3

14.22 (a) $CH_3\overset{O}{\overset{\|}{C}}CH_2CH_3$

(b) No reaction   (c) No reaction

(d)  phenyl—$CH_2CH=O$, then phenyl—CH_2CO_2H

(e) 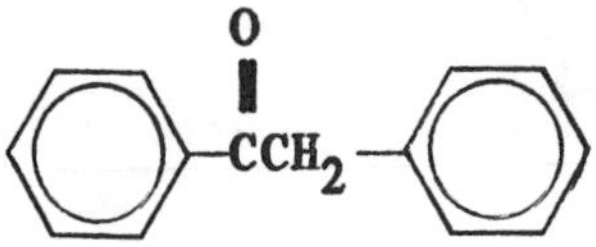 phenyl—$\overset{O}{\overset{\|}{C}}CH_2$—phenyl   (f) 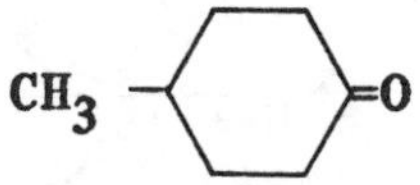 CH_3—cyclohexanone =O

14.23 (a) CH_3CH_2OH (b) $CH_3CH_2CH_2OH$ or $CH_3\underset{OH}{CH}CH_3$

(c) $HOCH_2\underset{CH_3}{CH}CH_3$ or $CH_3\underset{\underset{OH}{|}}{\overset{\overset{CH_3}{|}}{C}}CH_3$ (d) $CH_3\underset{OH}{CH}CH_2CH_3$

14.24 (a) $CH_3CH_2CH_2OH$ (b) $CH_3\underset{OH}{CH}CH_2CH_2CH_3$

(c) 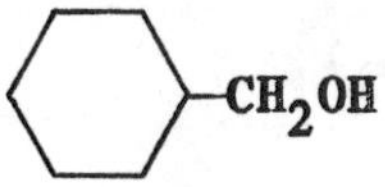 cyclohexyl—CH_2OH (d) $CH_3\underset{CH_3}{CH}CH_2OH$

14.25 Butyl alcohol. It is a 1^O alcohol that can be oxidized:

$$CH_3CH_2CH_2CH_2OH \; + \; (O) \; \longrightarrow \; CH_3CH_2CH_2CO_2H$$

The oxidizing agent – (O) – is the permanganate ion that changes to manganese dioxide, MnO_2. The 3° alcohol could not be oxidized in this manner.

14.26 2-Butanol. It, but not the alkane (pentane), can be oxidized by the dichromate ion:

$$\overset{\displaystyle OH}{\underset{\displaystyle |}{CH_3CH_2CHCH_3}} \; + \; (O) \; \longrightarrow \; \overset{\displaystyle O}{\underset{\displaystyle \|}{CH_3CH_2CCH_3}}$$

The orange-colored dichromate ion changes to the green-colored Cr^{3+} ion.

14.27 (a) $\underset{\displaystyle \underset{|}{SH}}{CH_3CHCH_3}$

(b) $CH_3CH_2CH_2CH_2SH$

(c) $CH_3 - S - S - CH_3$

(d) $HS - CH_2CH_2 - SH$

14.28 (a) $CH_3 - S - S - CH_3$

(b) $2(CH_3)_2CH - SH$

(c) ⬠—S — S—⬠

(d) $CH_3SH \; + \; HSCH_2CH_2CH_3$

14.29 Hydrogen bonding in the thioalcohol family does exist, but the hydrogen bonds are not as strong as those in the alcohol family.

14.30 Phenol is a strong enough acid to be able to neutralize sodium or potassium hydroxide:

$$\bigcirc\!\!-O - H \; + \; OH^- \; \longrightarrow \; \bigcirc\!\!-O^- \; + \; H_2O$$

14.31

o-Cresol

m-Cresol

p-Cresol

14.32 It is compound A, the phenol, because it reacts with sodium hydroxide to form an ion that, like the carboxylate ion, is soluble in water.

$$CH_3CH_2CH_2CH_2 - \underset{}{\bigcirc} - OH + OH^- \longrightarrow CH_3CH_2CH_2CH_2 - \underset{}{\bigcirc} - O^- + H_2O$$

Insoluble in water   Soluble in water

14.33 The unknown is I. II is a phenol that would react with aqueous KOH to form a water-soluble salt, but I is just a 3° alcohol and is not acidic enough to react with KOH.

14.34 Ether functions occur in structures a, d, e, g, and h.

14.35 (a) $(CH_3)_2CH - O - CH(CH_3)_2$ (b) $(CH_3)_2CHCH_2 - O - CH_2CH(CH_3)_2$

(c)   (d) $\bigcirc - CH_2 - O - CH_2 - \bigcirc$

14.36 (a) $2CH_3OH$ (b) $2CH_3CH_2CH_2CH_2OH$

14.37 $CH_3 - O - CH_3 + CH_3 - O - CH_2CH_3 + CH_3CH_2 - O - CH_2CH_3$

14.38 Nothing

14.39 (a) $(CH_3)_2C=O$ (b) No reaction

(c) No reaction   (d) No reaction

(e) $CH_3 - O - CH_3$ (f) $CH_3CH_2\overset{\overset{O}{\|}}{C}CH_2CH_3$

(g) No reaction   (h) $CH_3CH_2\underset{\underset{OH}{|}}{C}HCH_3$

(i) $CH_3CH=CH_2$ (j) $CH_3\overset{\overset{O}{\|}}{C}CH_2CH_3$

14.40 (a) $CH_3CH_2CH_3$ (b) No reaction

(c)

(d) No reaction

(e) ⬡ — CH=CH₂

(f) No reaction

(g) CH₃ — O — CH₂CH₂CH₃

(h) No reaction

(i) $$CH_3\overset{\overset{\displaystyle CH_3}{|}}{C}=CH_2$$

(j) CH₃CH₂CH=O

(k) $$CH_3 - O - CH_2\overset{\overset{\displaystyle O}{\|}}{C}CH_3$$

CHAPTER 15

ALDEHYDES AND KETONES

Review Exercises

15.1 Two carbons are always directly attached to the carbon atom of the carbonyl group in a ketone, whereas in aldehydes only one carbon is attached to this carbon (together with the aldehydic H atom).

15.2 (a) $CH_3CH_2\overset{\overset{\displaystyle CH_3}{|}}{C}HCH_2CHO$

(b)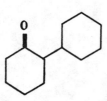

(c) $\langle\!\!\!\bigcirc\!\!\!\rangle - \overset{\overset{\displaystyle O}{\|}}{C}CH_2CH_2CH_3$

(d) $CH_3CH_2CH_2CH_2\overset{\overset{\displaystyle O}{\|}}{C}CH_2CH_2CH_2CH_3$

(e) $CH_3\overset{\overset{\displaystyle O}{\|}}{C} - \overset{\overset{\displaystyle O}{\|}}{C} - CH_3$

15.3 (a)

(b) $CH_3CH_2\overset{\overset{\displaystyle |}{C}}{H}CHO$
       $\quad\quad\quad\underset{\displaystyle CH_3CH_2}{}$

(c) 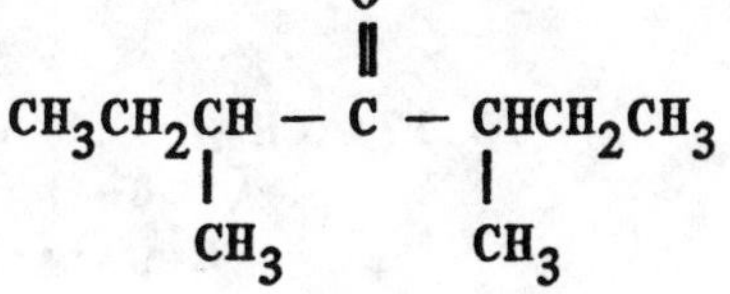 $CH_3CH_2\underset{\underset{CH_3}{|}}{CH}-\overset{\overset{O}{\|}}{C}-\underset{\underset{CH_3}{|}}{CH}CH_2CH_3$ (d) 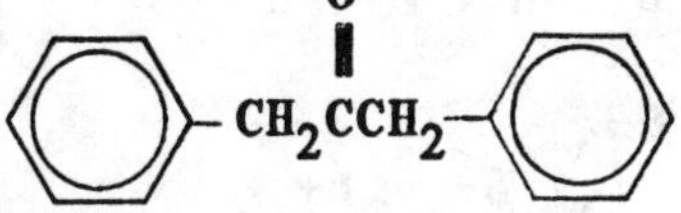 $\text{Ph}-CH_2\overset{\overset{O}{\|}}{C}CH_2-\text{Ph}$

(e) 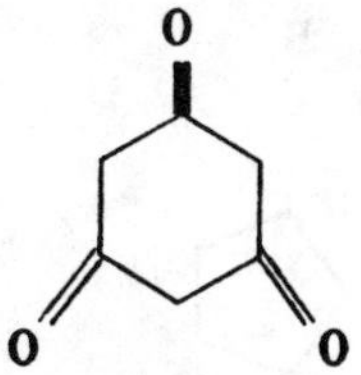 cyclohexane ring with O at top (=O) and two O (=O) at bottom positions

15.4 (a) 2-Methylcyclohexanone   (b) Propionic acid
      (c) 2-Pentanone   (d) Propanal
      (e) 2-Methylpentanal

15.5 (a) 3-Methyl-2-ethylpentanal   (b) 1-Phenylethanone
      (c) 2,3-Dimethylcyclopentanone   (d) Acetic acid
      (e) Butanal

15.6 2,3-Dimethyl-4-ketopentanal

15.7 4-Formylbenzoic acid. (Also acceptable: p-formylbenzoic acid.)

15.8 (a) 2-Bromo-3-methylpentanal   (b) 3-Methyl-2-pentanone
      (c) 2,3-Dimethylcyclohexanone   (d) 1-Phenyl-2-propanone
      (e) 2-Methyl-1-phenyl-1-propanone

15.9 (a) 5-Methyl-3-isobutyl-2-heptanone   (c) 2-Phenylethanal
      (b) 5-Methyl-3-t-butylhexanal   (d) 1-Phenylethanone
                    (e) 3,3-Dimethylcyclopentanone

15.10 Valeraldehyde

15.11 Anisic acid

15.12 C < B < D < A

15.13 B < A < D < C

15.14 C < B < D < A

15.15 B < A < D < C

15.16

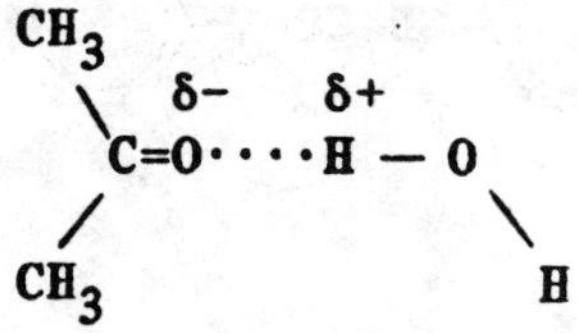

$$CH_3 \underset{CH_3}{\overset{\quad}{\diagup}} C=O \overset{\delta-}{\cdots} \overset{\delta+}{H} - O \diagdown H$$

15.17

$$CH_3 \underset{H}{\overset{\quad}{\diagup}} C=O \overset{\delta-}{\cdots} \overset{\delta+}{H} - O \diagdown CH_3$$

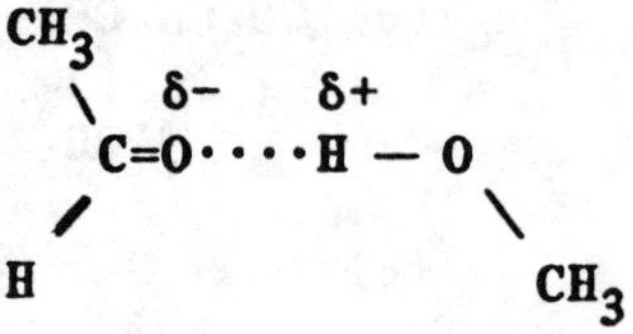

15.18 (a) $CH_3\overset{\overset{\displaystyle O}{\|}}{C}CH_2CH_3$

2-Butanone

(b) 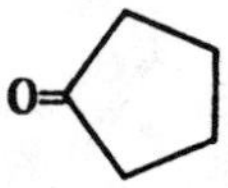 $O=\Bigl\langle\ \text{pentagon}\ \Bigr\rangle$

Cyclopentanone

(c) 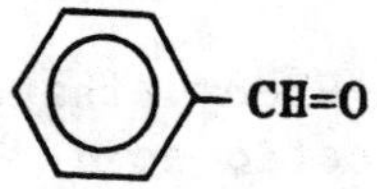 $\langle\!\!\langle\bigcirc\rangle\!\!\rangle - CH=O$

Benzaldehyde

(d) $(CH_3)_3CCH=O$

2,2-Dimethylpropanal

15.19 (a) $CH_3CH=O$
Ethanal

(b) This compound cannot exist in pure form because it has two — OH groups on the same carbon.

(c) This, a 3° alcohol, cannot be oxidized to an aldehyde or a ketone.

(d) $CH_3\overset{\overset{\displaystyle O}{\|}}{C}CH_2CO_2H$

3-Ketobutanoic acid

15.20 C_3H_6O is $CH_3CH_2CH=O$
$C_3H_6O_2$ is $CH_3CH_2CO_2H$

15.21 $C_3H_6O_2$ is $CH_3\overset{\overset{\displaystyle OH}{|}}{CH}CH=O$

$C_3H_4O_3$ is $CH_3\overset{\overset{\displaystyle O}{\|}}{C}CO_2H$

15.22 Positive Benedict's tests are given by (a) and (b).

15.23 Positive Benedict's tests are given by (a) and (c).

15.24 To keep the silver ion in solution as the ion: $Ag(NH_3)_2^+$

15.25 To keep the copper(II) ion in solution

15.26 Cu_2O

15.27 To test for the presence of glucose in urine

15.28 Silvering mirrors

15.29
$$\overset{\displaystyle OH}{\underset{\displaystyle}{CH_3CHCO_2^-}}$$

15.30
$$\overset{\displaystyle O}{\underset{\displaystyle}{CH_3CCH_2CO_2^-}}$$

15.31 $H:^- \ + \ H-O-CH_3 \ \longrightarrow \ H-H \ + \ ^-O-CH_3$

15.32 It is a very strong base because we know that its conjugate acid, $CH_3 - O - H$ (methyl alcohol), like all alcohols, is a very weak acid.

15.33 (a) $CH_3CH_2O^-$

(b) $CH_3CH_2O^- \ + \ H-OH \ \longrightarrow \ CH_3CH_2O-H \ + \ ^-OH$

(c) Ethanol

15.34 (a)
$$\overset{\displaystyle O^-}{\underset{\displaystyle}{CH_3 - CH - CH_3}}$$

(b)
$$\overset{\displaystyle O^-}{\underset{\displaystyle}{CH_3 - CH - CH_3}} \ + \ H-OH \ \longrightarrow \ \overset{\displaystyle OH}{\underset{\displaystyle}{CH_3 - CH - CH_3}} \ + \ ^-OH$$

(c) 2-Propanol

15.35   I   = $^+NH_3 - CH - CO_2^-$ II   = $^+NH_3 - CH - CO_2^-$

$\qquad\qquad\qquad\qquad CH_2CH_2O^- \qquad\qquad\qquad\qquad\qquad CH_2CH_2OH$

15.36   I   = $CH_3\overset{O^-}{\underset{}{C}}HCH_2\overset{O}{\underset{}{C}} - S - $ enzyme   II   = $CH_3\overset{OH}{\underset{}{C}}HCH_2\overset{O}{\underset{}{C}} - S - $ enzyme

15.37   (a)

     (b) $CH_3\overset{O}{\underset{}{C}}CH_3$

    (c)

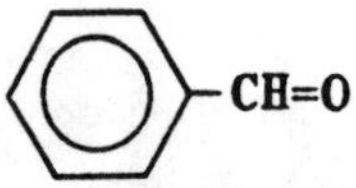

$-CH=O$

        (d)

$-CH=O$

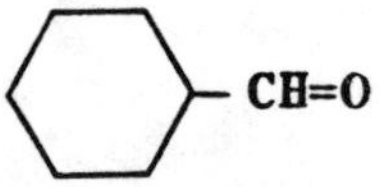

15.38   (a) Hemiacetal   (b) Acetal
        (c) Something else (a di-ether)   (d) Ketal

15.39   (a) Something else (an ether-alcohol)
        (b) Hemiacetal
        (c) An acetal with an alcohol group besides
        (d) An acetal

15.40   (a) 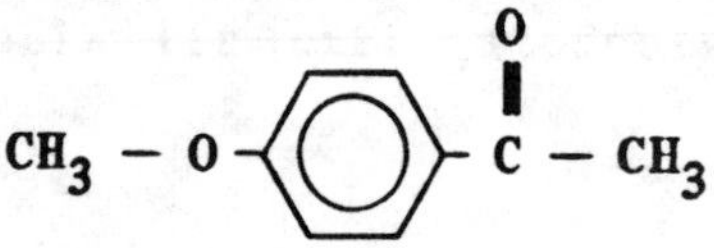 $CH_3 - O -$ $- \overset{O}{\underset{}{C}} - CH_3$ (b) $CH_3\overset{OH}{\underset{CH_3}{C}}CH_2\overset{O}{\underset{}{C}}CH_3$

    (c) $CH_3CH_2OCH_2CH=O$ (d) $O=CH\underset{CH_3}{CH}OCH_3$

15.41   (a) $CH_3\underset{OH}{CH} - O - CH_3$ (b) $CH_3\underset{OH}{CH} - O - CH_2CH_3$

$\qquad\qquad CH_3\underset{O-CH_3}{CH} - O - CH_3 \qquad\qquad\qquad CH_3\underset{O-CH_2CH_3}{CH} - O - CH_2CH_3$

15.42 (a)
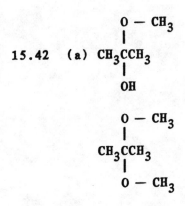

(b)
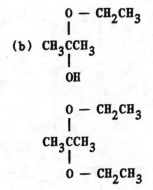

15.43
```
CH3
   \
    CH — O — H
   /
  CH2            CH=O
    \           /
     CH2 — CH2
```

15.44
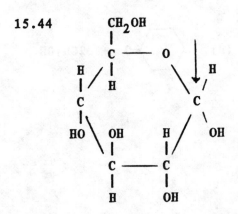   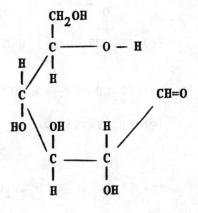

15.45
```
      O — H        CH3
     /              |
CH2               C=O
   \             /
    CH2 — CH2
```

15.46 (a)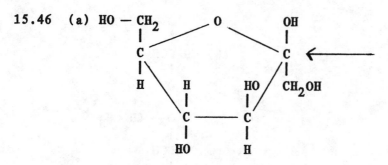

(b)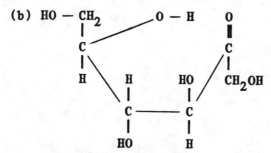

15.47 (a) $CH_3CH=O$ + $2CH_3OH$ (b) No hydrolysis

(c) $CH_3\overset{O}{\overset{\|}{C}}CH_3$ + $2CH_3OH$ (d) ⬡=O + $2CH_3OH$

15.48 (a) $CH_3CH=O$ + 2 HO—⬠

(b) No hydrolysis

(c) $CH_3\overset{O}{\overset{\|}{C}}CH_3$ + $HOCH_2CH_2OH$

(d) $CH_3CH_2CH=O$ + CH_3OH + CH_3CH_2OH

15.49 (a) $CH_3CH_2CH_2CH_3$ (b) $CH_3\overset{O}{\overset{\|}{C}}CH_3$

(c) $CH_3CH=O$ + $2CH_3OH$ (d) CH_3CH_2OH

(e) $CH_3CH_2CO_2H$ (f) No reaction

$$(g) \quad CH_3CH_2\overset{\overset{\displaystyle OH}{|}}{CH} - O - CH_3$$ (h) No reaction

(i) No reaction $$(j) \quad CH_3\overset{\overset{\displaystyle }{|}}{CH} - O - CH_3 \atop | \quad O - CH_3$$

15.50 (a) 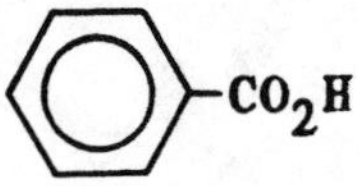 ⬡—CO_2H (b) No reaction

$$(c) \quad CH_3\overset{\overset{\displaystyle OH}{|}}{CH} - O - CH_2CH_2CH_3$$ (d) No reaction

$$(e) \quad CH_3\overset{\overset{\displaystyle O}{\|}}{C}CH_3 \ + \ CH_3CH_2OH$$ (f) $CH_3CH_2CH_2OH$

$$(g) \quad CH_3CH_2\overset{\overset{\displaystyle O}{\|}}{C}CH_3 \ + \ 2CH_3CH_2OH$$ $(h) \quad$ 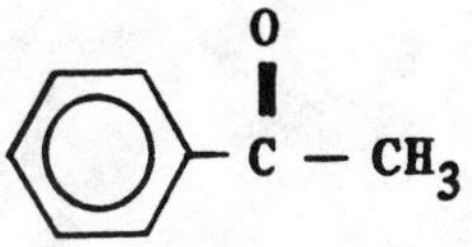 ⬡—$\overset{\overset{\displaystyle O}{\|}}{C} - CH_3$

(i) $CH_3 - O - CH_2CH_2CH_3$ $(j) \quad CH_3CH_2\overset{\overset{\displaystyle OCH_3}{|}}{CH} - O - CH_3$

15.51 A is $CH_3CH_2CH{=}O$

B is $CH_3CH_2CH_2OH$

C is $CH_3CH{=}CH_2$

$$D \text{ is } CH_3\overset{\overset{\displaystyle }{|}}{CH}CH_3 \atop | \quad OH$$

$$E \text{ is } CH_3\overset{\overset{\displaystyle O}{\|}}{C}CH_3$$

15.52   F is $(CH_3)_2CHCH_2OH$

G is $(CH_3)_2CHCH=O$

H is $(CH_3)_2CHCO_2H$

I is $(CH_3)_2C=CH_2$

J is $(CH_3)_3COH$

CHAPTER 16

CARBOXYLIC ACIDS AND ESTERS

Review Exercises

16.1 $\overset{\overset{\text{O}}{\|}}{-\text{C}}-\text{O}-\text{H}$ (which is often written as $-CO_2H$ and sometimes as $-COOH$)

For a compound to be an alcohol its molecules have the $-OH$ group attached to a carbon that has only <u>single</u> bonds. A compound is a ketone if its molecules have a carbonyl group attached on both sides to carbon atoms.

16.2 From fats and oils in the diet

16.3 In vinegar – acetic acid. In sour milk – lactic acid.

16.4 (a) $CH_3CH_2CH_2CO_2H$ (b) CH_3CO_2H

(c) HCO_2H (d) 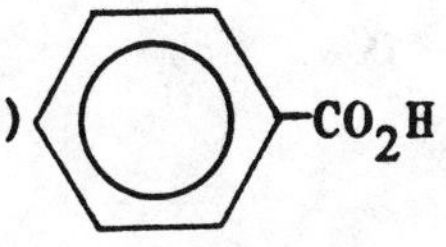 ⬡$-CO_2H$

16.5 (a) $CH_3\overset{\overset{\displaystyle CH_3}{|}}{\underset{\underset{\displaystyle CH_3}{|}}{C}}CO_2^-$ (b) $CH_3CH_2\overset{\overset{\displaystyle CH_3}{|}}{C}H\overset{\underset{\displaystyle Br}{|}}{C}HCH_2CO_2H$

(c) $HO_2CCH_2CH_2CO_2H$ (d) $CH_3CH=CHCO_2^-$

16.6 (a) 2,3-Dimethylpentanoic acid
(b) 2,3-Dimethyl-3-isopropylhexanoic acid
(c) Potassium 2-methylpropanoate

491

(d) Sodium 2,2-dimethylbutanoate

16.7 (a) 3-Bromo-2-methylbutanoic acid
(b) Sodium 2-hydroxypropanoate
(c) 3-Hydroxybutanoic acid
(d) 4,5,7-Trimethyl-2-isopropyl-4-<u>sec</u>-butyloctanoic acid

16.8 Sodium 3-ketobutanoate or sodium acetoacetate

16.9 2-Ketopentanedioic acid

16.10

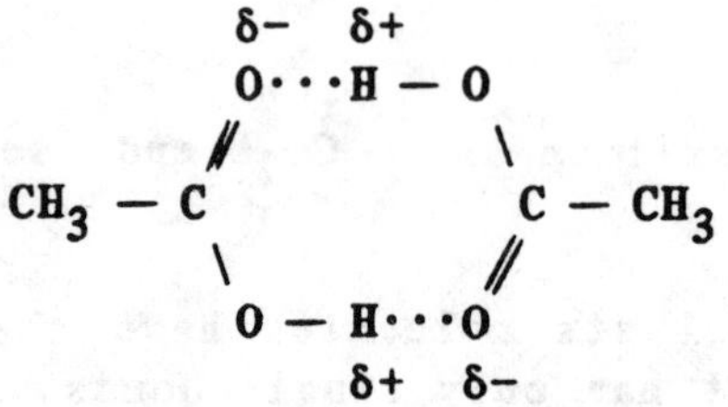

$$
\begin{array}{c}
\overset{\delta-}{O}\cdots\overset{\delta+}{H}-O \\
CH_3-C \qquad\qquad C-CH_3 \\
O-H\cdots O \\
\overset{\delta+}{} \quad \overset{\delta-}{}
\end{array}
$$

16.11

$$
\begin{array}{c}
H \\
\backslash \\
C=O\cdots etc. \\
/ \\
H \qquad H-O \\
\backslash \qquad \overset{\delta-}{} \quad \overset{\delta+}{} \\
C=O\cdots H-O \\
H \qquad / \quad \overset{\delta-}{} \quad \overset{\delta+}{} \\
\backslash \\
C=O\cdots H-O \\
/ \quad \overset{\delta-}{} \quad \overset{\delta+}{} \\
etc.\cdots H-O
\end{array}
$$

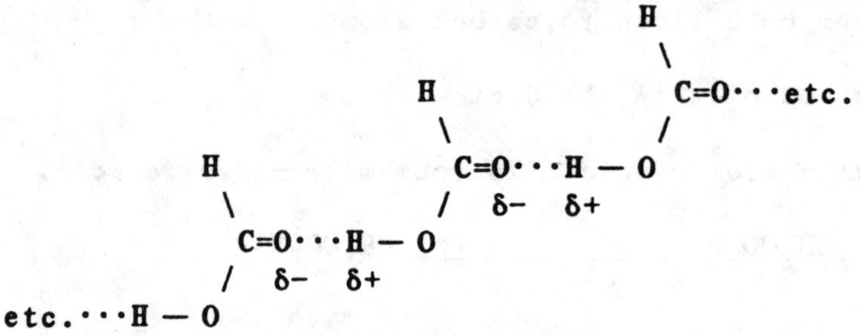

16.12 C < B < A

16.13 D < C < B < A

16.14 $CH_3CO_2H + H_2O \rightleftharpoons CH_3CO_2^- + H_3O^+$

16.15 $HCO_2H + H_2O \rightleftharpoons HCO_2^- + H_3O^+$

16.16 C < D < A < B

16.17 A < B < C < D

16.18 (a) $CH_3CH_2CO_2H + OH^- \longrightarrow CH_3CH_2CO_2^- + H_2O$

(b) 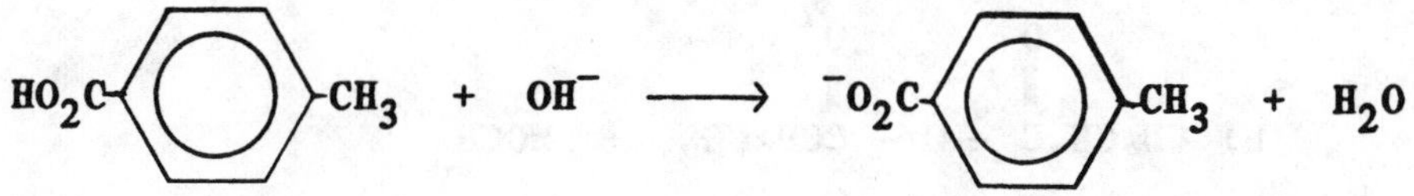 HO_2C⟨◯⟩CH_3 + OH^- ⟶ ^-O_2C⟨◯⟩CH_3 + H_2O

(c) No reaction

16.19   (a)

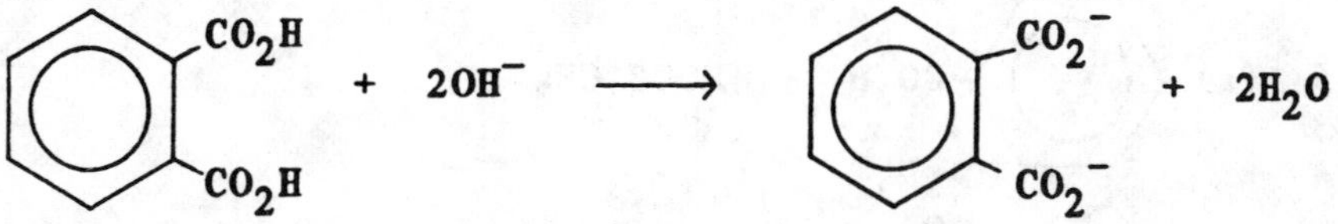

⟨◯⟩ with CO_2H / CO_2H + $2OH^-$ ⟶ ⟨◯⟩ with CO_2^- / CO_2^- + $2H_2O$

$$O$$
(b) $HOCH_2CCH_2CH_2CO_2H$ + OH^- ⟶ $HOCH_2CCH_2CH_2CO_2^-$ + H_2O

(c) 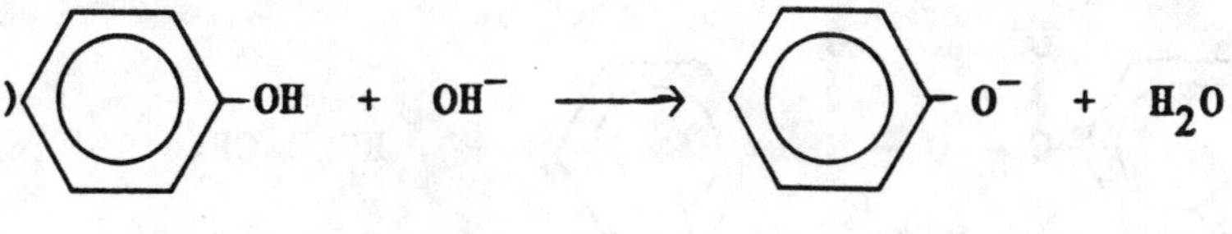 ⟨◯⟩$-OH$ + OH^- ⟶ ⟨◯⟩$-O^-$ + H_2O

16.20   B.   Sodium salts of acids are much more soluble in water than their parent acids.   The ions are strongly solvated.

16.21   A.   Ether is a relatively nonpolar solvent and it won't dissolve an ionic compound such as B.

16.22   (a) 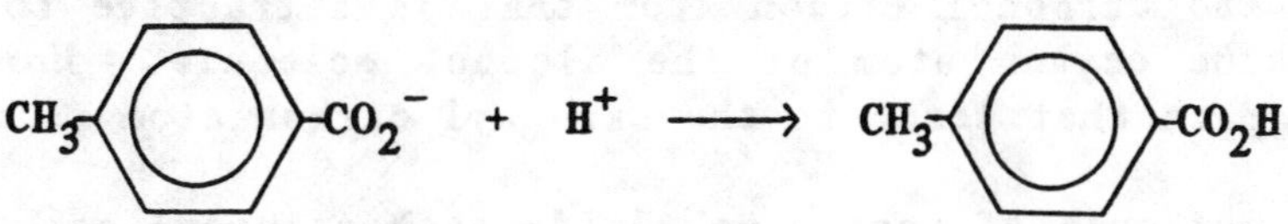 CH_3⟨◯⟩$-CO_2^-$ + H^+ ⟶ CH_3⟨◯⟩$-CO_2H$

(b) No reaction

(c) $^-O_2CCH_2CH_2CO_2^-$ + H^+ ⟶ $HO_2CCH_2CH_2CO_2^-$

16.23   (a) No reaction

(b) $CH_3CH_2CH_2CO_2^-$ + H^+ ⟶ $CH_3CH_2CH_2CO_2H$

(c) $HOCH_2CH_2CH_2CO_2^-$ + H^+ ⟶ $HOCH_2CH_2CH_2CO_2H$

$$O$$
16.24   (a) $CH_3CH_2C - Cl$ + $HOCH_3$

(b) $CH_3CH_2\overset{O}{\overset{\|}{C}} - O - \overset{O}{\overset{\|}{C}}CH_2CH_3$ + $HOCH_3$

(c) $CH_3CH_2CO_2H$ + $HOCH_3$

16.25 (a) 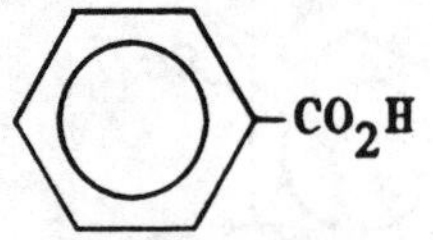 ⟨phenyl⟩$-CO_2H$ + $HOCH_2CH_3$

(b) 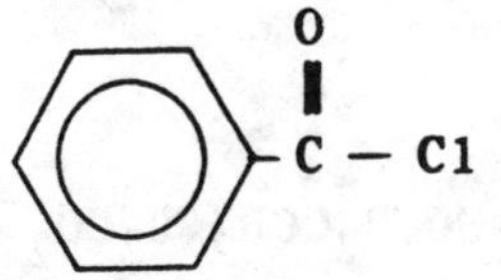 ⟨phenyl⟩$-\overset{O}{\overset{\|}{C}} - Cl$ + $HOCH_2CH_3$

(c) 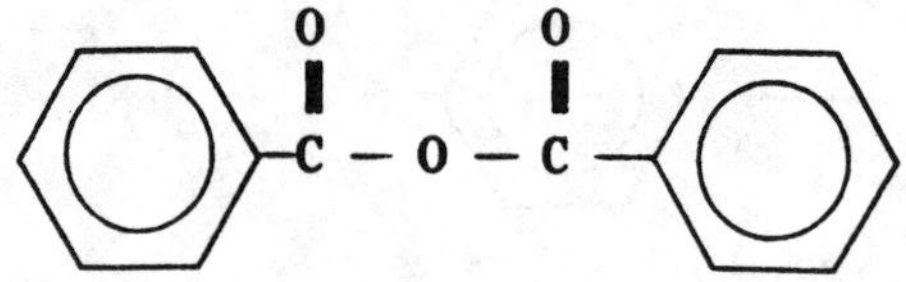 ⟨phenyl⟩$-\overset{O}{\overset{\|}{C}} - O - \overset{O}{\overset{\|}{C}}-$⟨phenyl⟩ + $HOCH_2CH_3$

16.26 (a) $CH_3\overset{O}{\overset{\|}{C}} - O - CH_3$

(b) In a molecule of acetyl chloride there is a large δ+ on the carbonyl carbon atom that is attractive to the δ− on the oxygen atom of the alcohol molecule. Moreover, the path that leads to the carbonyl carbon atom is wide open.

16.27 In a molecule of acetic anhydride each carbonyl carbon atom has a large δ+ which is attractive to the δ− on the oxygen atom of the alcohol molecule. In addition, the paths that lead to the carbonyl carbon atoms are unhindered by groups in the way.

16.28 (a) HCO_2CH_3

(b) $CH_3CH_2\overset{CH_3}{\overset{\|}{C}H}CO_2CH_3$

(c) 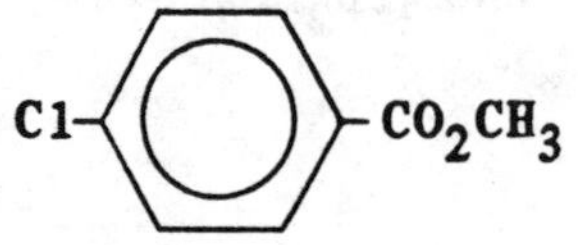 $Cl-$⟨phenyl⟩$-CO_2CH_3$

(d) $CH_3 - O - \overset{O}{\overset{\|}{C}} - \overset{O}{\overset{\|}{C}} - O - CH_3$

16.29 (a) $CH_3CH_2CO_2CH_2CH_3$

(b) 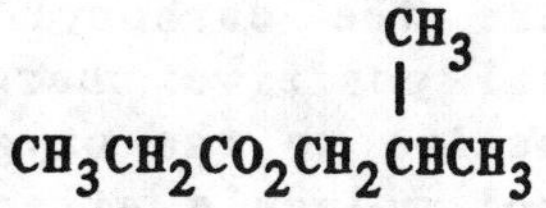 $CH_3CH_2CO_2CH_2\overset{\displaystyle CH_3}{\underset{|}{C}}HCH_3$

(c) $CH_3CH_2CO_2-$⟨benzene ring⟩

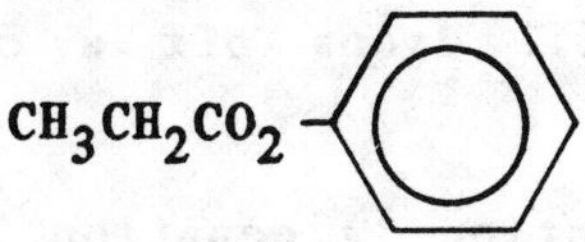

(d) $CH_3CH_2CO_2CH_2CH_2O_2CCH_2CH_3$

16.30

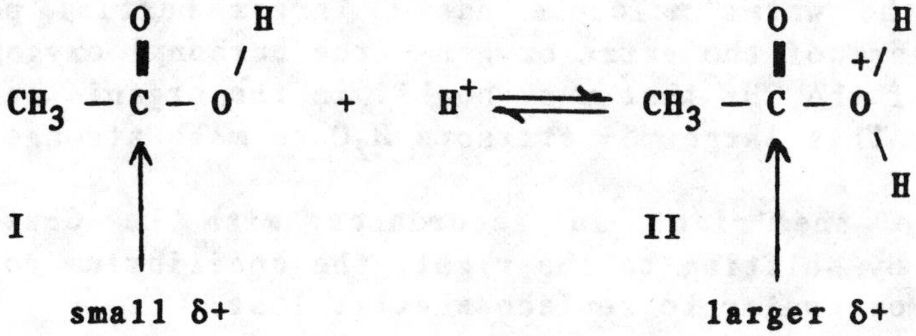

$$CH_3 - \overset{O}{\overset{\|}{C}} - \overset{H}{\overset{/}{O}} \quad + \quad H^+ \rightleftharpoons CH_3 - \overset{O}{\overset{\|}{C}} - \overset{+/\,H}{O\!\diagdown\!_H}$$

I
↑
small δ+

II
↑
larger δ+

Then:

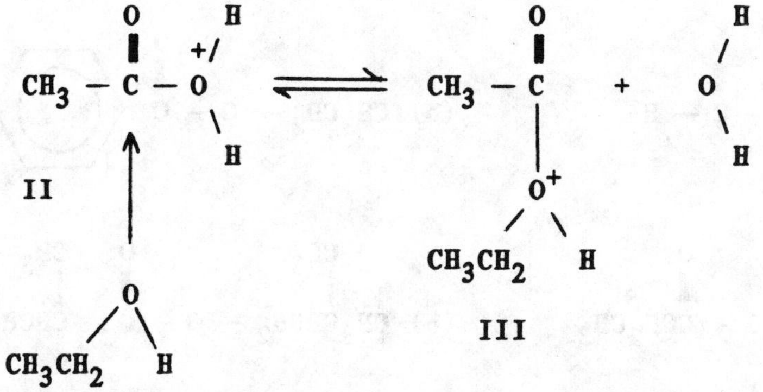

$$CH_3 - \overset{O}{\overset{\|}{C}} - \overset{+/\,H}{O} \rightleftharpoons CH_3 - \overset{O}{\overset{\|}{C}} \quad + \quad \overset{H}{\overset{/}{O}\diagdown_H}$$

II
↑
$$\overset{O}{\overset{\diagup\,\diagdown}{CH_3CH_2 \qquad H}}$$

$$\overset{O^+}{\overset{\diagup\,\diagdown}{CH_3CH_2 \quad H}}$$

III

Then:

$$CH_3 - \overset{O}{\overset{\|}{C}} - \overset{+/\,H}{O}\diagdown_{CH_2CH_3} \rightleftharpoons CH_3 - \overset{O}{\overset{\|}{C}} - O - CH_2CH_3 \quad + \quad H^+$$

III

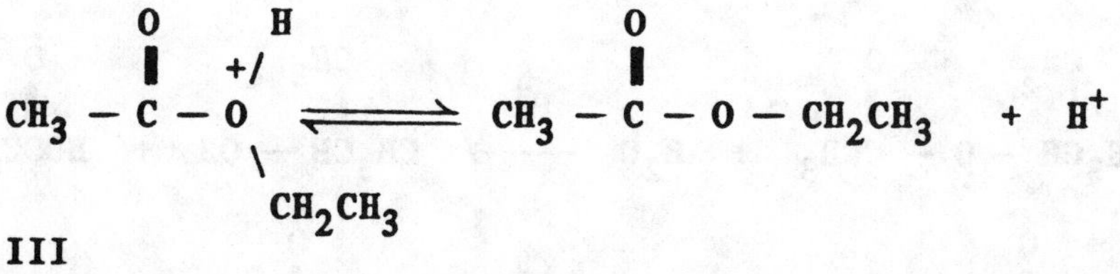

In the first step, the catalyst, H^+, forms a bond to oxygen in I. This does two things. It sets up a stable leaving group (H_2O), and it weakens the bond that holds this leaving group.

Because the carbonyl carbon atom in II has a much larger partial positive charge than it does in I, it is much more attractive to the oxygen atom of the alcohol. This helps the alcohol molecule attack and expel the water molecule to form III. Then in the last step, III drops off a proton (the catalyst is thus recovered).

16.31 The carbon atom of the organic cation in equation 2 that must be hit by the water molecule has a larger partial positive charge (because of the extra oxygen – the carbonyl oxygen) than the carbon of the CH_2 that must be hit in the organic cation of equation 1. This larger $\delta+$ attracts H_2O to make stronger hits.

16.32 Shift it to the right in accordance with Le Chatelier's principle. By shifting to the right, the equilibrium continues to produce more water to replace what is lost.

16.33 Ester molecules have smaller partial positive charges on their carbonyl carbon atoms than do acid chloride molecules.

16.34 (a) $CH_3 - O - \overset{\overset{\displaystyle O}{\|}}{C} - H$　　(b) $CH_3CH_2 - O - \overset{\overset{\displaystyle O}{\|}}{C} \bigcirc$

16.35 (a) $CH_3\overset{\overset{\displaystyle CH_3}{|}}{CH} - O - \overset{\overset{\displaystyle O}{\|}}{C}CH_2CH_3$　　(b) $CH_3\overset{\overset{\displaystyle CH_3}{|}}{CH}CH_2 - O - \overset{\overset{\displaystyle O}{\|}}{C} - \overset{\overset{\displaystyle CH_3}{|}}{CH}CH_2CH_3$

16.36 B < C < D < A

16.37 D < C < B < A

16.38 (a) $CH_3\overset{\overset{\displaystyle CH_3}{|}}{CH} - O - \overset{\overset{\displaystyle O}{\|}}{C}CH_3 + H_2O \overset{H^+}{\longrightarrow} CH_3\overset{\overset{\displaystyle CH_3}{|}}{CH} - OH + HO\overset{\overset{\displaystyle O}{\|}}{C}CH_3$

(b) $CH_3\underset{\underset{\displaystyle CH_3}{|}}{\overset{\overset{\displaystyle O}{\|}}{C}}HC - O - CH_3 + H_2O \overset{H^+}{\longrightarrow} CH_3\underset{\underset{\displaystyle CH_3}{|}}{\overset{\overset{\displaystyle O}{\|}}{C}}HC - OH + HO - CH_3$

(c) No reaction

(d) $HOCH_2\overset{\overset{\displaystyle O}{\|}}{C} - O - CH_3 + H_2O \overset{H^+}{\longrightarrow} HOCH_2\overset{\overset{\displaystyle O}{\|}}{C} - OH + HO - CH_3$

16.39 (a)

$CH_3CH_2 - O - \overset{\overset{\displaystyle O}{\|}}{C}\bigcirc + H_2O \overset{H^+}{\longrightarrow} CH_3CH_2 - OH + HO - \overset{\overset{\displaystyle O}{\|}}{C}\bigcirc$

(b)

$CH_3CH_2\overset{\overset{\displaystyle O}{\|}}{C} - O - \bigcirc + H_2O \overset{H^+}{\longrightarrow} CH_3CH_2\overset{\overset{\displaystyle O}{\|}}{C} - OH + HO - \bigcirc$

(c) No reaction

(d)

$CH_3 - O - \overset{\overset{\displaystyle O}{\|}}{C}CH_2CH_2\overset{\overset{\displaystyle O}{\|}}{C} - O - CH_3 + 2H_2O \overset{H^+}{\longrightarrow} 2CH_3OH + HO\overset{\overset{\displaystyle O}{\|}}{C}CH_2CH_2\overset{\overset{\displaystyle O}{\|}}{C}OH$

16.40
$CH_3(CH_2)_{10}CO_2H + CH_3(CH_2)_{16}CO_2H + CH_3(CH_2)_8CO_2H + HOCH_2\underset{\underset{\displaystyle OH}{|}}{CH}CH_2OH$

16.41 $HOCH_2CH_2CH_2CH_2CO_2H$

16.42   (a) $(CH_3)_2CHOH + CH_3CO_2Na$

(b) $(CH_3)_2CHCO_2Na + CH_3OH$

(c) No saponification, but $CH_3OCH_2CO_2Na$ does form.

(d) $HOCH_2CO_2Na + CH_3OH$

16.43   (a) $\bigcirc\!-CO_2K + CH_3CH_2OH$

(b) $CH_3CH_2CO_2K + \bigcirc\!-OK$

(c) No reaction

(d) $KO_2CCH_2CH_2CO_2K$ + $2CH_3OH$

16.44 (a)
$CH_3(CH_2)_{10}CO_2Na$ + $CH_3(CH_2)_{16}CO_2Na$ + $CH_3(CH_2)_8CO_2Na$ + $HOCH_2CHCH_2OH$
$$\qquad\qquad\qquad\qquad\qquad\qquad\qquad\qquad\qquad\qquad\qquad\qquad\underset{\displaystyle OH}{|}$$

16.45 $HOCH_2CH_2CH_2CH_2CO_2^-$

16.46 Use a large excess of ethanol to drive the following equilibrium to the right and use up the acid.

$$RCO_2H \; + \; CH_3CH_2OH \; \overset{H^+}{\rightleftharpoons} \; RCO_2CH_2CH_3 \; + \; H_2O$$

16.47

$$CH_3 - \overset{\displaystyle O}{\overset{\|}{C}} - \underset{\displaystyle CH_3}{\overset{\displaystyle}{O}} \; + \; H^+ \; \rightleftharpoons \; CH_3 - \overset{\displaystyle O}{\overset{\|}{C}} - \underset{\displaystyle CH_3}{\overset{\displaystyle +/H}{O}}$$

Then

$$CH_3 - \overset{\displaystyle O}{\overset{\|}{C}} - \underset{\displaystyle CH_3}{\overset{+/H}{O}} \; + \; \underset{\displaystyle H}{\overset{/H}{O}} \; \rightleftharpoons \; CH_3 - \overset{\displaystyle O}{\overset{\|}{C}} - \underset{\displaystyle H}{\overset{+/H}{O}} \; + \; \underset{\displaystyle CH_3}{\overset{/H}{O}}$$

Then

$$CH_3 - \overset{\displaystyle O}{\overset{\|}{C}} - \underset{\displaystyle H}{\overset{+/H}{O}} \; \rightleftharpoons \; CH_3 - \overset{\displaystyle O}{\overset{\|}{C}} - \underset{\displaystyle H}{\overset{\displaystyle}{O}} \; + \; H^+$$

16.48 (a) $CH_3CH_2 - O - \overset{\displaystyle O}{\underset{\displaystyle OH}{\overset{\|}{P}}} - OH$ (b) $CH_3 - O - \overset{\displaystyle O}{\underset{\displaystyle OH}{\overset{\|}{P}}} - O - \overset{\displaystyle O}{\underset{\displaystyle OH}{\overset{\|}{P}}} - OH$

(c) $CH_3CH_2CH_2 - O - \overset{\overset{\displaystyle O}{\|}}{\underset{\underset{\displaystyle OH}{|}}{P}} - O - \overset{\overset{\displaystyle O}{\|}}{\underset{\underset{\displaystyle OH}{|}}{P}} - O - \overset{\overset{\displaystyle O}{\|}}{\underset{\underset{\displaystyle OH}{|}}{P}} - OH$

16.49 To improve their solubility in water

16.50 The phosphoric acid anhydride system: $- \overset{\overset{\displaystyle O}{\|}}{\underset{\underset{\displaystyle |}{|}}{P}} - O - \overset{\overset{\displaystyle O}{\|}}{\underset{\underset{\displaystyle |}{|}}{P}} -$

The P — O bonds in this system are the bonds that break when ATP or other high energy phosphoric acid anhydride systems participate in endothermic processes.

16.51 The phosphorus atom in the P=O group is harder for an attacking molecule to reach than is the carbon atom in the C=O group.

16.52 (a) CH_3CO_2H (b) No reaction

(c) $CH_3CO_2CH_3$ (d) $CH_3CH_2\overset{\overset{\displaystyle O}{\|}}{C}CH_3$

(e) $CH_3CH=CH_2$ (f) CH_3OH + $HO_2CCH(CH_3)_2$

(g) $CH_3CH_2\underset{\underset{\displaystyle Cl}{|}}{C}HCH_3$ (h) CH_3CHO + $2CH_3OH$

(i) $(CH_3)_2CHCO_2Na$ + $HOCH_3$ (j) No reaction

(k) $CH_3CH_2CO_2H$ (l) $CH_3CH_2CH_2CO_2Na$

16.53 (a) $CH_3CH_2CO_2H$

(b) CH_3CO_2H +

(c) No reaction

(d) $CH_3CH_2CH_2OH$ + $HO_2CCH_2CH_2OH$ + HO_2CCH_3

(e) 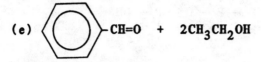—CH=O + $2CH_3CH_2OH$

(f) No reaction

(g) $CH_3CH_2CO_2CH(CH_3)_2$

(h) No reaction

(i) CH_3OH + $NaO_2CCH_2CH_2OH$ + $NaO_2CCH_2CH_3$

(j) $CH_3(CH_2)_6CO_2Na$

(k) $HO_2CCH_2CH_2CH_2CO_2H$

(1) $CH_3O_2CCH_2CH_2CH_2CO_2CH_3$

CHAPTER 17

AMINES AND AMIDES

Review Exercises

17.1   (a) Aliphatic amide (with an ether group)
        (b) Aliphatic amine (with an ester group)
        (c) Aliphatic and heterocyclic amine
        (d) Aromatic amine

17.2   (a) Aliphatic and heterocyclic amide
        (b) Aliphatic and heterocyclic amide
        (c) Aliphatic amine with a ketone group; also heterocyclic
        (d) Aromatic amine

17.3   (a) 1 = aliphatic and heterocyclic amine

        (b) 1 = aliphatic amine
            2 = ester
            3 = aromatic amine

        (c) 1 = heterocyclic amine   (This ring is also an aromatic
                                              system.)
            2 = aliphatic amine; also heterocyclic

        (d) 1 = 2° alcohol
            2 = aliphatic amine

17.4   (a) 1 = alkene
            2 = ester
            3 = aliphatic and heterocyclic amine

        (b) 1 = aliphatic and heterocyclic amine
            2 = 1° alcohol
            3 = ester

(c) 1 = alkene
2 = 2° alcohol
3 = ether
4 = heterocyclic amine

(d) 1 = amide
2 = aliphatic and heterocyclic amine
3 = alkene
4 = aliphatic and heterocyclic amine

17.5 (a) p-Chloroaniline
(b) Diethylisopropylamine
(c) Cyclopentylcyclohexylamine
(d) Isobutyl-t-butylammonium ion

17.6 (a) Tri-t-butylamine   (b) 2,4,6-Tribromoaniline
(c) Isopropylbutylamine   (d) Triethylammonium chloride

17.7 (a) No reaction

(b) 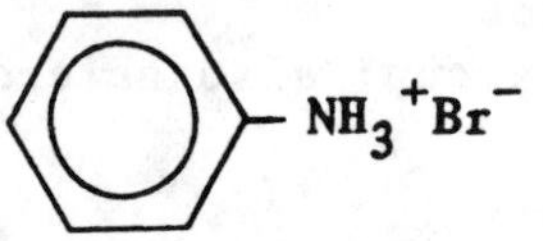 ⬡—$NH_3^+Br^-$ (Also acceptable: ⬡—NH_3Br)

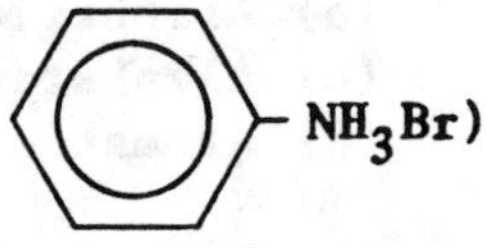

(c) NH_3 (+ H_2O + NaCl)

(d) 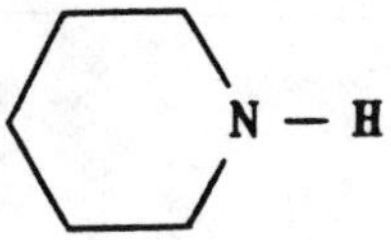 [piperidine ring]—N—H

(e) CH_3NH_2 (f) $NH_2CH_2CO_2^-$

17.8 (a) 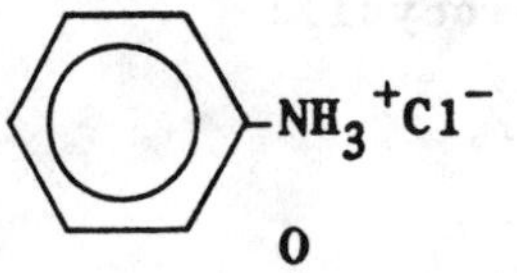 ⬡—$NH_3^+Cl^-$ (Also acceptable: ⬡—NH_3Cl)

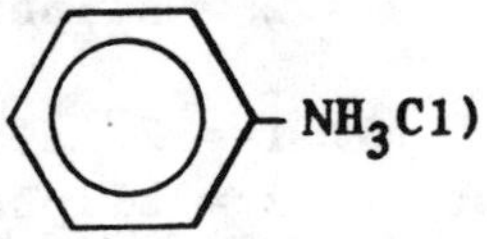

(b) 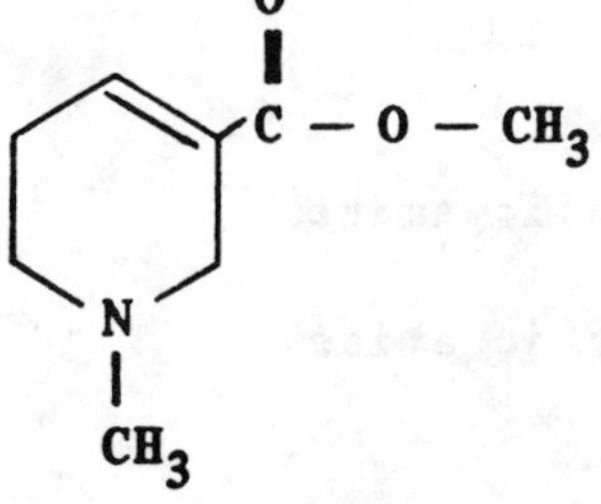 [ring structure with]
$$\overset{O}{\overset{\|}{C}} - O - CH_3$$
N
CH₃

(c)

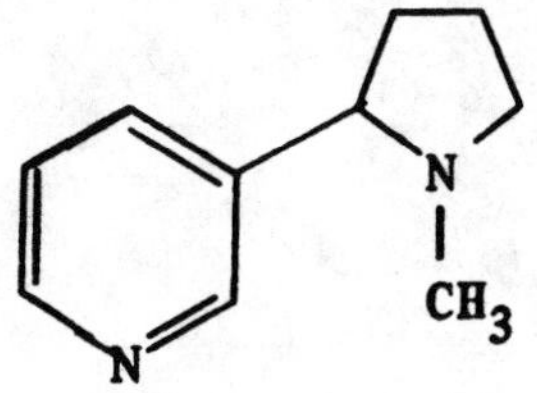

(d)

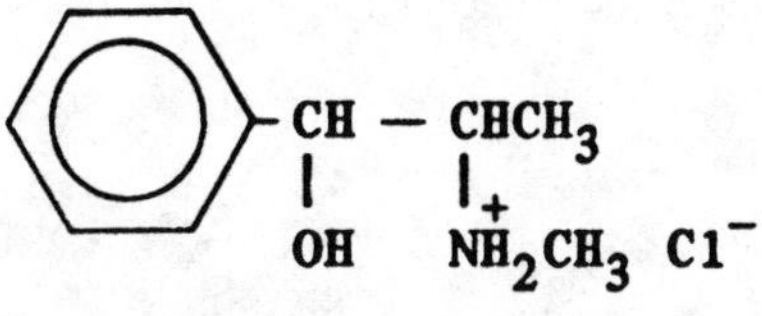

(e) 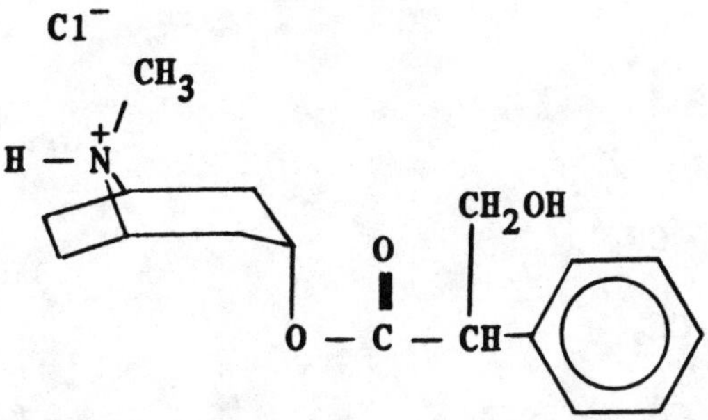 Cl^-

17.9   B. A is an amide, but B is an amine (as well as a ketone).

17.10   A. A is an amine, but B has no unshared pair of electrons on N for accepting a proton.

17.11   (a) 2,2-Dimethylpropanamide
   (b) 3-Bromo-2-methylbutanamide

17.12   Caproamide (sometimes written as capramide)

17.13

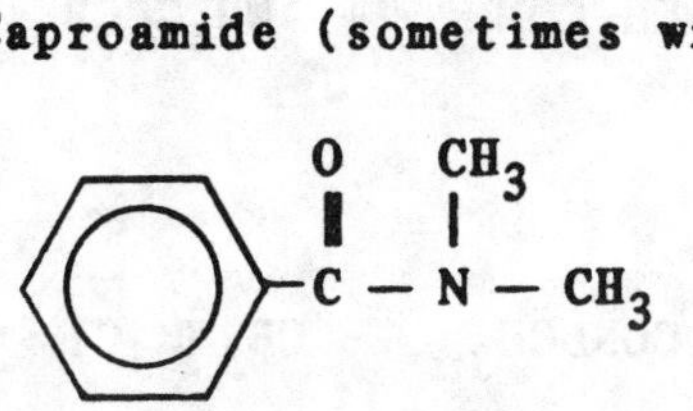

17.14 $NH_2 - \overset{\overset{\text{O}}{\|}}{C}CH_2CH_2\overset{\overset{\text{O}}{\|}}{C} - NH_2$

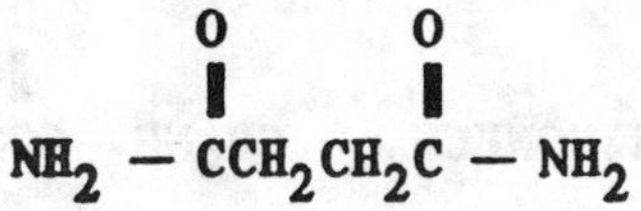

17.15

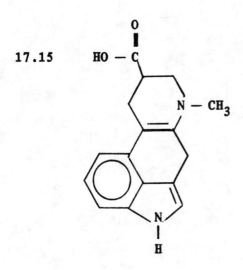

17.16

$$\text{C}_6\text{H}_5 - \underset{\underset{\text{HO}}{|}}{\text{CH}} - \underset{\underset{\underset{\underset{\text{O=C}-\text{CH}_3}{|}}{\text{N}-\text{CH}_3}}{|}}{\text{CH}} - \text{CH}_3$$

17.17 $\quad \text{CH}_3\overset{\text{O}}{\underset{\|}{\text{C}}} - \text{Cl} + 2\text{NH}_3 \longrightarrow \text{CH}_3\text{CONH}_2 + \text{NH}_4\text{Cl}$

$\quad \text{CH}_3\overset{\text{O}}{\underset{\|}{\text{C}}} - \text{O} - \overset{\text{O}}{\underset{\|}{\text{C}}}\text{CH}_3 + 2\text{NH}_3 \longrightarrow \text{CH}_3\text{CONH}_2 + \text{NH}_4\overset{\text{O}}{\underset{\|}{\text{O}}}\text{CCH}_3$

17.18 $\quad \text{CH}_3\overset{\text{O}}{\underset{\|}{\text{C}}} - \text{Cl} + 2\text{NH}_2\text{CH}_3 \longrightarrow \text{CH}_3\text{CONHCH}_3 + \text{CH}_3\text{NH}_3\text{Cl}$

$\quad \text{CH}_3\overset{\text{O}}{\underset{\|}{\text{C}}} - \text{O} - \overset{\text{O}}{\underset{\|}{\text{C}}}\text{CH}_3 + 2\text{NH}_2\text{CH}_3 \longrightarrow \text{CH}_3\text{CONHCH}_3 + \text{CH}_3\overset{+}{\text{NH}_3}{}^{-}\overset{\text{O}}{\underset{\|}{\text{O}}}\text{CCH}_3$

17.19 (a) $NH_2CH_2\overset{O}{\overset{\|}{C}}NHCH\overset{O}{\overset{\|}{C}} -$ (b) Two
$\underset{CH_3}{|}$

17.20 $CH_3\overset{O}{\overset{\|}{C}}NH_2$ + $CH_3CH_2\overset{O}{\overset{\|}{C}}NH_2$ (Ammonia can attack at random at either of the carbonyl carbons.)

17.21 (a) CH_3NH_2 + $CH_3CH_2CO_2H$ (b) No reaction

(c) CH_3CO_2H + $(CH_3)_2NH$ (d) $(CH_3)_2CHCO_2H$ + CH_3NH_2

17.22 (a) $2NH_2CH_2CO_2H$

(b) $(CH_3)_2NCH_2CO_2H$ + $NH_2CH_2CH_2CO_2H$ + CH_3NH_2

(c) $NH_2CH_2CH_2CH_2CH_2CO_2H$

(d) $(CH_3)_2NCH_2CH_2NH_2$ + $HO_2CCH_2CH_2CO_2H$ + CH_3NH_2

17.23 (a) The addition of water to a double bond to give an alcohol:

$$\overset{\diagdown \diagup}{\underset{\diagup \diagdown}{C}}=C + H_2O \xrightarrow{H^+} -\overset{|}{\underset{|}{C}} - \overset{|}{\underset{|}{C}} -$$
$$\qquad\qquad\qquad\qquad H \quad OH$$

The hydrolysis of an acetal or a ketal to give an aldehyde or a ketone plus two molecules of alcohol:

$$\overset{O-R}{\underset{O-R}{-\overset{|}{\underset{|}{C}}-}} + H_2O \xrightarrow{H^+} -\overset{O}{\overset{\|}{C}} - + 2HOR$$

The hydrolysis of an ester to give an acid and an alcohol:

$$\overset{\displaystyle O}{\underset{\displaystyle \|}{R - C - O - R'}} \quad + \quad H_2O \quad \longrightarrow \quad R - CO_2H \quad + \quad HOR'$$

The hydrolysis of an amide to give an acid and an amine (or ammonia):

$$\overset{\displaystyle O}{\underset{\displaystyle \|}{R - C - \overset{\displaystyle |}{N} -}} \quad + \quad H_2O \quad \longrightarrow \quad R - CO_2H \quad + \quad H - \overset{\displaystyle |}{N} -$$

(b) The reduction of a disulfide to two molecules of mercaptan:

$$R - S - S - R \quad + \quad 2(H) \quad \longrightarrow \quad 2RSH$$

The reduction of an alkene to give an alkane:

$$\overset{\diagdown}{\underset{\diagup}{C}} = \overset{\diagup}{\underset{\diagdown}{C}} \quad + \quad H_2 \quad \xrightarrow[\text{pressure}]{\text{Ni, heat}} \quad - \overset{|}{\underset{|}{C}} - \overset{|}{\underset{|}{C}} -$$

The reduction of an aldehyde or a ketone to give an alcohol:

$$\overset{\displaystyle O}{\underset{\displaystyle \|}{- C -}} \quad + \quad H_2 \quad \xrightarrow[\text{pressure}]{\text{Ni, heat}} \quad \overset{\displaystyle OH}{\underset{\displaystyle |}{- CH -}}$$

(c) The oxidation of a 1° alcohol or a 2° alcohol to an aldehyde or a ketone:

$$\overset{\displaystyle OH}{\underset{\displaystyle |}{- CH -}} \quad + \quad (O) \quad \longrightarrow \quad \overset{\displaystyle O}{\underset{\displaystyle \|}{- C -}} \quad + \quad H_2O$$

The oxidation of a mercaptan to a disulfide:

$$2RSH \quad + \quad (O) \quad \longrightarrow \quad R-S-S-R \quad + \quad H_2O$$

The oxidation of an aldehyde to a carboxylic acid:

$$RCH=O \quad + \quad (O) \quad \longrightarrow \quad RCO_2H$$

(Note: Strong oxidizing agents also attack side chains on benzene rings and double bonds.)

17.24 The acetal (or ketal) group in carbohydrates; the ester group in fats and oils; and the amide group in proteins.

17.25 (a) $CH_3CH_2CO_2Na$ (b) No reaction

OH
|
(c) CH_3CHCH_3 (d) CH_3OH + CH_3CO_2H

O
‖
(e) $CH_3C - NH(CH_3)_2$ + $(CH_3)_2CHNH_3^+Cl^-$ (f) No reaction

(g) $CH_3CH_2CH=O$ + $2CH_3OH$ (h) $- CO_2H$

OCH$_3$
|
(i) $CH_3CH_2CO_2Na$ + CH_3CH_2OH (j) $CH_3CH - OCH_3$

(k) No reaction   (1) No reaction

(m) $CH_3CH_2CO_2CH_3$ (n) $CH_3CH_2CH_2NH_3^+Cl^-$

(o) HCO_2H + NH_3 (p) $CH_3 - S - S - CH_3$

17.26 (a)$-CO_2Na$ + CH_3CH_2OH (b)$- CH - OCH_3$ (with OCH$_3$ above CH)

(c) No reaction   (d) No reaction

(e) $NaO_2CCH_2CH_2CH_3$

(f) $CH_3\overset{\overset{\displaystyle O}{\|}}{C}CH_3$ + $2HOCH_2CH_3$

(g) No reaction

(h) $2CH_3OH$ + $HO_2CCH_2CH_2CO_2H$

(i) $Cl^-\overset{+}{N}H_3 - CH_2CH_2CH_2 - NH_3^+Cl^-$

(j) $CH_3 - O - CH_2CH_2CO_2H$

(k) $CH_3(CH_2)_4CO_2CH_2CH_3$

(l) $2CH_3CH_2OH$ + $CH_3CH=O$

(m) $CH_3CH_2\overset{\overset{\displaystyle OH}{|}}{C}HCH_2CH_3$

(n) $CH_3\overset{\overset{\displaystyle O}{\|}}{C}NHCH_3$ + $CH_3NH_3^+ {}^-\overset{\overset{\displaystyle O}{\|}}{O}CCH_3$

(o) $2CH_3NH_2$ + $HO_2CCH_2CH_2CO_2H$

(p) $2CH_3SH$

CHAPTER 18

OPTICAL ISOMERISM

Review Exercises

18.1 $CH_3CH_2CH_2Cl$ and CH_3CHCH_3
                                               |
                                              Cl

18.2
$$
\begin{array}{c}
CH3 \qquad\qquad Cl \\
C\!=\!C \\
H \qquad\qquad H
\end{array}
\quad \text{and} \quad
\begin{array}{c}
CH_3 \qquad\qquad H \\
C\!=\!C \\
H \qquad\qquad Cl
\end{array}
$$

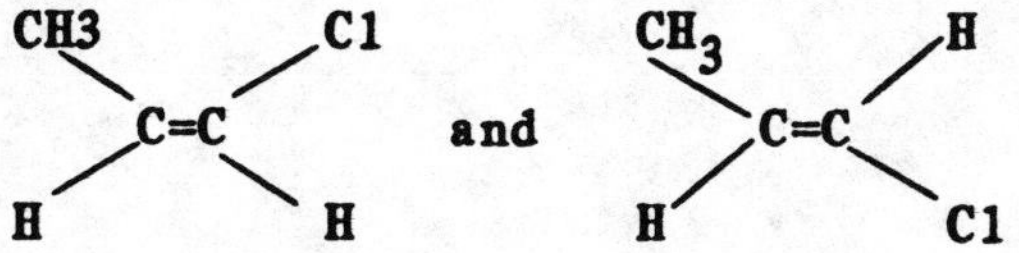

18.3   (a) Stereoisomers   (b) Structural isomers

18.4   Each substance has a unique structure.

18.5   (a) $HOCH_2 - \overset{*}{C}H - \overset{*}{C}H - \overset{*}{C}H - \overset{*}{C}H - CH\!=\!0$
                                | | | |
                               OH OH OH OH

       (b) All four
       (c) 16 (2^4) optical isomers   (d) 8 pairs of enantiomers

18.6   (a)

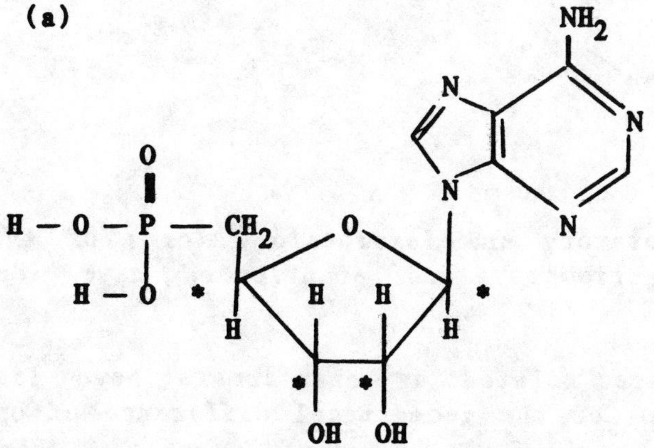

510

(b) 16

18.7   No, it has no chiral carbons.

18.8   No, it has no chiral carbons.

18.9   (a)

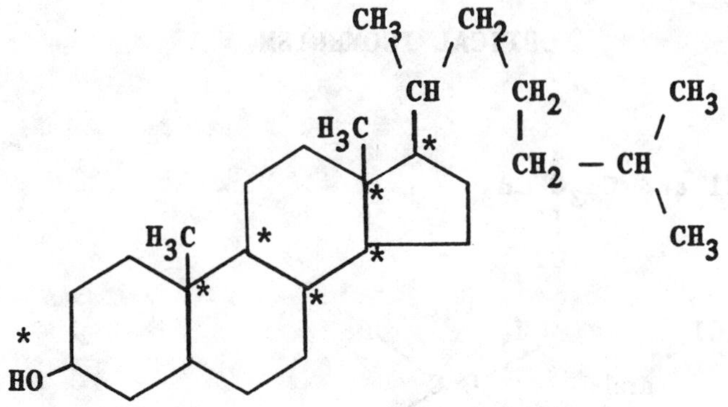

(b) 256 (2^8)

18.10   (a)

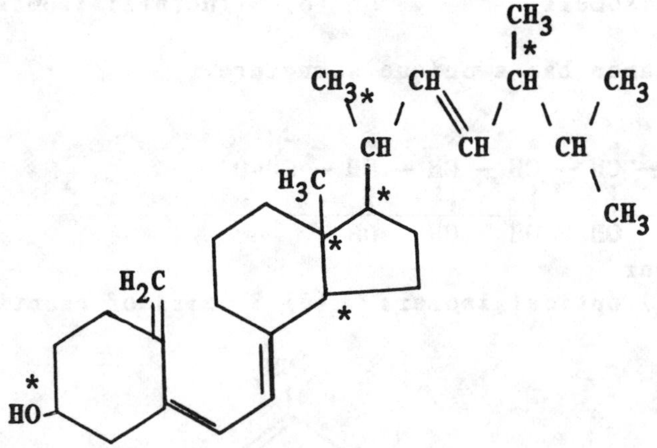

(b) 64 (2^6)

18.11   148.5 °C. Dextrotatory and levorotatory forms of the same compound are enantiomers, and enantiomers have identical melting points.

18.12   Molecules that are related as enantiomers have identical structures, except for the geometrical difference of opposite chirality. They have identical bond angles, bond distances, and identical molecular polarities. All properties related to these factors are therefore identical.

18.13 Usually, a reaction between the molecules of two reactants occurs only if their molecules somehow fit to each other. Just as either a left hand or a right hand interacts equally well with, say, a cube or a sphere (achiral objects), so the enantiomers of a chiral substance can react equally well with any reactant whose molecules are achiral.

18.14 Usually, a reaction between the molecules of two reactants occurs only if their molecules somehow fit to each other. Just as a left hand fits best to a left-hand glove, so a molecule of an enantiomer will fit best only to a molecule of just one of the two enantiomers of a potential reactant.

18.15 -37.5°

18.16 -21°

18.17 The amount of rotation is directly proportional to the populations of chiral molecules encountered by the polarized light.

18.18 The optical activity is lost. The effect on the polarized light by the chiral molecules of one enantiomer is cancelled by the opposite effect exerted by the molecules of the other enantiomer.

18.20 4.86 g/dL

18.22 $[\alpha] = +209^{\circ}$. The unknown is cortisone.

CHAPTER 19

CARBOHYDRATES

Review Exercises

19.1   Materials, energy, and information

19.2   Carbohydrates, lipids, and proteins

19.3   Enzymes and nucleic acids, the latter carrying genetic blueprints.

19.4   Organization

19.5   (a) C   (b) D
        (c) A   (d) B and D

19.6   (a) 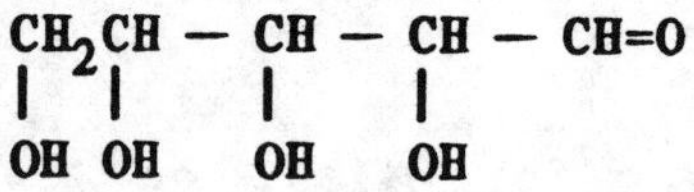 $CH_2CH - CH - CH - CH=0$
             $|\ \ |\ \ \ \ \ |\ \ \ \ \ |$
             $OH\ OH\ \ \ OH\ \ \ OH$

        (b) 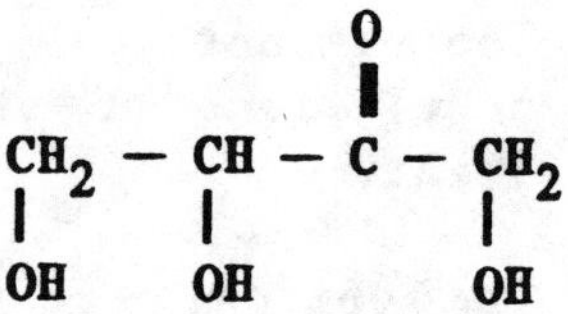 $CH_2 - CH - \overset{\displaystyle O}{\overset{\|}{C}} - CH_2$
            $\ |\ \ \ \ \ \ |\ \ \ \ \ \ \ \ \ \ \ \ |$
            $\ OH\ \ \ \ OH\ \ \ \ \ \ \ \ OH$

19.7 $CH_2 - CH - CH=0$ Glyceraldehyde
        $\ |\ \ \ \ \ \ |$
        $\ OH\ \ \ \ OH$

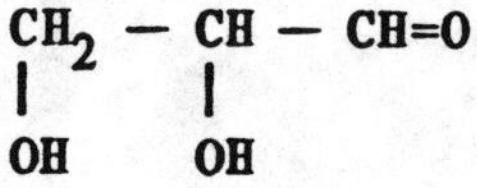

19.8 $CH_2 - \overset{\displaystyle O}{\overset{\|}{C}} - CH2$ Dihydroxyacetone
        $\ |\ \ \ \ \ \ \ \ \ |$
        $\ OH\ \ \ \ \ \ OH$

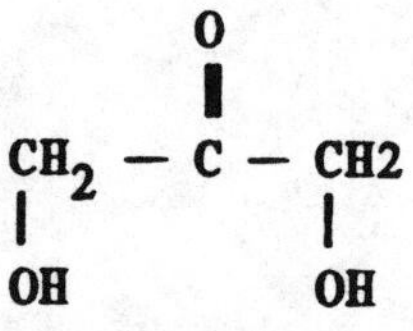

19.9   Polysaccharide

19.10   Nonreducing carbohydrate

512

19.11 (a) Glucose
(b) Glucose

19.12 Fructose

19.13 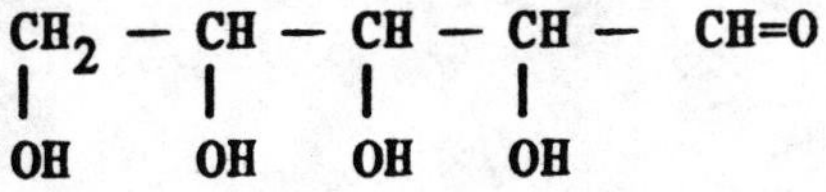
CH2 — CH — CH — CH — CH=O
OH OH OH OH

19.14 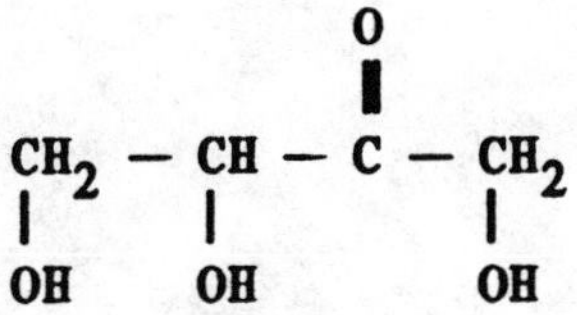
O
CH2 — CH — C — CH2
OH OH OH

19.15 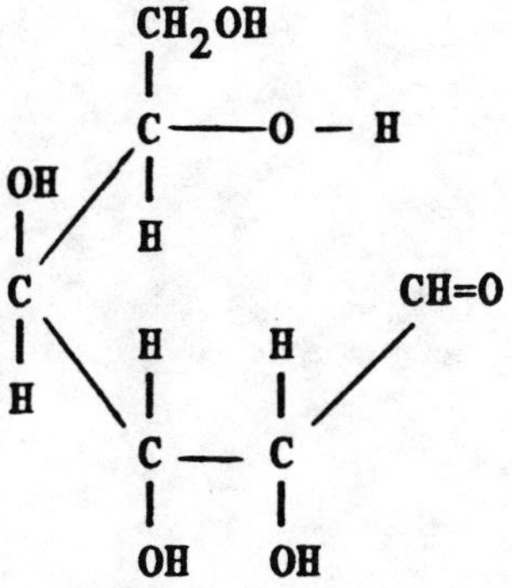
CH2OH
C—O — H
OH
H
C
CH=O
H H
H
C — C
OH OH

19.16 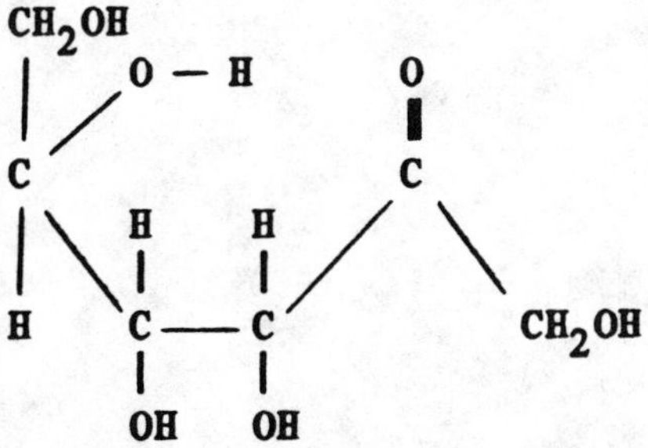
CH2OH
O — H
O
C
C
H H
H
C — C
CH2OH
OH OH

19.17

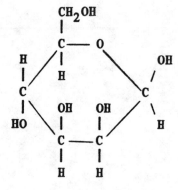

β–Mannose

Open-form of mannose

α–Mannose

19.18

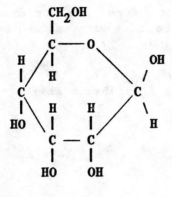

β-Allose

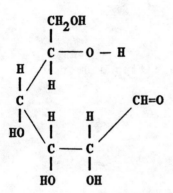

Open-form of allose

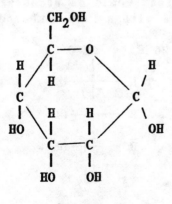

α-Allose

19.19   As molecules of the open-chain form are oxidized, the equilibrium continuously shifts to make more from the cyclic forms.

19.20   As the beta form is used, molecules of the other forms continuously change into it as the equilibrium shifts.

19.21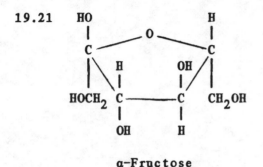

α-Fructose

19.22   The structure is that of β-fructose tipped over.

19.23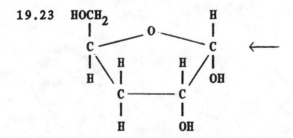

19.24   No, it has no — OH on carbon-4 that would be necessary to the formation of the cyclic hemiacetal with a five-membered ring.

19.25   19.26

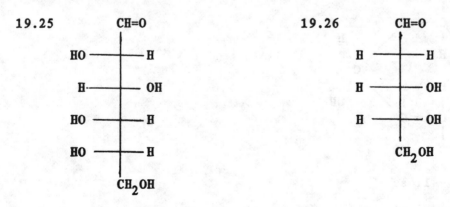

19.27

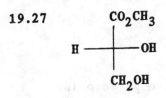

D-family, because the — OH group has the same projection as in D-glyceraldehyde. Moreover, no bonds change their orientation in the conversion to the methyl ester.

19.28 L-family

19.29

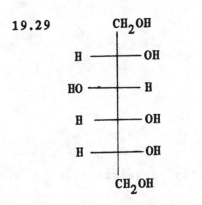

19.30

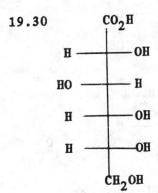

19.31

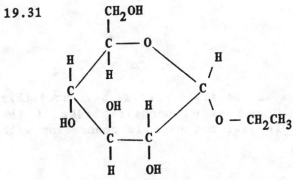

Ethyl α-glucoside

Ethyl β-glucoside

These two are not enantiomers, just stereoisomers.

518

19.32

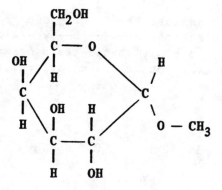

Methyl α-galactoside

Methyl β-galactoside

These are cis-trans isomers in the sense that we have used this concept, because the OCH$_3$ group is on opposite sides of the rings. However, the term is just not used in connection with glycosides.

19.33 Maltose, lactose, and sucrose

19.34 A 50:50 mixture of glucose and fructose

19.35 Its molecules have no hemiacetal or hemiketal group at which ring-opening and ring-closing can occur.

19.36 (a) and (b)

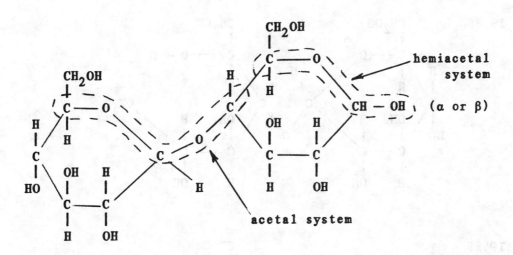

(c) Yes, it has the hemiacetal system.

(d) The bridge between the two glucose units is β(1——>4)
instead of α(1——>4).

(e) Two molecules of glucose:

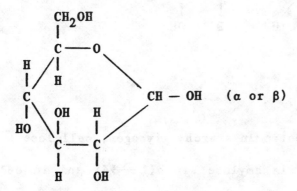

19.37 (a) No, it has no hemiacetal or hemiketal system.

(b) No, for the same reason given in (a) this cannot give a
Tollens' or a Benedict's test.

(c) Two molecules of glucose

520

19.38

```
        CH₂OH                          CH₂OH
          |                              |
          C — O                          C — O — H
     H  / |            H  H        H  / |
      | /  H            |  |          | /  H
      C                 C    C        C
      |                   \ O /       |
   HO \ |  OH   H          O       OH |    H       CH=O
       \ |   |   |                   |    |      /
        C — C                        C — C
        |   |                        |    |
        H   OH                       H    OH
```

19.39

```
                                    CH₂OH
                                      |
              CH₂OH                   C — O — H
                |              H   / |
                C — O           |  /  H
           HO / |               C
            | /  H         O   / |  OH   H          CH=O
            C              |  /  |    |   |        /
            |              C —  C     C — C
         H \ |  OH   H          |     |   |
            \ |   |   |         H     H   OH
             C — C
             |   |
             H   OH
```

19.40 Amylose and amylopectin in starch; glycogen; cellulose

19.41 The oxygen bridges in amylose are $\alpha(1\longrightarrow4)$ and in cellulose they are $\beta(1\longrightarrow4)$.

19.42 They are polymers of α-glucose.

19.43 Amylose is an entirely linear polymer of α-glucose in which all of the oxygen bridges are $\alpha(1\longrightarrow4)$, and amylopectin has branching.

19.44 Humans lack the enzyme for catalyzing this reaction.

19.45 A test for starch. The reagent is a solution of iodine in aqueous potassium iodide. It is used to test for starch, and a positive test is the immediate appearance of a blue-black color.

19.46 They are very similar except that glycogen is more branched.

19.47 To store glucose units

Review Exercises

20.1   It is not obtainable from living plants or animals.

20.2   It is extractable from animal and plant sources by relatively nonpolar solvents.

20.3   It is soluble in water, and it isn't present in plant or animal sources.

20.4   It is present in undecomposed plant or animal materials and is extractable by relatively nonpolar solvents.

20.5 $CH_3(CH_2)_{14}CO_2(CH_2)_{17}CH_3$

20.6   C. A is ruled out because <u>both</u> the acid and alcohol portions of the wax molecule are usually long-chain. B is ruled out because both of these portions are likely to have an even number of carbons.

20.7   Palmitic acid, $CH_3(CH_2)_{14}CO_2H$
        Stearic acid, $CH_3(CH_2)_{16}CO_2H$

20.8 $CH_3(CH_2)_7$ ⟍ $(CH_2)_7CO_2H$, oleic acid
               $C=C$
             H H

        $CH_3(CH_2)_4$ ⟍ CH_2 ⟍ $(CH_2)_7CO_2H$, linoleic acid
              $C=C$ $C=C$
            H H H H

$$CH_3CH_2 \overset{\displaystyle CH_2}{\underset{\displaystyle H}{C=C}} \overset{\displaystyle CH_2}{\underset{\displaystyle H \quad H}{C=C}} \overset{\displaystyle CH_2}{\underset{\displaystyle H \quad H}{C=C}} \overset{\displaystyle (CH_2)_7CO_2H,}{\underset{\displaystyle H}{}} \quad \text{linolenic acid}$$

20.9 (a) $CH_3(CH_2)_{14}CO_2H + NaOH \longrightarrow CH_3(CH_2)_{14}CO_2Na + H_2O$

(b) $CH_3(CH_2)_{14}CO_2H + CH_3OH \overset{H^+}{\longrightarrow} CH_3(CH_2)_{14}CO_2CH_3 + H_2O$

20.10 (a)

$CH_3(CH_2)_7CH=CH(CH_2)_7CO_2H + Br_2 \longrightarrow CH_3(CH_2)_7\underset{\displaystyle Br}{CH} - \underset{\displaystyle Br}{CH}(CH_2)_7CO_2H$

(b)

$CH_3(CH_2)_7CH=CH(CH_2)_7CO_2H + KOH \longrightarrow CH_3(CH_2)_7CH=CH(CH_2)_7CO_2K + H_2O$

(c)

$CH_3(CH_2)_7CH=CH(CH_2)_7CO_2H + H_2 \xrightarrow[\text{heat,pressure}]{Ni} CH_3(CH_2)_{16}CO_2H$

(d)

$CH_3(CH_2)_7CH=CH(CH_2)_7CO_2H + CH_3CH_2OH \xrightarrow[\text{heat}]{H^+} CH_3(CH_2)_7CH=CH(CH_2)_7CO_2CH_2CH_3$
$+ H_2O$

20.11 B. Molecules of fatty acids rarely have branched chains or an uneven number of carbon atoms.

20.12 Fatty acids whose molecules have five-membered rings to which long side chains are joined.

20.13

$CH_2 - O - \overset{\displaystyle O}{\overset{\|}{C}}(CH_2)_7CH=CHCH_2CH=CHCH_2CH=CHCH_2CH_3$

$CH \ - O - \overset{\displaystyle O}{\overset{\|}{C}}(CH_2)_7CH=CH(CH_2)_7CH_3$

$CH_2 - O - \overset{\displaystyle O}{\overset{\|}{C}}(CH_2)_{12}CH_3$

20.14

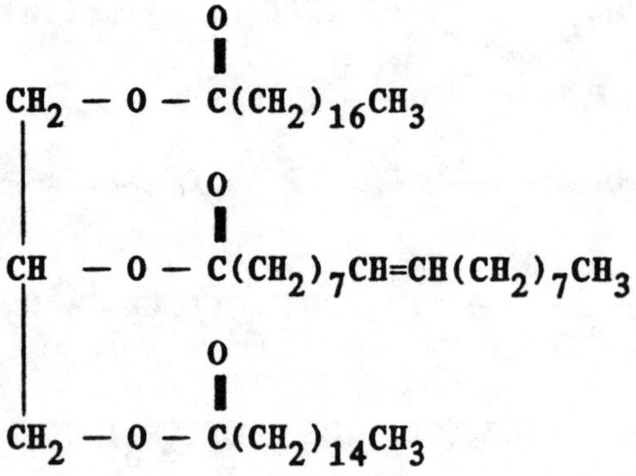

$$CH_2 - O - \overset{\overset{\displaystyle O}{\|}}{C}(CH_2)_{16}CH_3$$

$$CH \ - O - \overset{\overset{\displaystyle O}{\|}}{C}(CH_2)_7CH=CH(CH_2)_7CH_3$$

$$CH_2 - O - \overset{\overset{\displaystyle O}{\|}}{C}(CH_2)_{14}CH_3$$

20.15

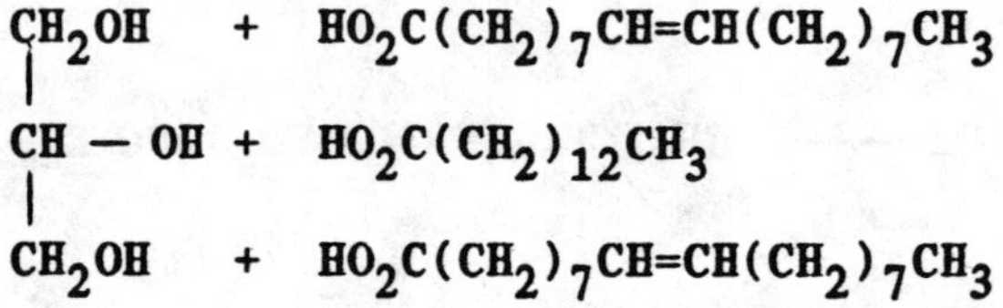

$CH_2OH \quad + \quad HO_2C(CH_2)_7CH=CH(CH_2)_7CH_3$

$CH - OH \ + \quad HO_2C(CH_2)_{12}CH_3$

$CH_2OH \quad + \quad HO_2C(CH_2)_7CH=CH(CH_2)_7CH_3$

20.16

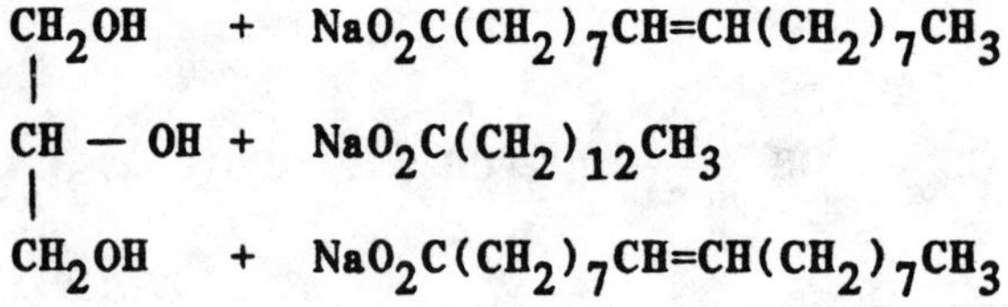

$CH_2OH \quad + \quad NaO_2C(CH_2)_7CH=CH(CH_2)_7CH_3$

$CH - OH \ + \quad NaO_2C(CH_2)_{12}CH_3$

$CH_2OH \quad + \quad NaO_2C(CH_2)_7CH=CH(CH_2)_7CH_3$

20.17

$$CH_2 - O - \overset{\overset{\displaystyle O}{\|}}{C}(CH_2)_{10}CH_3$$

$$CH \ - O - \overset{\overset{\displaystyle O}{\|}}{C}(CH_2)_7CH=CHCH_2CH=CH(CH_2)_4CH_3$$

$$CH_2 - O - \overset{\overset{\displaystyle O}{\|}}{C}(CH_2)_7CH=CH(CH_2)_7CH_3$$

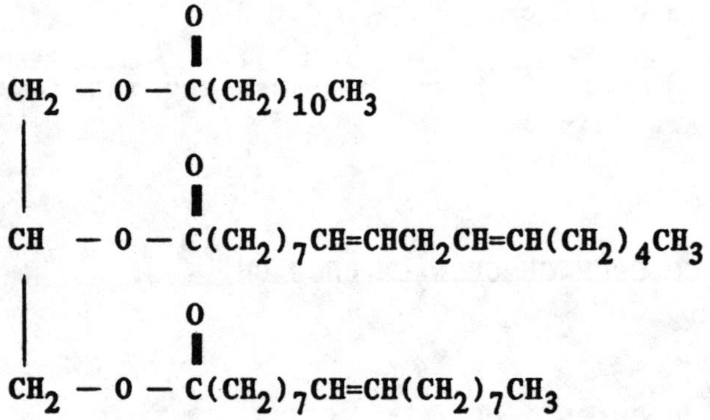

This structure shows just one possibility. The fatty acyl
units can be joined in different orders to the glycerol unit.

20.18

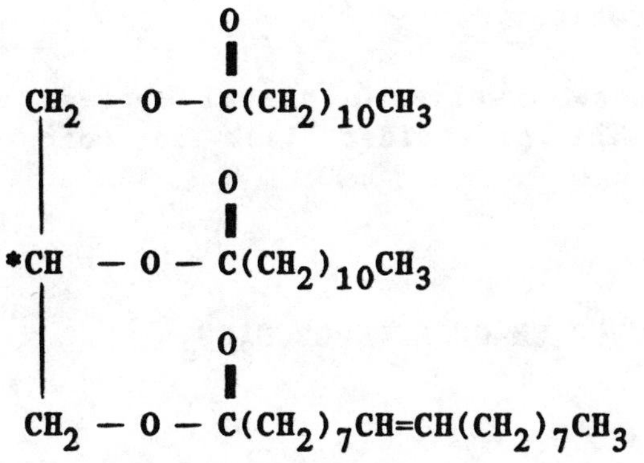

$$CH_2 - O - \overset{\overset{\displaystyle O}{\|}}{C}(CH_2)_{10}CH_3$$

$$*CH - O - \overset{\overset{\displaystyle O}{\|}}{C}(CH_2)_{10}CH_3$$

$$CH_2 - O - \overset{\overset{\displaystyle O}{\|}}{C}(CH_2)_7CH=CH(CH_2)_7CH_3$$

The asterisk (*) marks the chiral carbon. This is the only arrangement that can have a chiral carbon.

20.19 There are more alkene double bonds per molecule in vegetable oils than in animal fats.

20.20 The triacylglycerol molecules that are present have several alkene units per molecule, so the substances are more "polyunsaturated" than animal fats.

20.21 Hydrogenation

20.22 Butter melts on the tongue; lard and tallow do not.

20.23 Glycerol-based (phosphoglycerides and plasmalogens) and sphingosine-based (sphingolipids).

20.24 Phosphoglyceride molecules have a fatty acyl unit joined to the glycerol unit where plasmalogens have an unsaturated ether link.

20.25 The cerebrosides are glycolipids in which a sugar unit is joined to the sphingosine unit where, in the sphingomyelins, a phosphate diester is joined.

20.26 A sugar unit such as galactose or glucose

20.27 Their molecules have electrically charged sites.

20.28 Cell membranes

20.29 Sphingomyelins and cerebrosides

20.30 By glycosidic links. These involve an acetal system, not an ordinary ether group. The glycosidic link is more easily hydrolyzed.

20.31 (a)

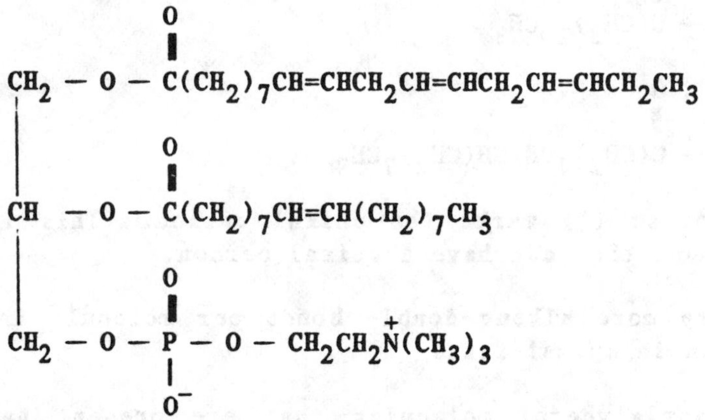

$$CH_2 - O - \overset{\overset{\displaystyle O}{\|}}{C}(CH_2)_7CH=CHCH_2CH=CHCH_2CH=CHCH_2CH_3$$

$$CH \ - O - \overset{\overset{\displaystyle O}{\|}}{C}(CH_2)_7CH=CH(CH_2)_7CH_3$$

$$CH_2 - O - \overset{\displaystyle O}{\underset{\displaystyle O^-}{\overset{\|}{P}}} - O - CH_2CH_2\overset{+}{N}(CH_3)_3$$

(b) Phosphoglyceride, because glycerol is one of the products of hydrolysis.

(c) Yes. The middle carbon of the glycerol unit is chiral.

(d) Lechithin, because it gives choline when hydrolyzed.

20.32 (a)

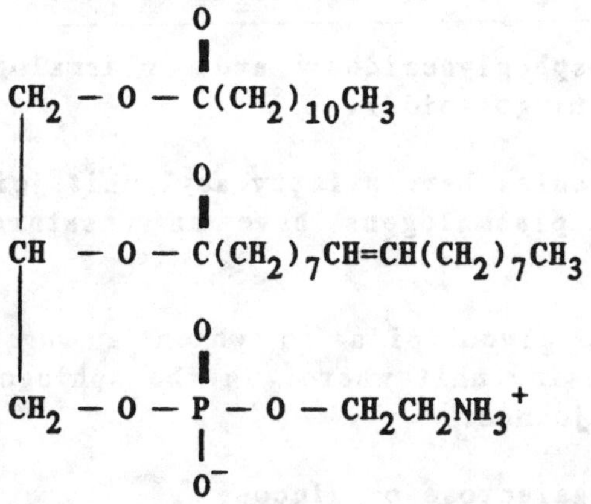

$$CH_2 - O - \overset{\overset{\displaystyle O}{\|}}{C}(CH_2)_{10}CH_3$$

$$CH \ - O - \overset{\overset{\displaystyle O}{\|}}{C}(CH_2)_7CH=CH(CH_2)_7CH_3$$

$$CH_2 - O - \overset{\displaystyle O}{\underset{\displaystyle O^-}{\overset{\|}{P}}} - O - CH_2CH_2NH_3^+$$

(b) Phosphoglyceride, because glycerol is one of the products of hydrolysis.

(c) Yes, the middle carbon of the glycerol unit is chiral.

(d) Cephalin, because aminoethanol is a hydrolysis product.

20.33   The anion of cholic acid

20.34   Vitamin D_3

20.35   Estradiol, progesterone, testosterone, and androsterone

20.36   Cholesterol

20.37   The membrane consists mostly of two layers of phospholipid molecules whose hydrophobic parts intermingle with each other between the layers and whose hydrophilic parts face toward the aqueous solutions whether they are inside the cell or outside.

20.38   The hydrophobic tails intermesh with each other between the two layers of the bilayer.

20.39   Proteins

20.40   The water-avoiding properties of the hydrophobic units and the water-attracting properties of the hydrophilic units.

20.41   Yes, there is a concentration gradient because the sugar concentration is not uniform throughout. In time, the process of diffusion makes the concentration uniform, and there is no chance that this uniformity will spontaneously change to reestablish a gradient.

20.42   Plasma

20.43   Cell fluid

20.44   Cell fluid

20.45   Energy-consuming reactions carry various chemical species through the membrane.

20.46   It is a membrane-bound mechanism for carrying sodium and potassium ions through a membrane against their individual concentration gradients.

20.47   Their molecules act as receptor sites for molecules that must either enter the cell or cause the cell to do something. They also serve as channels and pumps.

CHAPTER 21

PROTEINS

Review Exercises

21.1　$^+NH_3CH_2CO_2H$

21.2　$NH_2\underset{\underset{CH_3}{|}}{C}HCO_2^-$

21.3　$NH_2\underset{\underset{CH_3}{|}}{C}HCO_2CH_2CH_3$

The polarity of this molecule is much, much less than the polarity of the dipolar ionic structure of alanine, so the ester molecules stick together with weaker forces and the compound has a lower melting point.

21.4　The ester has an $-NH_2$ as the proton acceptor, whereas alanine itself has the $-CO_2^-$ group as the acceptor. And this group is made a weak acceptor by the electron withdrawal of the adjacent $-NH_3^+$ group in the dipolar ion. (The withdrawal of electron density from a proton-accepting site renders this site less able to take and hold a proton.)

21.5　B, because of the polar sites at the end of the side chain

21.6　A, because the side chain is purely hydrocarbonlike

21.7　(a)　pH 1, because at a relatively high concentration of H^+ the molecule tends to be fully protonated at all of its proton accepting sites
　　　(b)　Cathode

21.8   (a) At a pH of 10, because at a relatively high concentration
            of OH⁻ few proton-accepting sites can retain protons.
       (b) Anode

21.9

$$
\begin{array}{c}
CO_2^- \\
| \\
^+NH_3 \!-\!\!-\!\!-\!\!-\! H \\
| \\
CH_2OH
\end{array}
$$

21.10 $^+NH_3\overset{*}{C}HCO_2^-$ Isoleucine
              |
              $^*CHCH_3$
              |
              CH_2CH_3

        $^+NH_3\overset{*}{C}HCO_2^-$ Threonine
              |
              $CH_3\overset{*}{C}HOH$

21.11 $^+NH_3CH_2CONHCHCO_2^-$ and $^+NH_3CHCONHCH_2CO_2^-$
                    | |
                  $(CH_2)_4$ $(CH_2)_4$
                    | |
                  NH_2 NH_2

21.12 $^+NH_3CHCONHCHCO_2^-$ and $^+NH_3CHCONHCHCO_2^-$
              | | | |
            CH_2SH CH_2 CH_2 CH_2SH
                      | |
                    CH_2CO_2H CH_2CO_2H

21.13   Lys·Glu·Ala   Glu·Ala·Lys   Ala·Lys·Glu

        Lys·Ala·Glu   Glu·Lys·Ala   Ala·Glu·Lys

21.14   Cys·Gly·Ala   Gly·Cys·Ala   Ala·Gly·Cys

        Cys·Ala·Gly   Gly·Ala·Cys   Ala·Cys·Gly

21.15 $^+NH_3CHCONHCHCONHCHCO_2^-$
        | | |
       $CHCH_3$ $CHCH_3$ CH_2
       | | |
       CH_3 CH_2 C_6H_5
               |
               CH_3

21.16 $^+NH_3CHCONHCHCO_2CH_3$
        | |
       CH_2 CH_2
       | |
       CO_2H C_6H_5

21.17 $^+NH_3CHCONHCHCONHCHCONHCH_2CONHCHCO_2^-$
        | | | |
       CH_3 $CHCH_3$ CH_2 CH_2
               | | |
              CH_3 C_6H_5 $CHCH_3$
                              |
                              CH_3

21.18 $^+NH_3CHCONHCHCONHCHCONHCHCONHCHCO_2^-$
        | | | | |
       CH_2 $CHOH$ $(CH_2)_4$ CH_2 CH_2
       | | | |
      CO_2H CH_3 NH_2 CH_2
                          CO_2H

21.19 (a) A

      (b) B, because its side chains are hydrophilic

21.20 D, because its side chains are all hydrophobic

21.21 $^+NH_3CH_2CONHCHCONHCHCO_2^-$ Or: Gly·Cys·Ala

$\qquad\qquad\quad$ | $\quad$ |

$\qquad\qquad\quad CH_2 \quad CH_3$ $\qquad\qquad\qquad$ Gly·Cys·Ala

$\qquad\qquad\qquad$ |

$\qquad\qquad\qquad$ S

$\qquad\qquad\qquad$ |

$\qquad\qquad\qquad$ S

$\qquad\qquad\qquad$ |

$\qquad\qquad\qquad CH_2$

$\qquad\qquad\qquad$ |

$\qquad\quad ^+NH_3CH_2CONHCHCONHCHCO_2^-$

$\qquad\qquad\qquad\qquad\qquad$ |

$\qquad\qquad\qquad\qquad\qquad CH_3$

21.22   Two molecules of $^+NH_3CH_2CONHCHCONHCHCO_2^-$

$\qquad\qquad\qquad\qquad\qquad\qquad\qquad\qquad CH_2SH \quad CH_3$

21.23   The hydrogen bond: $C=O\cdots H-N$ 21.24   The hydrogen bond

21.25   Three polypeptide strands are wrapped into a triple helix, and these helices wrap into a right-handed super helix.

21.26   It is needed to make collagen.

21.27   The water-avoiding and the water-seeking characteristics of hydrophobic and hydrophilic side chains

21.28   A force of attraction between a (+) and a (−) charge on different side chains as in:

$-$ NHCHCO $-$ (Polypeptide backbone)

$\qquad$ |

$\qquad (CH_2)_4$ $\quad$ <——— Lysine side chain

$\qquad$ |

$\qquad NH_3^+$

$\qquad\qquad\qquad$ <——— Region in which the force of attraction

$\qquad$ CO_2^- $\qquad\qquad\qquad\qquad\qquad\qquad$ occurs

$\qquad$ |

$\qquad (CH_2)_2$ $\quad$ <——— Glutamic acid side chain

$-$ NHCHCO $-$ (Polypeptide backbone)

21.29 It involves the association of four polypeptide molecules (each bearing a heme unit).

21.30 (a) Hemoglobin has four associated polypeptide units. Myoglobin has just one.
(b) Hemoglobin is in the bloodstream. Myoglobin is in heart muscle.
(c) The same
(d) Hemoglobin transports oxygen. Myoglobin temporarily stores oxygen.

21.31

$$^{+}NH_3CHCO_2^{-} \ + \ ^{+}NH_3CHCO_2^{-} \ + \ ^{+}NH_3CHCO_2^{-} \ + \ ^{+}NH_3CHCO_2^{-} \ + \ ^{+}NH_3CH_2CO_2^{-}$$

with side chains:
- CH_3
- CH_2OH
- CH_2—(phenyl ring)
- CH_2—S—S—CH_2—$^{+}NH_3CHCO_2^{-}$

21.32 (a), (c), and (d) could form, but (b) could not; (b) has an amide bond that is not joined to the <u>alpha</u> amino group of lysine but is joined to its side chain amino group, instead.

21.33 At this pH, the protein molecules are not like-charged, so they cannot repel each other. Yet they are very polar, so they can attract each other and collect into a much larger particle that precipitates from the solution.

21.34 Digestion is the hydrolysis of a protein into its amino acid units. Denaturation is the disorganization of the secondary, tertiary, or quaternary structure of a protein so that it loses its ability to function biologically.

21.35 Fibrous proteins are insoluble in water but globular proteins do dissolve.

21.36 Collagen changes to gelatin when boiled in water.

21.37   They both have strengthening functions in tissue; both are fibrous proteins. The action of hot water on collagen turns it to gelatin, but elastin is unaffected in this way.

21.38   The globulins are less soluble in water than the albumins and they need the presence of dissolved salts to dissolve.

21.39   The protein that forms a blood clot. Fibrinogen is changed to fibrin by the clotting mechanism.

21.40   A lipoprotein

CHAPTER 22

NUCLEIC ACIDS

Review Exercises

22.1 All of the contents of a cell

22.2 They are small organelles inside cells (but not inside cell nuclei), and they manufacture ATP.

22.3 The contents of a cell outside of the cell nucleus

22.4 A weblike protein network in the nucleus and studded with replisomes

22.5 Nucleic acid and histones

22.6 Deoxyribonucleic acid (DNA)

22.7 Chromosomes are discrete, microscopically visible bodies made of chromatin.

22.8 Replication

22.9 The cell nucleus

22.10 Nucleic acids

22.11 Nucleotides

22.12 Ribose and deoxyribose

22.13 (a) Adenine, A Thymine, T Guanine, G Cytosine, C
(b) Adenine, A Uracil, U Guanine, G Cytosine, C

22.14 The main chains all have the same phosphate-deoxyribose-phosphate–deoxyribose repeating system.

22.15   In the sequence of bases attached to the deoxyribose units of the main chain

22.16   The main chains all have the same phosphate–ribose–phosphate–ribose repeating system.

22.17   DNA occurs as a double helix, and it alone has the base thymine (T). RNA alone has the base uracil (U). (The remaining three bases, A, G, and C, are the same in both DNA and RNA.) DNA molecules have one less — OH group per pentose unit than RNA molecules.

22.18   A and T pair to each other, so they must be in a 1:1 ratio regardless of the species. Similarly, G and C pair to each other and must be in a 1:1 ratio.

22.19   Hydrogen bond

22.20   Bases that project from the twin spirals of DNA chains fit to each other on opposite strands by hydrogen bonds. The geometries and functional groups are such that only A and T can pair (or only A and U can pair when RNA is involved), and only G and C can pair.

22.21   Original strand (given):   AGTCGGA   5' ——> 3'
                                   ....
                                   ....
                                   ....
                                   ....
                                   ....
        Opposite strand: TCAGCCT   3' <—— 5'

22.22   The synthesis of two new DNA double helices under the direction of and identical with an original DNA double helix

22.23   The base pairings of A with T and G with C

22.24   It is a site of DNA replication, a small mostly protein package fixed to the nuclear matrix at a number of places.

22.25   In cells of higher animals, a series of segments of a DNA molecule comprise one gene, each segment called an <u>exon</u> and each separated by DNA segments called <u>introns</u>.

22.26   The introns are b, d, and f, because they are the longer segments.

22.27   A sequence of base triplets that comprise a gene corresponds to a specific sequence of amino acid residues of a polypeptide.

22.28   It consists of several proteins plus <u>r</u>RNA, and it serves as the assembly area for the synthesis of polypeptides.

22.29   It is heterogeneous nuclear RNA (sometimes called primary transcript RNA). Its sequence of bases is complementary to a sequence of bases on DNA – those of both the exons and introns. It is processed to make <u>m</u>RNA whose base sequence is complementary only to the exons of the DNA.

22.30   A codon is a specific triplet of bases that corresponds to a specific amino acid residue in a polypeptide, and <u>m</u>RNA consists of a continuous sequence of codons.

22.31   (a)   5'   3'
              UUUCUUAUAGAGUCCCCAACAGAU

        (b)   5'   3'
              UUUUCCACAGAU

22.32   A triplet of bases found on <u>t</u>RNA and which is complementary to a codon found on <u>m</u>RNA

22.33   ATA cannot be a codon because T does not occur in any type of RNA.

22.34   (a) Phenylalanine   (b) Serine
        (c) Threonine   (d) Aspartic acid

22.35   Writing them in the 5' to 3' direction:

        (a) AAA   (b) GGA   (c) UGU   (d) AUC

22.36   Translation is the <u>m</u>RNA-directed synthesis of a polypeptide. Transcription is the DNA-directed synthesis of <u>m</u>RNA.

22.37   (a)   A large number of sequences are possible because three of
the specific amino acid residues are coded by more than
one codon.   The possibilities are indicated by:

Met·Ala·Try·Ser·Tyr

AUG GCU UGG UCU UAU   (5' ——→ 3')
    GCC   UCC UAC
    GCA   UCA
    GCG   UCG

(b)   CAU   (5' ——→ 3')

22.38   A substance whose molecules can block DNA-directed polypeptide
synthesis.

22.39   It cancels the effect of a repressor.

22.40   Some block the expression of a gene by interfering with DNA-
directed polypeptide synthesis.

22.41   The four-letter DNA language and the twenty-letter amino acid
language

22.42   The codons specify the same amino acids in all organisms.   No.

22.43   By causing chemical changes in the substances involved with the
genetic apparatus so that cell division is no longer controlled

22.44   At high enough, well-focused doses they destroy the genetic
apparatus of a cancer cell and render it incapable of dividing.

22.45   A substance that mimics the effects of radiations in cells

22.46   All viruses have nucleic acid, and many also contain a protein.

22.47   The protein part of a virus catalyzes the digestion of part of
the cell's membrane.   This opens a hole for the virus particle
or its nucleic acid to enter the cell.

22.48   Circular molecule of super-coiled DNA found in bacteria

22.49   One source is the plasmid and another is new material added to the bacterium or yeast.

22.50 Polypeptides of critical value to human medicine or technology

22.51 A defect in a gene

22.52 A gene needed for the metabolism of phenylalanine is defective. This leads to an increase in the level of phenylpyruvic acid in the blood, which causes brain dmage. A diet very low in phenylalanine is prescribed.

CHAPTER 23

NUTRITION

Review Exercises

23.1   It identifies the nutrients needed for health, determines the amounts required, and finds foods that are good sources of the nutrients.

23.2   It is any compound needed for health.

23.3   Foods are complex mixtures of nutrients.

23.4   Dietetics is the application of the findings of nutrition to the feeding of individuals whether ill or well.

23.5   To allow for individual differences among people and to ensure that practically all people can thrive

23.6   1. People with chronic diseases
2. People who must take special medications
3. Prematurely born infants
4. Pregnant women
5. Lactating women
6. People involved in strenuous physical activity
7. People exposed for prolonged periods to high temperatures.

23.7   No food contains all of the essential nutrients, and there might still be nutrients yet to be discovered but which are routinely provided by a varied diet.

23.8   (a) The body must make its own glucose, which can lead to a buildup of harmful substances.

(b) It lacks the essential fatty acids and it makes the absorption of the fat-soluble vitamins more difficult.

23.9   The body can make several amino acids itself.

23.10 Not all of the protein is digested, and of what is digested, not all is absorbed into circulation.

23.11 It breaks them down and eliminates the products.

23.12 Coefficient of digestibility $= \dfrac{(\text{N in food eaten} - \text{N in feces})}{\text{N in food eaten}}$

The numerator, (N in food eaten − N in feces), is the food nitrogen that is actually absorbed into circulation.

23.13 From an animal source

23.14 Mill them into flour, but this also lowers their vitamin and mineral content.

23.15 The proportions of essential amino acids available from it

23.16 Human milk protein is the best, but whole egg protein is very close.

23.17 The essential amino acid most poorly supplied by the protein

23.18 It is low in tryptophan and lysine.

23.19 (a) 127 g   (b) 1.7×10^3 g   (c) 6.1×10^3 kcal
       (d) Very likely not, since 1.7×10^3 g is nearly 3 lb

23.20 (a) 79 g   (b) 3.0×10^2 g   (c) 6.5×10^2 kcal
       (d) Since 3.0×10^2 g of peanuts is about 2/3 lb, a child could probably get this much down.

23.21

Vitamin	Source(s)	Problems If Deficient
A	carrots	night blindness, deterioration of mucous membranes, blindness
D	dairy products, fatty fish	poor bone development, rickets
E	vegetable oils	edema and anemia (in infants),

		accelerated hemolysis, possible heart disease
K	green leafy vegetables	increased susceptibility to hemorrhages
C	citrus fruits, potatoes, leafy vegetables, tomatoes	scurvy, possibly increased susceptibility to colds
Choline	meats, egg yolk, cereals, legumes	none known in humans; fatty liver and kidney disease in animals
Thiamine	lean meats, whole grains, legumes	beri beri
Ribo-flavin	milk, meat	problems with tissue around the mouth, nose and tongue; skin scaling; impaired wound healing
Niacin	meat, whole grains	pellagra
Folacin	green, leafy vegetables; liver, kidneys	megaloblastic anemia
B_6	meat, wheat, yeast	possibly disturbances in the central nervous system; hypochromic microcytic anemia
B_{12}	meat and dairy products	pernicious anemia
Panto-thenic acid	liver, kidney, egg yolk, skim milk	not observed clinically in humans
Biotin	egg yolk, liver, tomatoes, yeast	seldom observed; anorexia, nausea, pallor, dermatitis, depression

23.22 They are needed in much more than trace amounts, and they come from proteins.

23.23 Vitamin B_{12}

23.24 No single vegetable source has a balanced supply of essential amino acids.

23.25 Vitamins A, D, E, and K

23.26 Vitamin D

23.27 Vitamin D

23.28 Vitamin A

23.29 Vitamin A

23.30 Vitamin K

23.31 Vitamin C, choline, thiamine, riboflavin, niacin, folacin, vitamin B_6, vitamin B_{12}, pantothenic acid, and biotin

23.32 Vitamin C

23.33 Vitamin C, thiamine, riboflavin, niacin, folacin

23.34 Thiamine

23.35 Niacin

23.36 The quantity needed per day. Minerals are needed in the amount of more than 100 mg/day and trace elements in the amounts of less than 20 mg/day.

23.37 Calcium, Ca^{2+}

Phosphorus, P_i (Chiefly, the mix of HPO_4^{2-} and $H_2PO_4^{-}$ plus some PO_4^{3-} that exists at body pH and in bone.)

Magnesium, Mg^{2+}

Sodium, Na^{+}

Potassium, K^{+}

Chlorine, Cl^-

23.38

Trace Element	Function
Fluorine (F^-)	sound teeth
Chromium (Cr^{3+})	glucose metabolism; possible cofactor for insulin
Manganese (Mn^{2+})	nerves, bones, reproduction
Iron (Fe^{2+})	heme, many enzymes
Cobalt (Co^{2+})	vitamin B_{12}
Copper (Cu^{2+})	in proteins and enzymes
Zinc (Zn^{2+})	enzymes, nucleic acids, bones; prevention of dwarfism; possible protection against heart disease
Selenium	possible protection against heart disease
Molybdenum	nucleic acid metabolism
Iodine (I^-)	synthesis of thyroid hormones

23.39 Goiter

CHAPTER 24

ENZYMES, HORMONES, AND NEUROTRANSMITTERS

Review Exercises

24.1 (a) A catalyst
      (b) It consists of a protein.

24.2 Each enzyme catalyzes a reaction for a specific substrate or a specific kind of reaction.

24.3 (a) An apoenzyme is the wholly polypeptide part of the enzyme.
      (b) A cofactor is a non-polypeptide molecule or ion needed to make the complete enzyme.
      (c) A coenzyme is one kind of cofactor, an organic molecule.

24.4 Nicotinamide

24.5 Riboflavin

24.6 $\underline{NADH}$ + $\underline{H}^+$

24.7 $\underline{H}^+$ + $\underline{NADH}$ + FMN $\longrightarrow$ NAD^+ + $\underline{FMNH}_2$

24.8 (a) Sucrose   (b) Glucose   (c) A protein   (d) An ester

24.9 (a) An oxidation (or a dehydrogenation)
      (b) The transfer of a methyl group from one molecule to another
      (c) The hydrolysis of some bond
      (d) An electron-transfer event

24.10 Lactose is a disaccharide and the substrate for the enzyme, lactase.

24.11 Hydrolysis is a kind of reaction catalyzed by the enzyme hydrolase.

24.12   Enzymes of identical function but slightly different in structure

24.13   CK(MM) in skeletal muscle; CK(BB) in brain; and CK(MB) in heart muscle

24.14   Active site

24.15   By the necessity of the fitting of the substrate molecule to the surface of the enzyme much as a key must fit to a particular lock

24.16   The substrate molecule induces a change in the shape of the enzyme that makes the fit between the two possible.

24.17   (a) See Figure 24.5 in the text
         (b) See Figure 24.4 in the text

24.18   The rate diminishes.

24.19   There probably is more than one site, and that they exist normally in an inactive state. Activating one site activates all sites.

24.20   It induces a fit at one site and simultaneously causes another site to assume an activated shape.

24.21   The effector does not form a complex with the enzyme at the same place that the normal substrate does.

24.22   Some enzymes are made from a larger polypeptide – the zymogen – by having a small polypeptide fragment split off. Trypsinogen is a zymogen for the digestive enzyme, trypsin.

24.23   The competitor molecule can occupy the same active site as the true substrate molecule.

24.24   A molecule of a product of the work of the enzyme combines with the enzyme to inactivate it.

24.25   The product molecule is the stimulus that shuts off a mechanism (the enzyme) for some process until a reaction removes the product and lets the process run again.

24.26   In competitive inhibition the enzyme's active site is occupied whereas in allosteric inhibition some other site is occupied with the result that the active site is rendered inactive.

24.27   (a) CN^- inactivates an enzyme involved in cellular respiration.

(b) Hg^{2+} reacts with HS- groups on cysteine side chains and precipitates the enzyme.

(c) These inactivate cholinesterase, an enzyme needed in the signal-sending activities in the nervous system.

24.28   Antimetabolites are compounds that interfere with the metabolism of disease-causing bacteria. Antibiotics are those antimetabolites that are made by microorganisms.

24.29   Feedback inhibition

24.30   Competitive inhibition

24.31   The levels of these enzymes rise in blood as the result of a disease or injury to particular tissues, which causes tissue cells to release their enzymes.

24.32   The CK(MB) band originates in the leakage of this isoenzyme only from damaged heart muscle.

24.33   CK(MM)

24.34   Of the five LD isoenzymes, LD_1 normally is less concentrated than LD_2. The "flip" is the reversal of this relationship. LD_1 shows up as <u>more</u> concentrated than LD_2. This flip is observed in patients who have suffered a myocardial infarction.

24.35   If glucose is present, it is acted upon by glucose oxidase in the test strip. This produces hydrogen peroxide, and the enzyme peroxidase (also in the strip) catalyzes a reaction between hydrogen peroxide and an aromatic compound in the test strip. The product is a dye whose appearance and intensity of color signals that glucose is present and its approximate concentration.

24.36   They are primary chemical messengers.

24.37   When activated by a primary chemical messenger, it catalyzes
        the formation of cyclic AMP (from ATP), which then activates an
        enzyme inside the target cell.

24.38   It is an enzyme activator.

24.39   The cyclic AMP becomes AMP.

24.40   (a) Endocrine glands
        (b) Axon ends of presynaptic nerve cells

24.41   It explains how molecules of hormones or neurotransmitters
        recognize their particular target cells.

24.42   (1) Activating an enzyme, such as the work of epinephrine
        (2) Activating a gene, such as the work of sex hormones
        (3) Altering the permeability of a cell membrane toward a
             specific substance, as in the work of insulin

24.43   It is hydrolyzed back to acetic acid and choline. The enzyme
        is cholinesterase. Nerve poisons inactivate this enzyme.

24.44   It inactivates the receptor protein for acetylcholine.

24.45   It blocks the receptor protein for acetylcholine

24.46   They catalyze the deactivation of neurotransmitters such as
        norepinephrine and thus reduce the level of signal-sending
        activity that depends on such neurotransmitters.

24.47   Iproniazid inhibits the monoamine oxidases and thus lets
        norepinephrine work at a higher level of activity.

24.48   They inhibit the reabsorption of norepinephrine by the
        presynaptic neuron and thus reduce the rate of its deactivation
        by the monoamine oxidases.

24.49   Norepinephrine, acting as a hormone, serves as a backup to its
        acting as a neurotransmitter should some injury disrupt the
        latter action.

24.50 Dopamine

24.51 They bind to dopamine receptors in the postsynaptic nerve and inhibit the action of dopamine.

24.52 They accelerate the release of dopamine from the presynaptic neuron.

24.53 Degenerated neurons can use L–DOPA to make dopamine.

24.54 GABA (gamma–aminobutyric acid), whose signal–inhibiting activity is enhanced by Valium and Librium

24.55 Enkephalin molecules enter pain–signalling neurons and inhibit the release of substance P, a neurotransmitter that helps to send pain signals. Thus enkephalin, like an opium–drug, inhibits pain.

CHAPTER 25

EXTRACELLULAR FLUIDS OF THE BODY

Review Exercises

25.1   Interstitial fluid and blood

25.2   Saliva, gastric juice, pancreatic juice, and intestinal juice

25.3   (a) α-Amylase
        (b) Pepsinogen and gastric lipase
        (c) α-Amylase, lipase, nuclease, trypsinogen, chymo-
             trypsinogen, and procarboxypeptidase.
        (d) None
        (e) Amylase, aminopeptidase, sucrase, lactase, maltase,
             lipase, nucleases, enterokinase

25.4   (a) Pepsin from its zymogen in gastric juice; trypsin and
             chymotrypsin from zymogens in pancreatic juice
        (b) Lipases provided in gastric juice, pancreatic juice, and
             intestinal juice
        (c) Amylases in saliva, pancreatic juice, and intestinal juice
        (d) Sucrase in intestinal juice
        (e) Carboxypeptidase from its zymogen in pancreatic juice;
             aminopeptidase from its zymogen in intestinal juice
        (f) Nucleases in pancreatic juice and intestinal juice.

25.5   (a) Amino acids
        (b) Glucose, fructose, and galactose
        (c) Glycerol and fatty acids

25.6   (a) Peptide (amide) bonds in proteins
        (b) Acetal systems in carbohydrates
        (c) Ester groups in triacylglycerols

25.7   It catalyzes the conversion of trypsinogen to trypsin.   Then

549

trypsin catalyzes the conversion of other zymogens to chymo-trypsin and carboxypeptidase. Thus enterokinase turns on enzyme activity for three major protein-digesting enzymes.

25.8 They would catalyze the digestion of proteins that make up part of the pancreas to the serious harm of this organ.

25.9 They are surface active agents that help to break up lipid globules, wash lipids from the particles of food, and aid in the absorption of fat-soluble vitamins.

25.10 (a) Lubricates the food
(b) Protects the stomach lining from gastric acid and pepsin

25.11 (a) HCl   (b) Enterokinase
(c) Trypsin   (d) Trypsin

25.12 It helps to coagulate the protein in milk so that this protein stays longer in the stomach where it can be digested with the aid of pepsin.

25.13 This enzyme is inactive at the high acidity of the digesting mixture in the adult stomach, but the acidity of this mixture in the infant's stomach is less.

25.14 Pancreatic juice delivers its zymogens and enzymes into the duodenum, whereas the enzymes of the intestinal juice work within cells of the intestinal wall.

25.15 Dilute sodium bicarbonate released from the pancreas. This raises the pH of the chyme to the optimum pH for the action of the enzymes that will function in the duodenum.

25.16 They recombine to molecules of triacylglycerol during their migration from the intestinal tract toward the lymph ducts.

25.17 The flow of bile normally delivers colored breakdown products from hemoglobin in the blood that give the normal color to feces. When no bile flows, no colored products are available to the feces.

25.18 These lipids are less in need of the surfactant activity of bile salts as an aid to their digestion.

25.19   The concentration of soluble proteins is greater in blood.

25.20   The serum-soluble proteins (albumins, mostly)

25.21   Fibrinogen is a protein in blood that is changed to fibrin, the insoluble protein of a blood clot, by the clotting mechanism.

25.22   Albumin molecules transport hydrophobic molecules such as fatty acids and cholesterol, and albumins contribute as much as 75–80% of the osmotic effect of the blood.

25.23   It protects the body against infections.

25.24 Na^+ is in blood plasma and other extracellular fluids; K^+ is chiefly in intracellular fluids.

25.25   Maintain osmotic pressure relationships; be part of the regulatory mechanisms for acid-base balance; and participate in the smooth working of the muscles and the nervous system.

25.26   145 meq/L

25.27   136 meq/L

25.28   Such injuries allow the contents of cells to spill out and enter circulation.

25.29   5.0 meq/L

25.30   Hyponatremia

25.31   Hypermagnesemia and cardiac arrest

25.32   The activation of several enzymes

25.33   It initiates contraction.

25.34   They lower the availability of calcium ion and thus diminish the pressure with which the heart pumps, so they are used in connection with heart disease.

25.35   Hypocalcemia

25.36   Hypomagnesemia

25.37   Chloride ion, Cl^-

25.38   Loss of Cl^- causes retention of HCO_3^-, a base.

25.39   100–106 meq/L   or   100–106 mmol/L

25.40   Blood pressure that tends to force blood fluids out and osmotic pressure that tends to force fluids back. The return of fluids to the blood from the interstitial compartment is overbalanced by the blood pressure so that the net effect is a diffusion of fluids from the blood.

25.41   Blood pressure and osmotic pressure. The natural return of fluids to the blood from the interstitial compartment is not balanced by the now reduced blood pressure, so fluids return to the blood from which they left on the arterial side.

25.42   Serum proteins are lost from the blood, which upsets the osmotic pressure of the blood. Water leaves the blood for the interstitial compartment, and the blood volume drops. Loss of blood delivery to the brain leads to the symptoms of shock.

25.43   (a)   Blood proteins leak out which allows water to leave the blood and enter interstitial spaces throughout various tissues.

        (b)   Blood proteins are lost to the blood by being consumed which also leads to the loss of water from the blood and its appearance in interstitial compartments.

        (c)   Capillaries are blocked at the injured site reducing the return of blood in the veins, so fluids accumulate at the site.

25.44   Oxygen and carbon dioxide

25.45   Hemoglobin

25.46   The first oxygen molecule to bind changes the shapes of other parts of the hemoglobin molecule and makes it much easier for the remaining three oxygen molecules to bind. This ensures that all four oxygen-binding sites of each hemoglobin molecule will leave the lungs fully loaded with oxygen.

25.47 $HHb + O_2 \rightleftharpoons HbO_2^- + H^+$

    (a) To the left   (b) To the left
    (c) To the right   (d) To the left
    (e) To the left   (f) To the right.

25.48 (a) $HHb + O_2 \longleftarrow HbO_2^- + H^+$

                    Isohydric shift in active tissue

        $H_2CO_3 \longrightarrow HCO_3^- + H^+$

    (b) $HHb + O_2 \longrightarrow HbO_2^- + H^+$

                    Isohydric shift in an alveolus

        $H_2CO_3 \longleftarrow HCO_3^- + H^+$

25.49 It generates H^+ needed to convert HCO_3^- to H_2CO_3 (and thence to CO_2 and H_2O) and to convert $HbCO_2^-$ to HHb and CO_2.

25.50 It combines with water to give H_2CO_3 which furnishes H^+ needed to react with HbO_2^- to form HHb and release O_2.

25.51 It helps to shift the following equilibrium to the left:

$$HHb + O_2 \rightleftharpoons HbO_2^- + H^+$$

25.52 It is found in red cells.

    (a) It catalyzes the decomposition of H_2CO_3.

    (b) It catalyzes the formation of H_2CO_3.

It can do both because it accelerates both the forward and the reverse reactions in the equilibrium:

$$CO_2 + H_2O \rightleftharpoons H_2CO_3$$

Other factors, such as the value of the partial pressure of carbon dioxide, determine whether the forward or the reverse reaction is favored.

25.53 The migration of DPG molecules out of hemoglobin molecules as the first oxygen molecules enter the hemoglobin helps the remaining oxygen molecules to bind more readily.

25.54 It migrates into a cavity within the hemoglobin molecule and helps to change the shapes of subunits so that oxygen molecules are more easily ejected.

25.55 The body makes more hemoglobin and red cells. This allows the body to pick up oxygen more readily. The body also makes more DPG, which helps the system release oxygen where needed.

25.56 The blood produced while at a high altitude has a higher concentration of hemoglobin and of DPG. This aids in their ability to use oxygen during a race.

25.57 For oxygenation:
$$HHb-DPG + O_2 + HCO_3^- \longrightarrow HbO_2^- + DPG + CO_2 + H_2O$$

For deoxygenation:
$$HbO_2^- + DPG + CO_2 + H_2O \longrightarrow HHb-DPG + O_2 + HCO_3^-$$

25.58 As HCO_3^- in the serum and as $HbCO_2^-$ (carbaminohemoglobin) in red cells

25.59 It is lowered. Where the partial pressure of CO_2 is relatively high (as in actively metabolizing tissue) there is a need for oxygen, so the effect of CO_2 on oxygen affinity helps to release O_2.

25.60 The exchange of a chloride ion for a bicarbonate ion between a red blood cell and blood serum. This brings Cl^- inside the red cell when it is needed to help deoxygenate HbO_2^-.

25.61 Myoglobin can take oxygen from oxyhemoglobin and thus ensure that the oxygen needs of myoglobin-containing tissue are met.

25.62 These animals can store more oxygen in heart muscle, which helps them to go longer without breathing.

25.63 Fetal hemoglobin can take oxygen from the oxyhemoglobin of the mother's blood and thus ensure that the fetus gets needed oxygen.

25.64

Condition	pH	pCO_2	$[HCO_3^-]$
Normal	7.35–7.45	35–40 mm Hg	25–30 meq/L
Metabolic acidosis	↓ 7.20	↓ 30	↓ 14
Metabolic alkalosis	↑ 7.45	↑ >45	↑ >29
Respiratory acidosis	↓ 7.10	↑ 68	↑ 40
Respiratory alkalosis	↑ 7.54	↓ 32	↓ 20

25.65 The pH of the blood decreases in both but both pCO_2 and $[HCO_3^-]$ increase in respiratory acidosis and both decrease in metabolic acidosis.

25.66 Hyperventilation is observed in metabolic acidosis, and HCO_3^- can be given intravenously to neutralize excess acid. Hyperventilation is also observed in respiratory alkalosis (because the patient can't help hyperventilating), and CO_2 is given (by rebreathing one's own air) to keep up the supply of H_2CO_3, which can neutralize excess base.

25.67 Hypoventilation is observed in metabolic alkalosis, and isotonic ammonium chloride can be given to neutralize the excess base. Involuntary hypoventilation is observed in respiratory acidosis, and isotonic sodium bicarbonate might be given to neutralize excess acid.

25.68 In metabolic acidosis, because it helps to blow out CO_2 and thereby to reduce the level of H_2CO_3 in the blood and simultaneously raise the pH.

25.69 In respiratory alkalosis. The involuntary loss of CO_2 reduces the level of H_2CO_3 in the blood and thereby reduces the level of H^+.

25.70 In metabolic alkalosis

25.71 In respiratory acidosis

25.72 The kidneys work to remove acids from the blood, but to remove them they must also remove water from the blood. If too much water is taken in this way, then the blood obtains more water by taking it from interstitial and intracellular compartments.

25.73

(a) Respiratory alkalosis	(b) Metabolic alkalosis
(c) Respiratory acidosis	(d) Respiratory acidosis
(e) Metabolic acidosis	(f) Respiratory alkalosis
(g) Metabolic acidosis	(h) Metabolic alkalosis
(i) Respiratory acidosis	(j) Respiratory acidosis

25.74

(a) Hyperventilation	(b) Hypoventilation
(c) Hypoventilation	(d) Hypoventilation
(e) Hyperventilation	(f) Hyperventilation
(g) Hyperventilation	(h) Hypoventilation
(i) Hypoventilation	(j) Hypoventilation

25.75 CO_2 is removed at an excessive rate, which removes carbonic acid, so the blood becomes more alkaline and the pH of the blood rises (alkalosis).

25.76 Hypoventilation in emphysema lets the blood retain carbonic acid, and the pH decreases.

25.77 The loss of acid with the loss of the stomach contents results in a loss of acid from the blood, which means a rise in the blood's pH – alkalosis.

25.78 The loss of alkaline fluids from the duodenum and lower intestinal tract leads to a loss of base frm the bloodstream, too. The result is a decrease in the blood's pH – acidosis.

25.79 The blood has become more concentrated in solutes.

25.80   It acts to prevent the loss of water via the urine by letting the hypophysis secrete vasopressin whose target cells are in the kidneys. This helps to keep the blood's osmotic pressure from rising further. The thirst mechanism is also activated, which leads to bringing in more water to dilute the blood.

25.81   Aldosterone is secreted from the adrenal cortex, and it instructs the kidneys to retain sodium ion in the blood.

25.82   The kidneys secrete a trace of renin into the blood. This catalyzes the conversion of angiotensinogen into angiotensin I, which catalyzes the formation of angiotensin II, a neurotransmitter and powerful vasoconstrictor. By constricting capillaries, the blood pressure has to increase to keep the delivery of blood going.

25.83   The rate of diuresis increases.

25.84   They transfer hydrogen ions into the urine and put bicarbonate ions into the bloodstream.

CHAPTER 26

BIOCHEMICAL ENERGETICS

Review Exercises

26.1   All of them, but chiefly fatty acids and carbohydrates.

26.2   Carbon dioxide and water

26.3   Combustion produces just heat. The catabolism of glucose uses about half of the energy to make ATP and the remainder appears as heat.

26.4   Carbon dioxide and water.

26.5   Adenosine $-$ O $-$ P $-$ O $-$ P $-$ O $-$ P $-$ O$^-$ (each P bearing a double-bonded O above and O$^-$ below)

26.6   ADP = adenosine $-$ O $-$ P $-$ O $-$ P $-$ O$^-$ (each P bearing a double-bonded O above and O$^-$ below)

AMP = adenosine $-$ O $-$ P $-$ O$^-$ (P bearing a double-bonded O above and O$^-$ below)

26.7   The singly and doubly ionized forms of phosphoric acid, $H_2PO_4^-$ + HPO_4^{2-}

26.8   The first two have phosphate group transfer potentials equal to or higher than that of ATP, whereas this potential is lower than that of ATP for glycerol 3-phosphate.

26.9   The left

26.10 (a) Yes   (b) Yes   (c) No   (d) No

26.11 The substrate is directly phosphorylated by the <u>transfer</u> of a phosphate group from an organic donor.

26.12 It stores phosphate group energy and transfers phosphate to ADP to remake the ATP consumed by muscular work.

26.13 As the supply of ATP decreases, the supplies of ADP and P_i increase, and this development launches events that make creatine phosphate so that more ATP can be made.

26.14 (a) The aerobic synthesis of ATP
       (b) The synthesis of ATP when a tissue operates anaerobically
       (c) The supply of metabolites for the respiratory chain
       (d) The supply of metabolites for the respiratory chain and for the citric acid cycle

26.15 The disappearance of ATP by some energy-demanding process and the simultaneous appearance of ADP + P_i

26.16 The citric acid cycle

26.17 (a) 4 2 5 1 3
       (b) 2 4 3 1

26.18 It starts with glycolysis and since this is aerobic as stated, it ends with the respiratory chain.

26.19 A pair of electrons on the left of the arrow

26.20 Respiratory chain

26.21
$$
\begin{array}{c}
CH_2CO_2^- \\
| \\
{}^-O_2CCH_2
\end{array}
\;+\; FAD \;\longrightarrow\;
\begin{array}{c}
CHCO_2^- \\
\| \\
{}^-O_2CCH
\end{array}
\;+\; FADH_2
$$

(a) $^-O_2CCH_2CH_2CO_2^-$ is oxidized.

(b) FAD is reduced.

26.22

$$CH_3\overset{\overset{\displaystyle OH}{|}}{C}HCO_2^- \qquad NAD^+$$

$$CH_3\overset{\overset{\displaystyle O}{\|}}{C}CO_2^- \longrightarrow NADH + H^+$$

26.23 NAD+   FMN   FeS-P   Cyt <u>a</u>

26.24 It catalyzes the reduction of oxygen to water.

26.25 It is a riboflavin-containing coenzyme that in its reduced form, $FADH_2$, passes electrons and H^+ into the respiratory chain.

26.26 Across the inner membrane of the mitochondrion. The value of $[H^+]$ is higher on the outer side of this inner membrane.

26.27 The flow of protons across the inner mitochondrial membrane

26.28 If the membrane is broken, then the simple process of diffusion defeats any mitochondrial effort to set up a gradient of H^+ ions across the membrane, but the chain itself can still operate.

26.29 $\underline{M}H_2$ + $6H^+$ from inside the inner membrane + $\frac{1}{2}$ O_2 $\longrightarrow$ $\underline{M}$ + $6H^+$ now on the outside of the inner membrane + H_2O

26.30 Oxidative phosphorylation is the kind made possible by the energy released from the operation of the respiratory chain. Substrate phosphorylation arises from the direct transfer of a phosphate unit from a higher to a lower energy phosphate.

26.31 A gradient of positive charge. The migration of <u>any</u> cation away from the region of higher positive charge density or the migration of <u>any</u> anion toward this region will be spontaneous.

26.32 In one theory, protons interact with PO_4^{3-} to make a very high energy form, PO_3^{-}, which can react directly with ADP to make ATP.

26.33 When the respiratory chain starts up, the citric acid cycle must start up to keep the chain going.

26.34 The acetyl unit, $CH_3\overset{\displaystyle O}{\overset{\|}{C}}-$, of acetyl coenzyme A

26.35 Two

26.36 Two

26.37 (a) $CH_3 - \overset{\displaystyle O}{\overset{\|}{C}} - H$, ethanal (acetaldehyde)

(b) $CH_3 - \overset{\displaystyle O}{\overset{\|}{C}} - OH$, acetic acid

(c) α-Ketoglutarate

26.38 Pyruvate ion

26.39 ADP, because when ADP appears in the cell it is time to make ATP, an outcome of the action of this enzyme

26.40 (a) 15   (b) 12

26.41 9

CHAPTER 27

METABOLISM OF CARBOHYDRATES

Review Exercises

27.1   Glucose, fructose, and galactose

27.2   After a few steps, the metabolic pathways of galactose and fructose merge with the pathway of glucose.

27.3   The concentration of reducing monosaccharides, chiefly glucose, in the blood is called the blood sugar level. The normal fasting level is the blood sugar level after several hours of fasting.

27.4   65–95 mg/dL   (Note: Various references give slightly different ranges of values, e.g., 70–110 mg/dL.)

27.5   (a) Glucose in urine
        (b) A low blood sugar level
        (c) A high blood sugar level
        (d) The conversion of glycogen to glucose
        (e) The synthesis of glucose from smaller molecules
        (f) The synthesis of glycogen from glucose

27.6   The lack of glucose means the lack of the one nutrient most needed by the brain.

27.7   It rises.

27.8   Muscle tissue

27.9   The enzyme adenylate cyclase

27.10   4 3 5 1 2

27.11   It partly activates phosphorylase kinase.

27.12   Glucose might be changed back to glycogen as rapidly as it is released from glycogen, and no glucose would be made available to the cell.

27.13   The activated form of protein kinase

27.14   Glucose 1-phosphate is the end product and phosphoglucomutase catalyzes its change to glucose 6-phosphate.

27.15   Liver, but not muscles, has the enzyme glucose 6-phosphatase that catalyzes the hydrolysis of glucose 6-phosphate. This frees glucose for release from the liver to the bloodstream.

27.16   It is a polypeptide hormone made in the alpha cells of the pancreas and released into circulation when the blood sugar level drops. At the liver it activates adenylate cyclase, which leads to glycogenolysis and the release of glucose into circulation.

27.17   Glucagon, because it works better at the liver than epinephrine in initiating glycogenolysis, and when glycogenolysis occurs at the liver there is a mechanism for releasing glucose into circulation.

27.18   It stimulates the release of glucagon, which leads to the release of glucose into circulation.

27.19   It is a polypeptide hormone released from the beta cells of the pancreas in response to a rise in the blood sugar level, and it acts most effectively at adipose tissue.

27.20   A rise in the blood sugar level

27.21   Too much insulin leads to a sharp drop in the blood sugar level and therefore a drop in the supply of the chief nutrient for the brain.

27.22   Somatostatin is a polypeptide hormone released by the hypothalamus, and it acts at the pancreas to inhibit the release of glucagon and slow down the release of insulin.

27.23 The body's ability to manage dietary glucose without letting the blood sugar level swing too widely from its normal fasting level

27.24 To test for the possibility of diabetes mellitus. An adult patient is given a drink that has 75 g of glucose. For children, 1.75 g of glucose per kilogram of body weight is given. Then the blood sugar level is measured at regular intervals.

27.25 (a) The blood sugar level initially rises rapidly, but then drops sharply and slowly levels back to normal.
  (b) The blood sugar level, already high to start, rises much higher and never sharply drops. It only very slowly comes back down.

27.26 An over-release of epinephrine (as in a stressful situation) that induces an over-release of glucose

27.27 Most of it enters gluconeogenesis. Some of it is further catabolized to supply the energy needed for this pathway.

27.28 (a) Little if any, because glucose is changed to glucose 6-phosphate upon entering muscle cells, and in this form it is trapped.
  (b) Yes, because the lactate made by glycolysis will have carbon-13, and lactate can circulate back to the liver.
  (c) Yes, because carbon-13 labeled lactate can be used in the liver to make glucose which can be released into the blood.

27.29 $C_6H_{12}O_6 + 2ADP + \underline{2P_i} \longrightarrow 2C_3H_5O_3^- + 2H^+ + \underline{2ATP}$

27.30 Glycolysis can operate and make ATP even when the oxygen supply is low, so a tissue in oxygen debt can continue to function.

27.31 Glyceraldehyde 3-phosphate is in the direct pathway of glycolysis, so changing dihydroxyacetone phosphate into it ensures that all parts of the original glucose molecule are used in glycolysis.

27.32 (a) It undergoes oxidative decarboxylation and becomes the acetyl group in acetyl CoA.

(b) Its keto group is reduced to a $2°$ alcohol group in lactate, which enables NADH to be reoxidized to NAD^+ and then reused for more glycolysis.

27.33 It is reoxidized to pyruvate, which then undergoes oxidative decarboxylation to the acetyl group in acetyl CoA. This enters the citric acid cycle.

27.34 (a) 17 ATP   (b) 18 ATP

27.35 NADPH forms, and the body uses it as a reducing agent to make fatty acids.

27.36 It makes glucose out of smaller molecules obtained by the catabolism of fatty acids and amino acids.

27.37 All are catabolized, and parts of some of their molecules are used to make fatty acids and, thence, fat.

27.38 (a) Alanine   (b) Aspartic acid

27.39 (a) 15 ATP   (b) 3 ATP
       (c) 3 Glucose (since 6 ATP are needed to make each molecule of glucose by gluconeogenesis)

CHAPTER 28

METABOLISM OF LIPIDS

Review Exercises

28.1 Fatty acids and glycerol

28.2 They become reconstituted into triacylglycerols.

28.3 The transport of lipids received from the digestive tract to the liver

28.4 They unload some of their triacylglycerol.

28.5 They are absorbed.

28.6 Some cholesterol has originated in the diet and some has been synthesized in the liver.

28.7 (a) Very low density lipoprotein complex
       (b) Intermediate density lipoprotein complex
       (c) Low density lipoprotein complex
       (d) High density lipoprotein complex

28.8 Triacylglycerol

28.9 Triacylglycerol

28.10 The liver

28.11 Cholesterol

28.12 The synthesis of steroids and the fabrication of cell membranes

28.13 IDL and LDL

28.14 When the receptor proteins are reduced in number, the liver cannot remove cholesterol from the blood, so the blood cholesterol level increases.

28.15 Return to the liver any cholesterol that extrahepatic tissue cannot use.

28.16 A low level of HDL means a low ability to carry cholesterol from extrahepatic tissue back to the liver for export via the bile.

28.17 The grams of a particular tissue or fluid needed to store one kilocalorie of reserve energy

28.18 3 1 2

28.19 Fasting and diabetes

28.20 Insulin suppresses the lipase needed to hydrolyze triacylglycerols in storage prior to the release of their fatty acids into circulation.

28.21 5 4 1 3 2

28.22 The operation of the fatty acid cycle feeds electrons and protons directly to the respiratory chain, and it makes acetyl CoA that the citric acid cycle catabolizes as it also sends electrons and protons into the respiratory chain.

28.23 The citric acid cycle processes the acetyl units manufactured by the fatty acid cycle and so fuels the respiratory chain.

28.24 Epinephrine and glucagon. They help to keep the blood sugar level up to normal.

28.25 A rise in the blood sugar level triggers the release of insulin which inhibits the release of fatty acids from adipose fat.

28.26 It is changed to dihydroxyacetone phosphate, which enters the pathway of glycolysis.

28.27 They are joined to coenzyme A as fatty acyl CoA.

28.28 (a)

$$CH_3CH_2CH_2CH_2CH_2\overset{\overset{\displaystyle O}{\|}}{C} - SCoA \ + \ FAD \ \longrightarrow \ CH_3CH_2CH_2CH{=}CH\overset{\overset{\displaystyle O}{\|}}{C} - SCoA \ + \ FADH_2$$

(b)

$$CH_3CH_2CH_2CH{=}CH\overset{\overset{\displaystyle O}{\|}}{C} - SCoA \ + \ H_2O \ \longrightarrow \ CH_3CH_2CH_2\overset{\overset{\displaystyle OH}{|}}{CH}CH_2\overset{\overset{\displaystyle O}{\|}}{C} - SCoA$$

(c)

$$CH_3CH_2CH_2\overset{\overset{\displaystyle OH}{|}}{CH}CH_2\overset{\overset{\displaystyle O}{\|}}{C} - SCoA + NAD^+ \ \longrightarrow \ CH_3CH_2CH_2\overset{\overset{\displaystyle O}{\|}}{C}CH_2\overset{\overset{\displaystyle O}{\|}}{C} - SCoA + NADH + H^+$$

(d)

$$CH_3CH_2CH_2\overset{\overset{\displaystyle O}{\|}}{C}CH_2\overset{\overset{\displaystyle O}{\|}}{C} - SCoA + CoASH \ \longrightarrow \ CH_3CH_2CH_2\overset{\overset{\displaystyle O}{\|}}{C} - SCoA + CH_3\overset{\overset{\displaystyle O}{\|}}{C} - SCoA$$

28.29 $\quad CH_3CH_2CH_2\overset{\overset{\displaystyle O}{\|}}{C} - SCoA \ + \ FAD \ \longrightarrow \ CH_3CH{=}CH\overset{\overset{\displaystyle O}{\|}}{C} - SCoA \ + \ FADH_2$

$$CH_3CH{=}CH\overset{\overset{\displaystyle O}{\|}}{C} - SCoA \ + \ H_2O \ \longrightarrow \ CH_3\overset{\overset{\displaystyle OH}{|}}{CH}CH_2\overset{\overset{\displaystyle O}{\|}}{C} - SCoA$$

$$CH_3\overset{\overset{\displaystyle OH}{|}}{CH}CH_2\overset{\overset{\displaystyle O}{\|}}{C} - SCoA \ + \ NAD^+ \ \longrightarrow \ CH_3\overset{\overset{\displaystyle O}{\|}}{C}CH_2\overset{\overset{\displaystyle O}{\|}}{C} - SCoA \ + NADH \ + \ H^+$$

$$CH_3\overset{\overset{\displaystyle O}{\|}}{C}CH_2\overset{\overset{\displaystyle O}{\|}}{C} - SCoA \ + \ CoASH \ \longrightarrow \ 2 \ CH_3\overset{\overset{\displaystyle O}{\|}}{C} -SCoA$$

No more turns of the fatty acid cycle are possible.

28.30 FADH$_2$ passes its hydrogen into the respiratory chain and is changed back to FAD.

28.31 NADH passes its hydrogen into the respiratory chain and is changed back to NAD$^+$.

28.32 Steps 1-3 succeed in oxidizing the beta position of the fatty acyl unit to a ketone group.

28.33 (a) 7   (b) 6   (c) 6
(d)

Intermediate	Maximum no. of ATP from each	Total number of ATP possible from each as acetyl CoA forms
6 FADH$_2$	2	12
6 NADH	3	18

$$\text{7 CH}_3\overset{\overset{\text{O}}{\|}}{\text{C}}\text{SCoA} \qquad 12 \qquad \underline{84}$$

$$\text{Sum} = 114$$

Deduct 2 $\underline{-2}$

Net ATP produced   112

28.34 (a) Inside mitochondria   (b) Cytosol

28.35

$$\text{CH}_3\overset{\overset{\text{O}}{\|}}{\text{C}} - \text{SCoA} + \text{HCO}_3^- + \text{ATP} \longrightarrow \ ^-\text{O}\overset{\overset{\text{O}}{\|}}{\text{C}}\text{CH}_2\overset{\overset{\text{O}}{\|}}{\text{C}}\text{SCoA} + \text{H}^+ + \text{ADP} + \text{P}_i$$

$\downarrow$ ACP

$$^-\text{O}\overset{\overset{\text{O}}{\|}}{\text{C}}\text{CH}_2\overset{\overset{\text{O}}{\|}}{\text{C}}\text{S} - \text{ACP} + \text{CoA}$$

$$\text{CH}_3\overset{\overset{\text{O}}{\|}}{\text{C}} - \text{SCoA} + \text{ACP} \longrightarrow \text{CH}_3\overset{\overset{\text{O}}{\|}}{\text{C}} - \text{ACP}$$

$$+ \ \text{CoA}$$

$$\overset{\text{OH} \quad \ \text{O}}{\underset{|}{\text{CH}_3}\text{CH}\text{CH}_2\overset{\overset{\text{O}}{\|}}{\text{C}} - \text{S} - \text{ACP}} \longleftarrow \overset{\text{O} \quad \ \text{O}}{\text{CH}_3\overset{\|}{\text{C}}\text{CH}_2\overset{\|}{\text{C}} - \text{S} - \text{ACP} + \text{CO}_2 + \text{ACP}}$$

$$\text{NADP}^+ \quad \overline{\text{NADPH} + \text{H}^+}$$

$$\longrightarrow \text{H}_2\text{O}$$

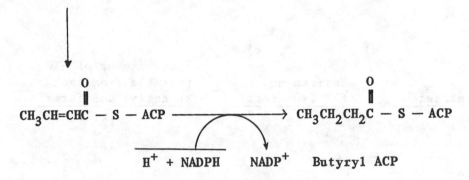

28.36   The pentose pathway of glucose catabolism

28.37   Mevalonate

28.38   Cholesterol inhibits the synthesis of HMG–CoA reductase.

28.39   Fatty acids

28.40   Acetyl CoA

28.41   Acetoacetic acid, $CH_3\overset{\displaystyle O}{\overset{\|}{C}}CH_2CO_2H$

28.42   Acetoacetate: $CH_3\overset{\displaystyle O}{\overset{\|}{C}}CH_2CO_2^-$

β–Hydroxybutyrate: $CH_3\overset{\displaystyle OH}{\overset{|}{C}H}CH_2CO_2^-$

Acetone: $CH_3\overset{\displaystyle O}{\overset{\|}{C}}CH_3$

28.43   An above normal concentration of the ketone bodies in the blood

28.44   An above normal concentration of the ketone bodies in the urine

28.45   Enough acetone vapor in exhaled air to be detected by its odor

28.46   Ketonemia   +   ketonuria   +   acetone breath

28.47   Metabolic acidosis brought on by a rise in the level of the ketone bodies in the blood

28.48   Acetoacetic acid

28.49   Diuresis is accelerated to remove ketone bodies from the blood, and their removal requires the simultaneous removal of water, so the urine volume rises.

28.50   Their over-production leads to acidosis.

28.51   Amino acids are catabolized at a faster than normal rate to participate in gluconeogenesis, and their nitrogen is excreted largely as urea.

28.52   Each Na^+ ion that leaves corresponds to the loss of one HCO_3^- ion, the true base, because HCO_3^- neutralizes acid generated as the ketone bodies are made. And for every negative ion that leaves with the urine, a positive ion, mostly Na^+, has to leave to ensure electrical neutrality.

CHAPTER 29

METABOLISM OF NITROGEN COMPOUNDS

Review Exercises

29.1   The entire collection of nitrogen compounds found anywhere in the body

29.2   To synthesize protein
        To synthesize nonprotein compounds of nitrogen
        To synthesize nonessential amino acids
        To contribute to the synthesis of ATP

29.3   Infancy

29.4   They are catabolized. Some are converted to fatty acids.

29.5 $^-O_2CCH_2CH_2\overset{\overset{\displaystyle O}{\|}}{C}CO_2^-$ + NH_4^+ + NADPH + H^+ $\longrightarrow$

$^-O_2CCH_2CH_2\overset{\overset{\displaystyle NH_3^+}{|}}{C}HCO_2^-$ + $NADP^+$ + H_2O

29.6 $-CH_2\overset{\overset{\displaystyle O}{\|}}{C}CO_2H$

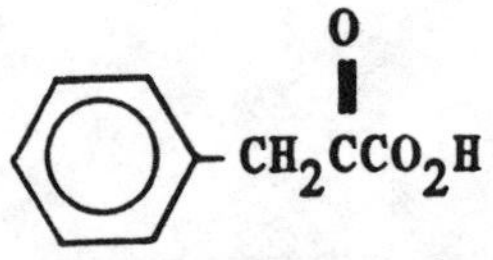

29.7 $CH_3\overset{\overset{\displaystyle CH_3}{|}}{C}H-\overset{\overset{\displaystyle O}{\|}}{C}CO_2H$

29.8
$CH_3\overset{\overset{\displaystyle NH_3^+}{|}}{C}HCO_2^-$ + $^-O_2CCH_2CH_2\overset{\overset{\displaystyle O}{\|}}{C}CO_2^-$ $\longrightarrow$ $CH_3\overset{\overset{\displaystyle O}{\|}}{C}CO_2^-$ + $^-O_2CCH_2CH_2\overset{\overset{\displaystyle NH_3^+}{|}}{C}HCO_2^-$

572

Then:

$$^-O_2CCH_2CH_2\overset{\overset{\displaystyle NH_3^+}{|}}{C}HCO_2^- + NAD^+ + H_2O \longrightarrow \ ^-O_2CCH_2CH_2\overset{\overset{\displaystyle O}{\|}}{C}CO_2^- + NADH + H+ + NH_4^+$$

29.9   4 1 5 2 3

29.10   5 2 1 3 4

29.11   Serine $\longrightarrow$ pyruvate $\longrightarrow$ acetyl CoA $\longrightarrow$ (lipigenesis) $\longrightarrow$ palmitic acid

29.12   Glucose $\longrightarrow$ (glycolysis) $\longrightarrow$ pyruvate $\longrightarrow$ (transamination) $\longrightarrow$ alanine

29.13 $CH_3CH_2\overset{\overset{\displaystyle O}{\|}}{C}CO_2^-$

29.14   HO—⟨benzene⟩—$CH_2CH_2NH_2$

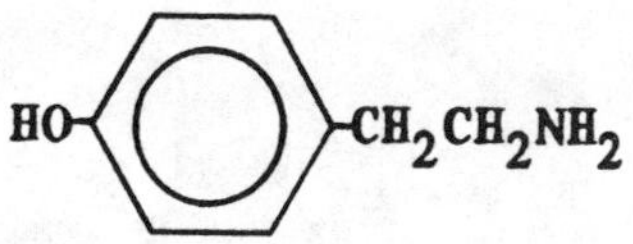

29.15

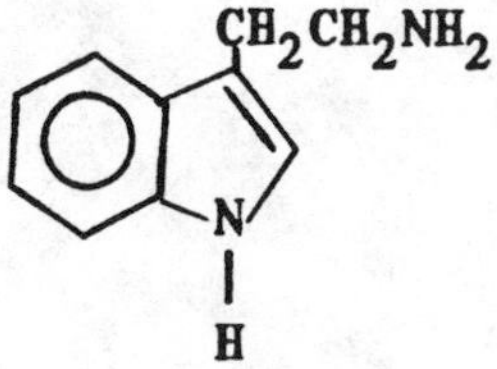

29.16   To make glucose via gluconeogenesis, and to generate ATP

29.17   (a) Originally, the amino groups of amino acids.
(b) Carbon dioxide

29.18   An above normal concentration of ammonium ion in the blood. Infants improve on a low-protein diet.

29.19 $2NH_3 + H_2CO_3 \longrightarrow NH_2\overset{\overset{\displaystyle O}{\|}}{C}NH_2 + 2H_2O$

29.20   The purine bases of nucleic acids, adenine and guanine

29.21   Sodium urate

29.22   3 2 1 4 5 7 6